Springer Series in

# MATERIALS SCIENCE 78

Springer Series in
# MATERIALS SCIENCE

*Editors:* R. Hull R. M. Osgood, Jr. J. Parisi H. Warlimont

The Springer Series in Materials Science covers the complete spectrum of materials physics, including fundamental principles, physical properties, materials theory and design. Recognizing the increasing importance of materials science in future device technologies, the book titles in this series reflect the state-of-the-art in understanding and controlling the structure and properties of all important classes of materials.

61 **Fatigue in Ferroelectric Ceramics and Related Issues**
By D.C. Lupascu

62 **Epitaxy**
Physical Principles and Technical Implementation
By M.A. Herman, W. Richter, and H. Sitter

63 **Fundamentals of Ion-Irradiated Polymers**
By D. Fink

64 **Morphology Control of Materials and Nanoparticles**
Advanced Materials Processing and Characterization
Editors: Y. Waseda and A. Muramatsu

65 **Transport Processes in Ion-Irradiated Polymers**
By D. Fink

66 **Multiphased Ceramic Materials**
Processing and Potential
Editors: W.-H. Tuan and J.-K. Guo

67 **Nondestructive Materials Characterization**
With Applications to Aerospace Materials
Editors: N.G.H. Meyendorf, P.B. Nagy, and S.I. Rokhlin

68 **Diffraction Analysis of the Microstructure of Materials**
Editors: E.J. Mittemeijer and P. Scardi

69 **Chemical–Mechanical Planarization of Semiconductor Materials**
Editor: M.R. Oliver

70 **Applications of the Isotopic Effect in Solids**
By V.G. Plekhanov

71 **Dissipative Phenomena in Condensed Matter**
Some Applications
By S. Dattagupta and S. Puri

72 **Predictive Simulation of Semiconductor Processing**
Status and Challenges
Editors: J. Dabrowski and E.R. Weber

73 **SiC Power Materials**
Devices and Applications
Editor: Z.C. Feng

74 **Plastic Deformation in Nanocrystalline Materials**
By M.Yu. Gutkin and I.A. Ovid'ko

75 **Wafer Bonding**
Applications and Technology
Editors: M. Alexe and U. Gösele

76 **Spirally Anisotropic Composites**
By G.E. Freger, V.N. Kestelman, and D.G. Freger

77 **Impurities Confined in Quantum Structures**
By P.O. Holtz and Q.X. Zhao

78 **Macromolecular Nanostructured Materials**
Editors: N. Ueyama and A. Harada

N. Ueyama A. Harada (Eds.)

# Macromolecular Nanostructured Materials

With 187 Figures, 59 Schemes and 26 Tables

Kodansha

Springer

Professor Norikazu Ueyama
Professor Akira Harada

Osaka University
Graduate School of Science
Department of Macromolecular Science
Toyonaka, Osaka 560-0043
Japan
E-mails: ueyama@chem.sci.osaka-u.ac.jp
harada@chem.sci.osaka-u.ac.jp

*Series Editors:*

Professor Robert Hull
University of Virginia
Dept. of Materials Science and Engineering
Thornton Hall
Charlottesville, VA 22903-2442, USA

Professor Jürgen Parisi
Universität Oldenburg, Fachbereich Physik
Abt. Energie- und Halbleiterforschung
Carl-von-Ossietzky-Straße 9–11
26129 Oldenburg, Germany

Professor R. M. Osgood, Jr.
Microelectronics Science Laboratory
Department of Electrical Engineering
Columbia University
Seeley W. Mudd Building
New York, NY 10027, USA

Professor Hans Warlimont
Institut für Festkörper-
und Werkstofforschung,
Helmholtzstraße 20
01069 Dresden, Germany

ISSN 0933-033X

ISBN 4-06-211242-6 Kodansha Ltd., Tokyo
ISBN 3-540-22327-4 Springer Berlin Heidelberg New York

Library of Congress Control Number: 3540223274



Springer is a part of Springer Science+Business Media.

springeronline.com


Printed in Japan



Cover concept: eStudio Calamar Steinen
Cover production: *design & production* GmbH, Heidelberg

Printed on acid-free paper SPIN: 11017363 57/3141/di 5 4 3 2 1 0

## List of Contributors

Numbers in parentheses refer to the sections.


Aoshima, Sadahito (2.5)

Department of Macromolecular Science, Graduate School of Science, Osaka University, Machikaneyama, Toyonaka 560-0043, Japan

Celestre, Paul C. (1.1)

Department of Chemistry and Biochemistry University of California, Los Angeles, 405 Hilgard Avenue, Los Angeles CA 90095, USA

Chujo, Yoshiki (3.4)

Department of Polymer Chemistry, Graduate School of Engineering, Kyoto University, Katsura, Nishikyo-ku, Kyoto 615-8510, Japan

Fuhrhop, Jürgen-Hinrich (3.1)

Institut für Organische Chemie der Freien Universität Berlin, Takustrasse 3, D-14195 Berlin, Germany

Gronwald, Oliver (2.3)

Asahi Kasei Corporation, SIF Research Group, Central Technology Laboratory, Samejima, Fuji, Shizuoka 416-8501, Japan

Hanabusa, Kenji (2.4)

Graduate School of Science and Technology, Shinshu University, Tokida, Ueda 386-8567, Japan

Harada, Akira (1.2,4.2)

Department of Macromolecular Science, Graduate School of Science, Osaka University, Machikaneyama, Toyonaka 560-0043, Japan

Hirao, Toshikazu (3.2)

Department of Applied Chemistry, Graduate School of Engineering, Osaka University, Yamada-oka, Suita 565-0871, Japan

Khodabakhsh, Saghar (1.3)

Department of Chemistry, Imperial College of Science, Technology and Medicine, South Kensington, London SW7 2AY, UK

Kim, Kyung-Min (3.4)
Department of Polymer Science & Engineering, Chungju National University, 123 Geomdan-ri, Iryu-myeon, Chungju, Chungbuk, 380-702, Korea

Klyszcz, Andreas (3.1)
Institut für Organische Chemie der Freien Universität Berlin, Takustrasse 3, D-14195 Berlin, Germany

Koumoto, Kazuya (4.4)
Department of Chemical Processes and Environment, Faculty of Environmental Engineering, University of Kitakyushu, Wakamatsu-ku, Kitakyushu 808-0135, Japan

Krasia, Theodora (1.3)
Max Planck Institute for Colloids and Interfaces, Golm 14424 Potsdam, Germany

Liaw, Der-Jang (2.2)
Department of Chemical Engineering, National Taiwan University of Science and Technology, Taipei, Taiwan

Lauer, Matthias (3.1)
Institut für Organische Chemie der Freien Universität Berlin, Takustrasse 3, D-14195 Berlin, Germany

Li, Guangtao (3.1)
Institut für Organische Chemie der Freien Universität Berlin, Takustrasse 3, D-14195 Berlin, Germany

Michinobu, Tsuyoshi (2.1)
Department of Applied Chemistry, Waseda University, Shinjuku-ku, Tokyo 169-8555, Japan

Murata, Masaki (3.5)
Department of Chemistry, Graduate School of Science, The University of Tokyo, Bunkyo-ku, Tokyo 113-0033, Japan

Nishide, Hiroyuki (2.1)
Department of Applied Chemistry, Waseda University, Shinjuku-ku, Tokyo 169-8555, Japan

Nishihara, Hiroshi (3.5)

Department of Chemistry, Graduate School of Science, The University of Tokyo, Bunkyo-ku, Tokyo 113-0033, Japan

Okamura, Taka-aki (3.6)

Department of Macromolecular Science, Graduate School of Science, Osaka University, Machikaneyama, Toyonaka 560-0043, Japan

Osaki, Shigeyoshi (4.5)

Department of Chemistry, Faculty of Medicine, Nara Medical University, Shijo-cho, Kashihara 634-8521, Japan

Reedijk, Jan (4.1)

Leiden Institute of Chemistry, Gorlaeus Laboratories, Leiden University, P.O.Box 9502, 2300 RA Leiden, The Netherlands

Sakurai, Kazuo (4.4)

Department of Chemical Processes and Environment, Faculty of Environmental Engineering, University of Kitakyushu, Hibikino, Wakamatsu-ku, Kitakyushu 808-0135, Japan

Shinkai, Seiji (2.3,4.4)

Department of Chemistry and Biochemistry, Graduate School of Engineering, Kyushu University, Higashi-ku, Fukuoka 812-8581, Japan

Steinke, Joachim H.G. (1.3)

Department of Chemistry, Imperial College of Science, Technology and Medicine, South Kensington, London SW7 2AZ, UK

Stoddart, J. Fraser (1.1)

Department of Chemistry and Biochemistry, University of California, Los Angeles 405 Hilgard Avenue, Los Angeles CA 90095, USA

Toshima, Naoki (3.3)

Department of Materials Science and Environmental Engineering, Faculty of Science and Engineering, Tokyo University of Science, Yamaguchi [The former Science University of Tokyo in Yamaguchi], Daigaku-dori, Onoda 756-0884, Japan

Tseng, Hsian-Rong (1.1)

Department of Chemistry and Biochemistry, University of California, Los Angeles, 405 Hilgard Avenue, Los Angeles CA 90095, USA

Tuncel, Dönüs (1.3)

Department of Chemistry, Bilkent University, 06800 Bilkent, Ankara, Turkey

Ueyama, Norikazu (3.6,4.3)

Department of Macromolecular Science, Graduate School of Science, Osaka University, Machikaneyama, Toyonaka 560-0043, Japan

Yamaguchi, Hiroyasu (4.2)

Department of Macromolecular Science, Graduate School of Science, Osaka University, Machikaneyama, Toyonaka 560-0043, Japan

Yamamoto, Hitoshi (4.3)

Department of Macromolecular Science, Graduate School of Science, Osaka University, Machikaneyama, Toyonaka 560-0043, Japan

# Preface

This volume contains information on the most recent advances in the rapid development of nanostructured materials. The authors are world leaders in this area and a group of young scientists who have devoted themselves to the development of macromolecular nanostructured materials. The first development phase of this field was characterized by the construction of nanostructured materials. We have now entered the second phase, the aim of which has been the creation of a functional device for various applications. Polymeric materials have already established a nanosized structure for these infinitesimal purposes. However, these polymers are used via a top-down processing approach from the mass. The goal of creating new nanostructured materials is to produce instead a device which can be self-assembled and used in a bottom-up processing approach. Although the top-down processing approach has given rise to the problem of how to dispose of scraps generated by industrial and daily usage, the bottom-up processing approach will involve a specific reaction after usage, thereby eliminating this disposal problem.

The collected articles present important concepts for the precise synthesis of nanostructured materials. These small molecules are designed according to the molecular recognition results for their functions, such as molecular electronics, mesoscopic assemblies using organic building blocks or polymers with well-defined shape, size and a deeply buried functional group, and novel magnets consisting of organic radicals, sol-gel hybrids applicable for hardeners of fluids or liquid crystals. The second part mainly describes macromolecular metal complexes and their assemblies having high ability in redox and catalytic reactions and mechanical hardness as well as and the nanostructured assemblies of inorganic polymers that contain various metal complex units. The third part deals with biologically-relevant nanostructured assemblies of DNA-metal, DNA-organic polymer, fibrillate proteins and biomineralliztion.

The aim of this volume is to understand the synthetic foundations for the construction of macromolecular nanostructured materials and to describe the reviews and results of each research project supported as frontier topics. This volume endeavors to serve as a reference work not only for specialists and non-specialists but also for graduate students who are interested in learning about national projects conducted in countries all around the world. By including articles from various different scientific backgrounds, we hope that this will increase the knowledge of the readers and impress upon them the scope of the field of nanoscience.

This monograph started out simply as a paper presented at the Osaka University Macromolecular Symposium in Osaka (November 5–7, 2001). Because of the great strides made recently in nanoscience, it was considered appropriate to expand this paper into a complete book. Some of the concepts based on nanostructured material science have stimulated the main subjects of the 21st Century COE projects in a number of universities in Japan, for example, "Integrated EcoChemistry" in Osaka University.

The editors are grateful to all the authors for making time in their busy schedules to contribute a chapter to this work and finally express our grateful acknowledgment to the support by Grant-in-Aid for Publication of Scientific Research Results from Japan Society for the Promotion of Science (No.155291).

Norikazu Ueyama
Akira Harada
Osaka
December 2003

# Contents

# 2 Macromolecular Organic Assemblies

# 3 Macromolecular Inorganic Assemblies

# 1
# Precise Synthesis of Supramolecules

# 1.1 An Integrated Systems-oriented Approach to Molecular Electronics

**Hsian-Rong Tseng, Paul C. Celestre and J. Fraser Stoddart**

## 1.1.1 Introduction

Fundamental physical limitations and commercial considerations relating to spiraling costs mean that conventional lithographically constructed solid-state devices may well face increasingly formidable technological hurdles to overcome by the end of the first decade of the new millennium. Although necessity has long been hailed as the mother of invention, the time is also ripe now for innovation to thrive at the interfaces between chemical design and synthesis and the physics and engineering of devising computer architectures, fabricating devices, and integrating electronic circuits. The rapidly developing practice of molecular electronics[1)] at the nanoscale level – often nowadays referred to as nanoelectronics – offers a bottom-up alternative to the present silicon-based top-down construction of memory and logic circuits.

Molecular electronic devices can, in principle, be scaled down single-molecule junctions of a few cubic nanometers where switching of bistable molecules can be performed at voltages that we associate with flashlight batteries and currents that are routinely in the nanoampere range. Clearly, such a bottom-up approach to device construction opens up the prospect of a highly energy-efficient means of information processing and could create the opportunity for the further cost-saving manufacture of electronic circuits at the nanoscale level using redox-active molecules as one of their key components.

As a result of the combined effects of multidisciplinary teams of chemists, engineers and physicists, molecular electronics[1)] has evolved considerably over the past few years. A number of prototypical molecular nanoelectronic devices, incorporating molecular rectifiers and wires,[2)] carbon nanotubes[3)] and redox-active, bistable molecules[4)] have been successfully demonstrated.

During the past 20 years, mechanically interlocked molecules, known as catenanes and rotaxanes,[5)] many of them[6)] redox-active, have become readily accessible using template-directed protocols[7)] that rely upon the precepts of molecular recognition[8)] and self-assembly[9)] and the tenets of supramolecular assistance to covalent synthesis.[10)] By incorporating different recognition units with dissimilar redox properties into appropriate components, these compounds can often be induced to switch hysteretically between ground and metastable co-conformations[10] as a result of substantial relative mechanical movements of either a rotary or linear type. Such molecules may be viewed as nanometer-scale switches and can be activated by controlling the redox properties of the recognition units chemically,[11] electrochemically[12] and optically[13] to mimic the moving parts in macroscopic machines.[14] This section will highlight the design and synthesis as well as solution- and solid-state characterizations of various molecular switches that have been incorporated into crossbar electronic devices. Additionally, we will discuss the performance of these devices. The information gleaned from exploring the behavior of these molecules in the solution state, as well as in solid-state devices, leads to the creation of feedback loops that enable the improvement of the molecular design, affording a truly integrated systems-oriented approach for the construction of even better molecular electronic devices.

### 1.1.2 Catenane-based Devices

The first electronically configurable switch that we incorporated into a solid-state device[15] was based on the non-degenerate, bistable [2]catenane $\mathbf{4}^{4+}$. It contains a tetrathiafulvalene (TTF) unit and a 1,5-dioxynaphthalene (DNP) ring system in its π-electronic-rich macrocyclic polyether component which is catenated by a π-electron-deficient tetracationic cyclophane component. A protocol[16] involving self-assembly followed by template-direction was employed (Fig. 1.1) in the synthesis of $\mathbf{4}^{4+}$ as its tetrakis(hexafluorophosphate) salt. The bis(hexafluorophosphate) salt $\mathbf{1}\cdot 2PF_6$ was treated with α,α′-dibromo-*p*-xylene (**2**) in the presence of the macrocyclic polyether **3** to obtain $\mathbf{4}\cdot 4PF_6$ in 23% yield after counterion exchange. X-ray crystallography, performed on a single crystal of $\mathbf{4}\cdot 4PF_6$, revealed (Fig. 1.2) that the TTF unit in the macrocyclic polyether resides preferentially inside the cavity of the tetracationic cyclophane in the solid state. This observation is consistent with the solution-state absorption spectrum of $\mathbf{4}\cdot 4PF_6$ in MeCN. It exhibits a band at 850 nm (curve (a) in Fig. 1.3), corresponding to a

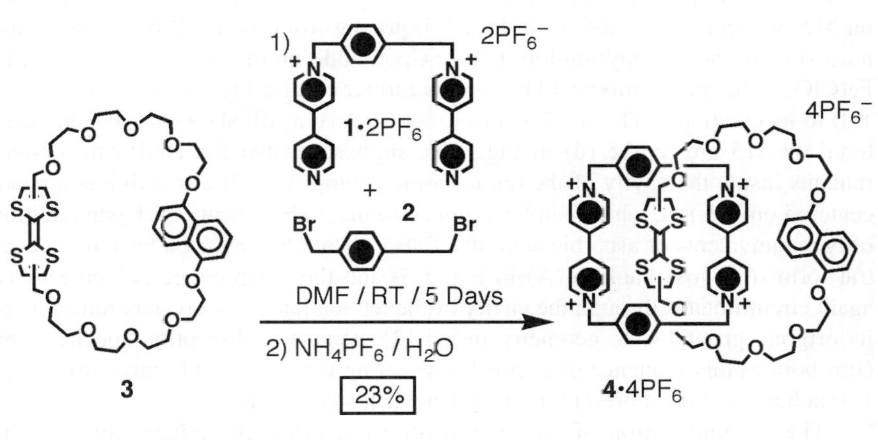


**Fig. 1.1** The template-directed synthesis of the [2]catenane $\mathbf{4}\cdot 4PF_6$.

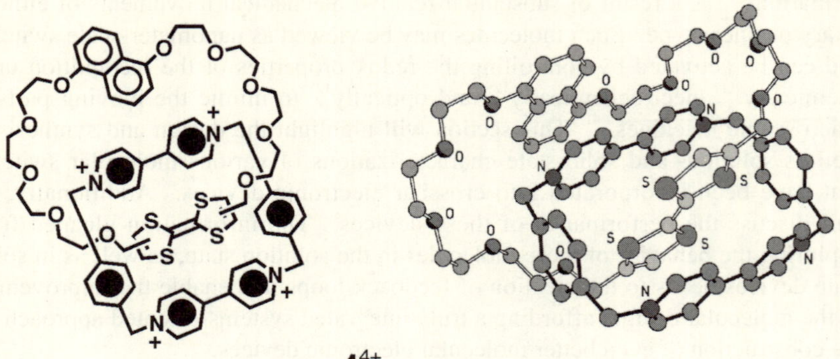

$4^{4+}$

**Fig. 1.2** The X-ray crystal structure of the [2]catenane $\mathbf{4}^{4+}$.

charge transfer (CT) interaction between the TTF unit and the bipyridinium units in the tetracationic cyclophane. Thus, the ground-state geometry of this potential molecular switch has been established in both the solid and solution states. The question remains − does the molecule switch under redox control, at least, in the solution state? Indeed, it does. The switching process has been demonstrated by monitoring the circumrotation of the macrocyclic polyether component through the cavity of the tetracationic cyclophane in the UV/visible spectrum. Switching can be activated reversibly by oxidation and subsequent reduction of the TTF unit. Upon addition of one equivalent of $Fe(ClO_4)_3$ to the [2]catenane, the TTF unit is oxidized ((A) goes to (B) in Fig. 1.3) to form a radical cation ($TTF^{+\bullet}$). Circumrotation of the macrocyclic polyether occurs as a result of the electrostatic repulsion between $TTF^{+\bullet}$ and the tetracationic cyclophane. The CT band, centered on 850 nm, gradually disappears and is replaced by a new CT band, centered on 515 nm (curves (b) and (c) in Fig. 1.3) arising from the DNP ring system and its two surrounding bipyridinium units. Upon addition of a second equivalent of $Fe(ClO_4)_3$, the monocationic $TTF^{+\bullet}$ unit is further oxidized (from (B) to (C) in Fig. 1.3) to its dication ($TTF^{2+}$). The absorption spectrum still shows a CT band, centered on 515 nm (curve (d) in Fig. 1.3), suggesting that the DNP ring system remains inside the cavity of the tetracationic cyclophane. It also exhibits a band, centered on 375 nm, which can be assigned to the $TTF^{2+}$ dication. Upon addition of two equivalents of ascorbic acid, the $TTF^{2+}$ dication is reduced back to its neutral form ((C) goes back to (A) in Fig. 1.3) and the macrocyclic polyether once again circumrotates through the cavity of the tetracationic cyclophane, returning to its original ground-state geometry in the [2]catenane. Absorption spectroscopy corroborates this sequence of events by revealing the same band (curve (e) in Fig. 1.3) at 850 nm as that present in the starting [2]catenane $\mathbf{4}\cdot 4PF_6$.

The circumrotation of the macrocyclic polyether component through the cavity of the tetracationic cyclophane can also be induced electrochemically. In

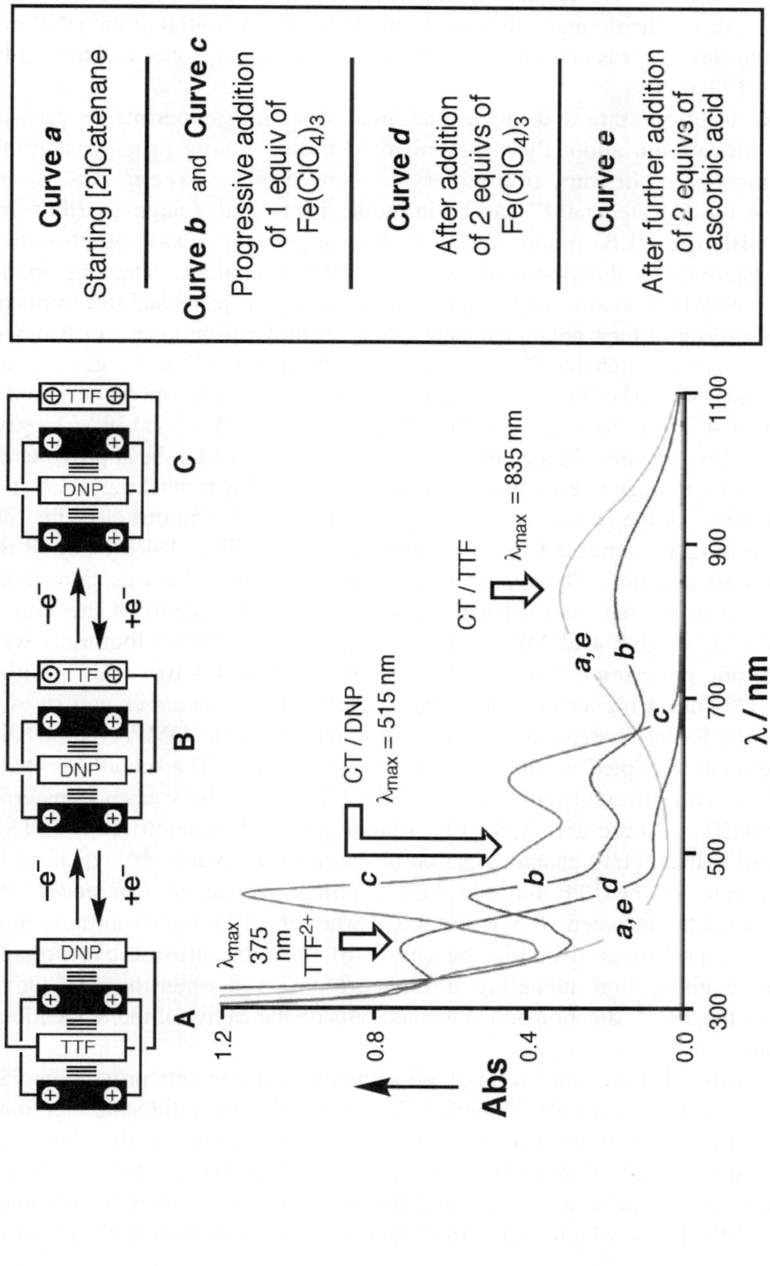


**Fig. 1.3** Absorption spectra of the redox processes of the [2]catenane $\mathbf{4}\cdot 4PF_6$.

cyclic voltammetry (CV) experiments[16] on **4•4PF**$_6$, a bielectronic oxidation process occurs that is associated with the TTF unit at +0.80 V *vs.* a saturated calomel electrode (SCE), while the oxidation of the DNP ring system is strongly displaced toward +1.60 V *vs.* SCE compared with that (+1.17 V *vs.* SCE) of the free crown ether. The dramatic increase in the oxidation potential of the DNP ring system indicates that it is encircled by the tetracationic cyclophane after the oxidation of the TTF unit.

These solution-state and electrochemical switching experiments provide detailed information about the mechanism of the switching processes in this bistable molecule. In some preliminary condensed-phase experiments, it was shown that the [2]catenane $\mathbf{4}^{4+}$ can be introduced into both Langmuir films and Langmuir-Blodgett (LB) monolayers[17] by exchanging the hexafluorophosphate ($PF_6^-$) counterions for dimyristoylphosphatidic ($DMPA^-$) anions. The amphiphilic nature of the $DMPA^-$ anions helps to prevent the films of [2]catenane salt from collapsing as a result of their polar character. Using a similar technique, an LB monolayer of the redox-switchable [2]catenane $\mathbf{4}^{4+}$ has been shown[18] to be stable at the air-water interface. The lower curve illustrated in Fig. 1.4 corresponds to the $\pi$-$A$ isotherm of $\mathbf{4}^{4+}$ plus four equivalents of the sodium salt of $DMPA^-$ to give **4•4DMPA**. The area per [2]catenane tetracation was estimated to be approximately 1.2 $nm^2$. This footprint resembles the two-dimensional $\pi$-$\pi$ stacking solid-state superstructure[16] of the [2]catenane, leading us to believe that, in this case, the four $DMPA^-$ anions are amassed in an amphiphilic layer (Fig. 1.4(a)) above the [2]catenane tetracations. The upper curve illustrated in Fig. 1.4 corresponds to the $\pi$-$A$ isotherm of the oxidized [2]catenane $\mathbf{4}^{6+}$ with six equivalents of the sodium salt of $DMPA^-$ to give **4•6DMPA**. Here, two distinct molecular footprints were observed, one covering 2.7 $nm^2$ at low pressures (Fig. 1.4(b)), and the other covering 1.5 $nm^2$ at higher pressures (Fig. 1.4(c)). These increases and steps in the molecular footprint areas can be interpreted in terms of the $DMPA^-$ anions dictating the areas occupied by the [2]catenane hexacations. The monolayers containing these two different redox states of the [2]catenane have been transferred onto a Au(III) substrate and studied by scanning tunneling microscopy (STM). The current voltage ($I/V$) characterization of the resulting materials is dictated by the redox states of the TTF unit and their resulting mechanical differences. No current is detected between –0.5 V and 0.0 V when the TTF unit is initially in its neutral state and located inside the cavity of the tetracationic cyclophane. However, relatively high tunneling currents are observed when the TTF unit is oxidized to the $TTF^{2+}$ dication and is located outside the cavity of the tetracationic cyclophane.

The results of these condensed-phase switching experiments provide us with the impetus to incorporate the bistable [2]catenane $\mathbf{4}^{4+}$ into solid-state electronic devices. In collaboration with members of the Heath group in the California NanoSystems Institute (CNSI) here at UCLA, the [2]catenane **4•4DMPA** was introduced[15] first of all into two-terminal devices to form molecular switch tunnel junctions (MSTJs) in which each MSTJ functions as an electronically reconfig-

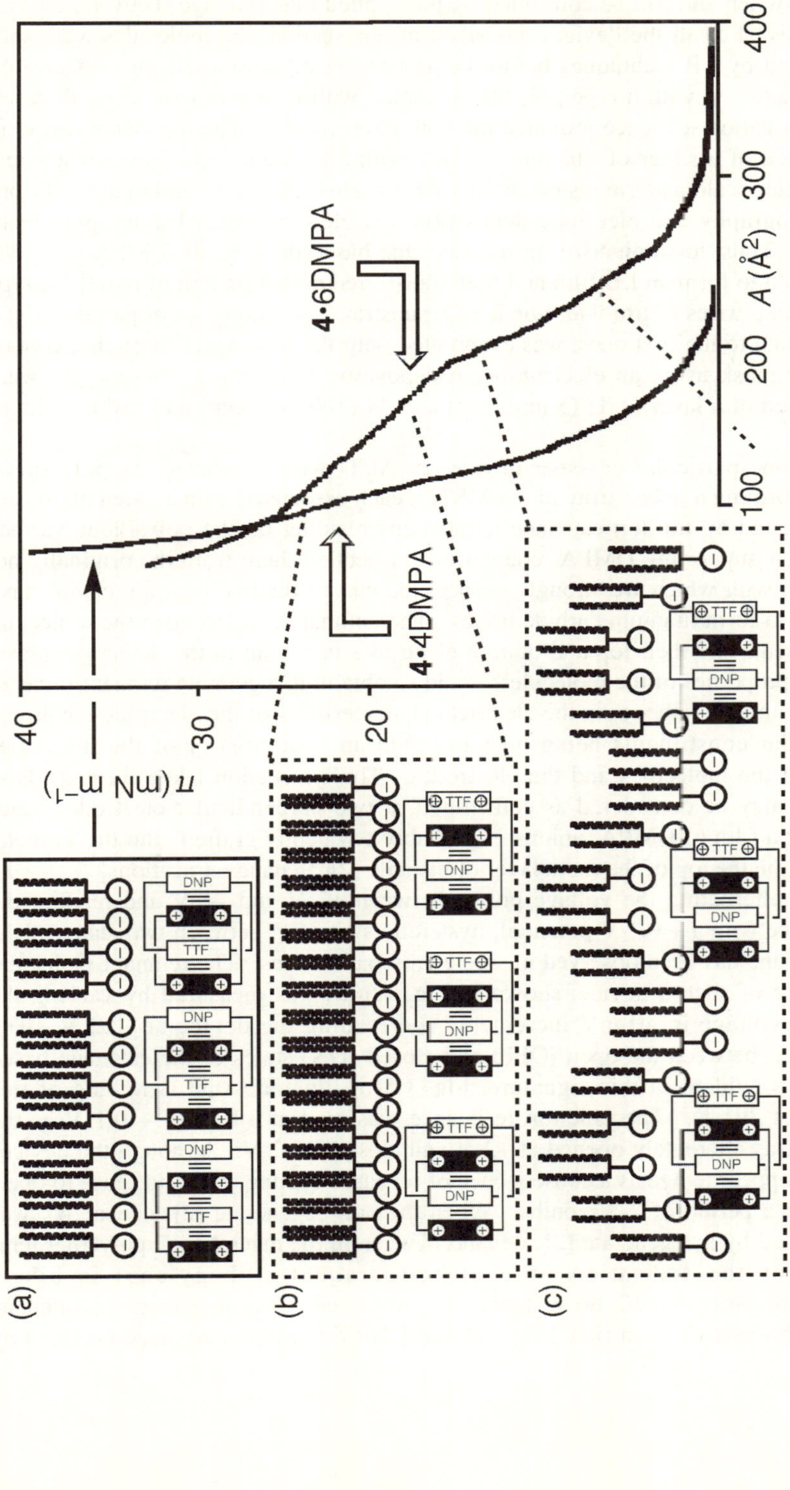


**Fig. 1.4** The isotherms of the [2]catenanes **4•4**DMPA and **4•6**DMPA at the air-water interface.

urable switch and can be controlled by the applied bias (voltage) between the two terminals. For all the devices described in this section, the molecules were self-organized by LB techniques before being transferred onto a silicon dioxide substrate patterned with n-type polysilicon coated with its native oxide. In all cases, this fabrication sequence provided the bottom electrodes. The top electrodes were composed of a sliver of titanium covered with aluminum. The micrometer- and nanometer-scale patternings of circuits were carried out by a combination of optical lithography and electrode-deposition and electron-beam lithographic techniques. Thus, to summarize, in the catenane-based devices, **4**•4DMPA was self-organized to form an LB film and then transferred onto a pattern of parallel n-type polysilicon wires (7 μm wide) on a $SiO_2$ substrate. A second set of parallel wires, orthogonal to the first ones, was evaporated onto the monolayer through a contact shadow mask using an electron-beam deposition technique. These wires were composed of a layer of Ti (5 nm deep) and Al (100 nm deep) and had a width of 10 μm.

In one particular crossbar device, an Al/Ti wire overlayed six polysilicon wires, forming a linear array of six MSTJs, each associated with an area of 70 $\mu m^2$ with around 50 million catenane tetracations residing on the polysilicon surface with their supporting $DMPA^-$ counterions protecting them from the originally hot titanium with which we strongly suspect the end of the hydrocarbon chains have reacted to form titanium carbide bonds. The compatibility between the molecular constituents and their top and bottom electrodes is crucial to the devices expressing the properties of the molecules – and so obtaining a genuine remnant molecular signature – rather than the electrical characteristics of the electrodes (with the molecular constituents being no more than an insulator) or of the interfaces between the molecules and the electrodes. Thus, a section taken through these MSTJs may be considered as constituted of two perpendicular electrodes sandwiching a film of $DMPA^-$ anions – probably covalently grafted onto the Ti metal – lying on the top of the monolayer of bistable [2]catenane tetracations.

By alternating the voltage pulses between +2 V and –2 V and reading the resistance with a + 0.1 V potential, hysteretic switching between two states of the [2]catenane has been observed in single crossbar devices. The remnant molecular signature of such a device, shown in Fig. 1.5(a), was measured by varying the writing voltage in 40-mV increments and reading the device at –0.2 V. The hysteresis between the open (OFF) and closed (ON) states of the catenane-based device is evident in this signature. Fig. 1.5(b) illustrates the switching of this device at 291 K. The junction resistance was read at a bias of + 0.1 V as the device was alternately opened at +2 V and closed at –2 V. Although the devices did not appear to age as a consequence of repeated cycling, they did cease to operate after a period of 2-3 months. When the non-degenerate [2]catenane $\mathbf{4}^{4+}$ was substituted by a degenerate [2]catenane,[19] wherein the macrocyclic polyether template for the formation of cyclobis(paraquat-*p*-phenylene) is bis-*p*-phenylene[34]crown-10, no switching was observed. This degenerate [2]catenane in which the π-electron rich TTF unit and DNP ring system are each replaced by

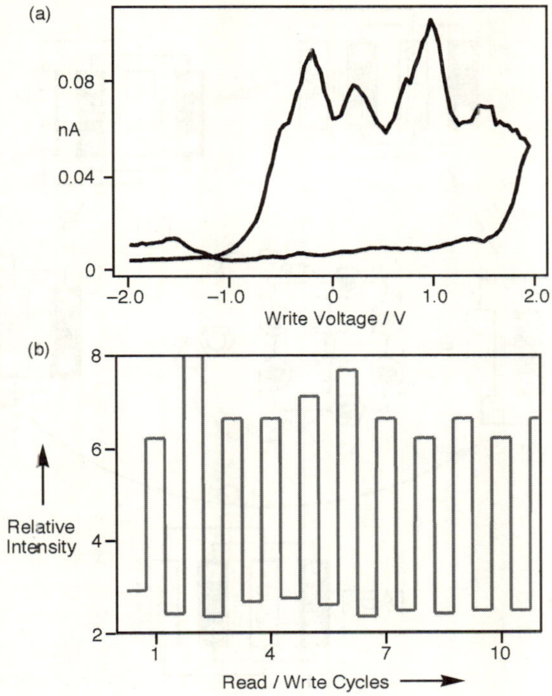


**Fig. 1.5** (a) The remnant molecular signature of the [2]catenane **4**•4DMPA device which was monitored at –0.1 V. (b) The switching cycles of this device.

hydroquinone rings, is a monostable compound which is only capable of inversion between enantiomeric forms, i.e., it is incapable of switching hysteretically between ground and metastable states. This control compound and others lend phenomenological support to an electromechanical switching mechanism.

The proposed mechanism is outlined in Fig. 1.6 : co-conformation $[A^0]$ corresponds to the "switch open" state and the metastable, co-conformation $[B^0]$ the "switch closed" state of the device. When the applied bias approaches +2 V, the [2]catenane $\mathbf{4}^{4+}$ is oxidized to state $[A^+]$ wherein the TTF unit is positively charged and so experiences an electrostatic repulsion inside the tetracationic cyclophane, resulting in circumrotation of the macrocyclic polyether and formation of its co-conformational state $[B^+]$. When the voltage is reduced to near-zero bias, the TTF $^{+\bullet}$ cation radical is reduced, affording the metastable state $[B^0]$ which does not return spontaneously to its stable co-conformation $[A^0]$ by further circumrotation of the macrocyclic polyether via state $[AB^\#]$ until the bipyridinium units in the tetracationic cyclophane are reduced. This electromechanical switching mechanism is consistent with temperature-dependent measurements,[15] which indicate

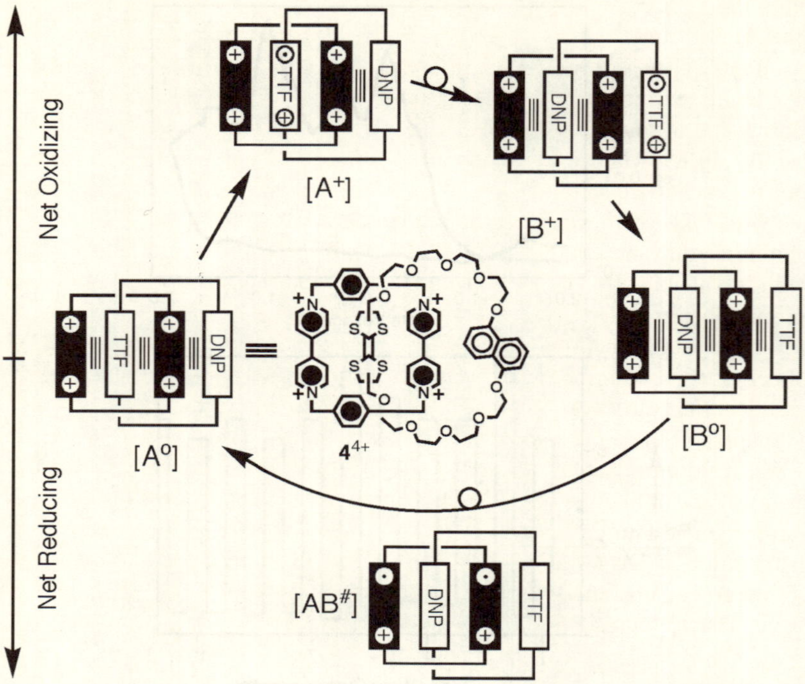


**Fig. 1.6** Proposed electromechanical mechanism for the operation of the device based on the bistable [2]catenane $\mathbf{4}^{4+}$.

that the overall cycling of the switches has at least one thermally activated process and is quenched below 200 K.

Although this electronically reconfigurable, bistable catenane-based switch might be useful in the generation of random access memory (RAM), the single crossbar devices had problems associated with them that were likely to limit their usefulness in increasingly complex devices. First, the current levels it was possible to pass through the [2]catenane MSTJs were low (in the tens of picoamperes for micrometer-level devices) in the switch closed state. Second, the switching ratio was about 2 : 1 (switch closed / switch open) and so would not be sufficient for use in more complicated circuits. We surmised that these deficiencies might find an explanation in the fact that the redox-active catenane switches are asymmetrically located within the LB film next the polysilicon electrode and distant from the Ti/Al electrode. The obvious way to construct less asymmetric MSTJs was to make the transition from bistable [2]catenanes to switchable [2]rotaxanes, rendering them amphiphilic in the process.

### 1.1.3 Amphiphilic Single-station Rotaxane-based Devices

With a view to increasing the tunneling current in MSTJ devices, a selection of amphiphilic [2]rotaxanes were designed and synthesized. In devices incorporating this kind of interlocked molecule, it is much easier to locate the redox-active components more symmetrically between the two electrodes. Consequently, higher levels of tunnel currents might be expected with bistable [2]rotaxanes straddling the electrodes. At the outset, an amphiphilic [2]rotaxane **6**•4PF$_6$ was prepared[20] in 8% yield using a template-directed protocol (Fig. 1.7). The reaction of the dicationic salt **1**•2PF$_6$ with α,α′-dibromo-*p*-xylene (**2**) was carried out in the presence of the dumbbell-shaped compound **5** as the template. This amphiphilic rotaxane contains only one formal recognition site, a monopyrrolotetrathiafulvalene (MP-TTF) for encirclement by the tetracationic cyclophane, an inter-relationship which gives rise to a CT band centered on 810 nm. The incorporation of a tetraarylmethane-based hydrophobic stopper at one end of the dumbbell-shaped component and a dendron-like hydrophilic stopper at the other end renders the [2]rotaxane $6^{4+}$ sufficiently amphiphilic so that the molecules will self-organize at the air-water interface as their tetrakis(hexafluorophosphate) salt into stable monolayers that can be transferred using the LB procedure into devices. Thus, unlike the bistable catenane described in Section 1.1.2, this rotaxane forms stable Langmuir films at the air-water interface without the addition of DMPA$^-$ anions. These monolayers have been transferred onto polysilicon electrode-patterned substrates (5 μm wide) as LB films prior to the deposition of the familiar Ti/Al electrodes that were either 0.5 or 10 μm wide in the different

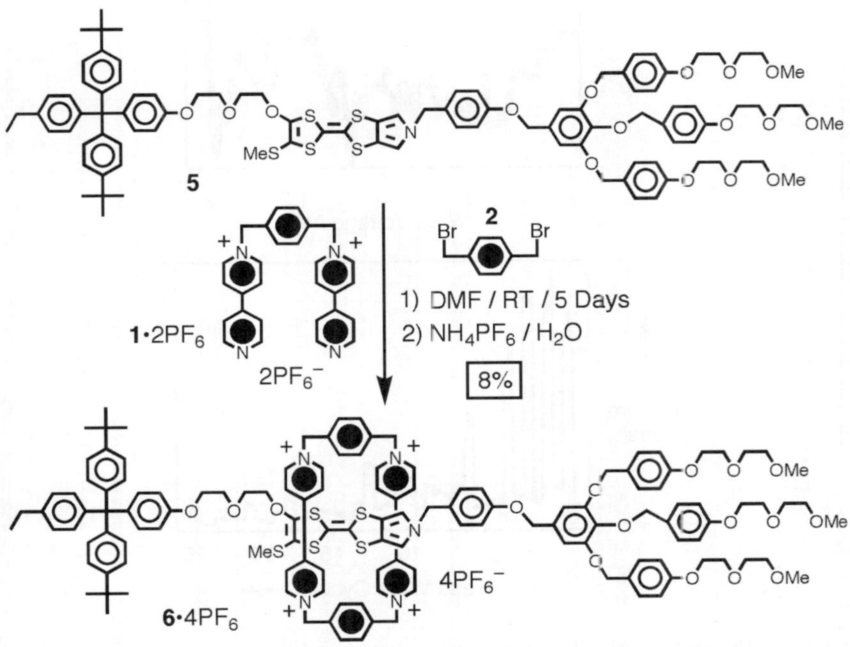


**Fig. 1.7** The template-directed synthesis of the "monostable" [2]rotaxane **6**•4PF$_6$.

devices that have been fabricated.[21] A very weak yet highly reproducible remnant molecular signature (Fig. 1.8(a)) was obtained from these devices by alternating the voltage pulses in increments between +2 V and –2 V and reading the resistance at a potential of + 0.2 V. However, a minimal switching response was only observed for a few device cycles (Fig. 1.8(b)), after which the remnant molecular signature collapsed. The lack of a genuine second recognition site, e.g., a DNP ring system, in $\mathbf{6}^{4+}$ could be the cause of this relatively unstable device behavior. It is not unlikely that the *p*-alkoxybenzyl group attached directly to the pyrrolo nitrogen atom on the MP-TTF unit fulfills the role of a weak second recognition site, albeit uncomfortably close to the MP-TTF unit to sustain a clear-cut hysteretic electromechanical switching mechanism. The investigation of an amphiphilic, single-station, rotaxane-based device constitutes an excellent control experiment and may be regarded as the logical starting point for the evaluation of bistable, amphiphilic [2]rotaxane-based devices. First of all, however, let us see what transpires at a supramolecular level when amphiphilic bistable pseudorotaxanes are incorporated into nanoelectronic devices.

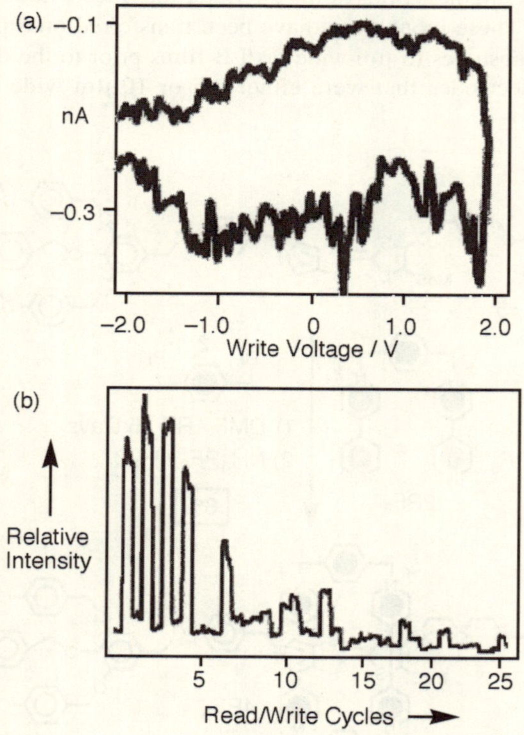


**Fig. 1.8** (a) The remnant molecular signature of the [2]rotaxane **6•4PF$_6$** device which was monitored at + 0.2 V. (b) The switching cycles of this device.

### 1.1.4 Amphiphilic Bistable Pseudorotaxane-based Devices

Mixing (Fig. 1.9) the semi-dumbbell compound **7** with the tetracationic cyclophane **8**•4PF$_6$ afforded[21,22] the two-station [2]pseudorotaxane [**8**⊃**7**]$^{4+}$. Since [**8**⊃**7**]$^{4+}$ contains two π-donor recognition sites – an MP-TTF unit and a DNP ring system – it is possible that it could exist as a mixture of two co-conformations – one with the tetracationic cyclophane encircling the MP-TTF unit and the other with the cyclophane residing around the DNP ring system. In fact, a 1 : 1 mixture of **7** and **8**•4PF$_6$ in MeCN exhibits two CT bands, centered around 540 nm and 785 nm, in keeping with two co-conformations being present in equilibrium in solution. A solution containing a mixture of the DMPA•Na salt, the tetracationic cyclophane **8**•4PF$_6$ and the semi-dumbbell compound **7**, respectively, in a molar ratio of 12 : 2 : 1 led to the [2]pseudorotaxane forming a stable Langmuir monolayer at the air-water interface. Brewster angle microscopy (BAM) performed on these Langmuir monolayers and AFM of the derived LB films reveal the influence that varying the nature of the subphase can have on our ability to control the characteristics (size, shape, etc.) of domains. A $CdCl_2$ subphase diminished the electrostatic repulsion between the head groups of the DMPA$^-$ anions, generating a more uniform thin film of the pseudorotaxane entities. BAM becomes an essential technique for collecting data on the nature of surface domains. It created an informed feedback loop which permits the evaluation of device performance in relation to the interfacial characteristics of the (supra)molecular switches.

The remnant molecular signature of an MSTJ fabricated from the [2]pseudorotaxane [**8**⊃**7**]$^{4+}$ is presented in Fig. 1.10(a). This device, which is based on a polysilicon bottom electrode and a Ti/Al top electrode, was fabricated in a manner similar to that already described in Section 1.1.3 for the single-station

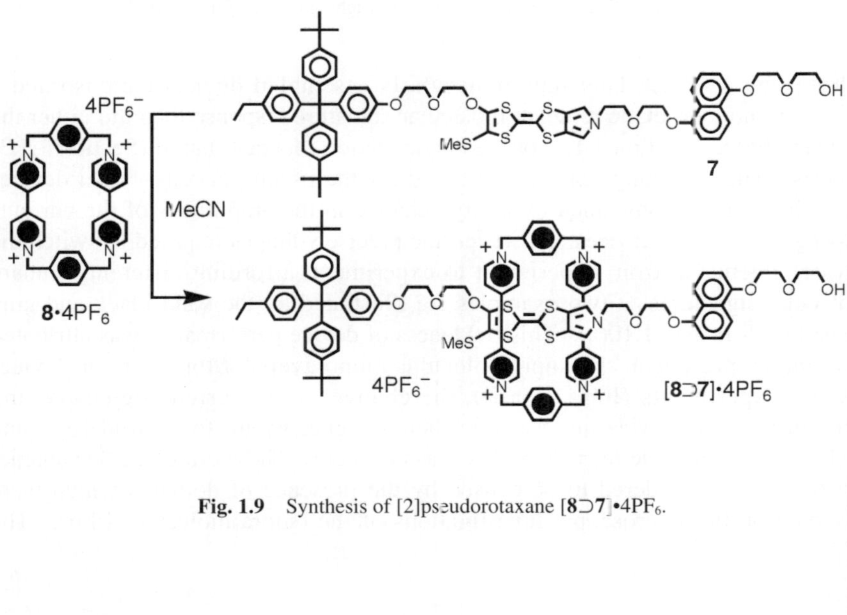


**Fig. 1.9** Synthesis of [2]pseudorotaxane [**8**⊃**7**]•4PF$_6$.

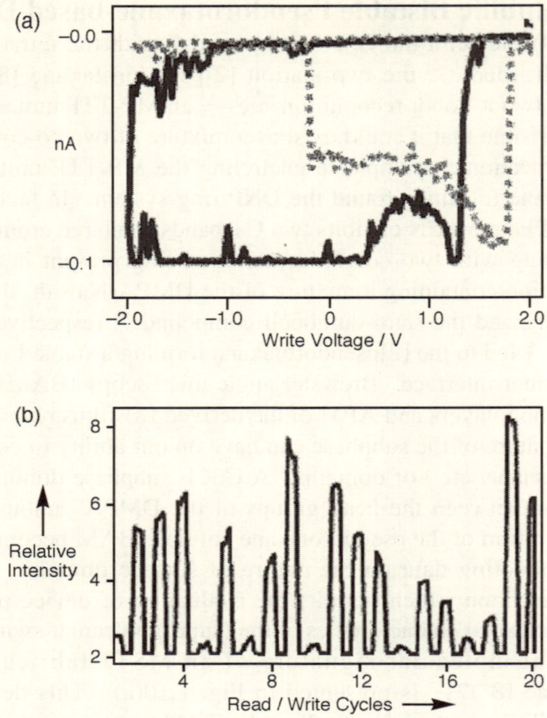


**Fig. 1.10** (a) The remnant molecular signature of the [2]pseudorotaxane [8⊃7]$^{4+}$ device which was monitored at +0.1 V. This device exhibited large-amplitude switching signatures that fluctuated from measurement to measurement. Only two representative measurements (solid and dashed lines) are plotted. (b) The switching cycles of this device.

[2]rotaxane **6**•4PF$_6$. This supramolecularly assembled device demonstrated a markedly more dramatic remnant molecular signature response than did either the catenane-based (Section 1.1.2) or the single-station rotaxane-based (Section 1.1.3) devices. The switching voltages observed for the pseudorotaxane-based devices were sharp and accompanied by a large change in the amplitudes of the currents passing through the devices. However, the precise voltages required to switch the devices fluctuated from experiment to experiment, affording different remnant molecular signatures – two examples are illustrated by the solid black and gray dashed plots in Fig. 1.10(a). This fickleness of device performance was attributed to domains present in the supramolecular monolayers. Moreover, in device-cycling experiments (Fig. 1.10(b)), the change in the current signal passing through the device was quite erratic, being higher by up to 10-fold for some cycles. Cycle-to-cycle reproducibility was also poor. These erratic device characteristics were considered to be caused by the presence of domains which were observed during microscopic investigations of the (supra)molecular films. The

lack of a sizable hydrophilic stopper in the dumbbell-shaped portion of the pseudorotaxane-based devices is possibly the root cause of the formation of domains during the self-organization process.

### 1.1.5 A Slow Amphiphilic Bistable Rotaxane-based Device

The preliminary results obtained with the pseudorotaxane-based devices encouraged us to investigate rotaxane-based devices in some considerable detail. Taking into account the potential ease of synthesis, we began this exercise by targeting the bistable amphiphilic [2]rotaxane **9**•4PF$_6$, shown as its two translational isomers in Fig. 1.11. The rotaxane was synthesized[23)] by using the dumbbell-shaped compound **10** as the template for the formation of the, by now, familiar tetracationic cyclophane from its usual precursor, namely, **1**•2PF$_6$ and the dibromide **2**. The molecular shuttle was isolated as an analytically pure brown solid after column chromatography on silica gel using a solution of $NH_4PF_6$ in $Me_2CO$ as the eluent. Spectroscopic studies indicated the presence of approximately equal amounts of the two stable translational isomers – **9**•4PF$_6$-Green and **9**•4PF$_6$-Red – in the isolated product. The visible absorption spectrum of the [2]rotaxane, recorded in $Me_2CO$ solution, showed a broad CT band centered on 805 nm, characteristic of a co-conformation (**9**•4PF$_6$-Green) containing the TTF unit located inside the cavity of the tetracationic cyclophane, as well as a CT band observed as a shoulder at 540 nm, which results from the DNP ring system encircled by the tetracationic cyclophane, in the other co-conformation (**9**•4PF$_6$-Red). The $^1$H NMR spectrum, recorded at 298 K in $(CD_3)_2CO$, also indicated the presence of two translational isomers which are, however, interconverting slowly on the $^1$H NMR timescale. The ratio of the two translational isomers was estimated by the integration of their two anisochronous SMe resonances to be approximately 1 : 1. We were able to isolate the translational isomer (**9**•4PF$_6$-Red) in which the tetracationic cyclophane encircles the DNP ring system by using preparative thin layer chromatography. If a solution of this red translational isomer is allowed to stand at room temperature for one day, the “original” brown solution containing a mixture of the two translational isomers is re-established. From a first order kinetic analysis of the equilibration process, a rate constant of $2 \times 10^{-5}$ s$^{-1}$ was obtained for the slow shuttling of the tetracationic cyclophane between the two recognition sites at room temperature. The corresponding free energy of activation for this isomerization is 24 kcal mol$^{-1}$. The MP-TTF unit plays a primary role in determining this energy barrier since the placement and bulk of the SMe group severely hinders the cyclophane from shuttling back and forth between the two recognition sites. Despite its sluggish performance as a molecular shuttle in solution, this amphiphilic, bistable [2]rotaxane was self-organized as LB monolayers which were subsequently incorporated into devices. Inferior remnant molecular signatures were obtained (e.g., Fig. 1.12(a)) for this rotaxane in devices similar to those previously described in Section 1.1.2 for the bistable catenane. We attributed this poor device performance to the slow switching of the rotaxane between its stable and metastable states – an explanation which is wholly consistent with the solution-state observa-

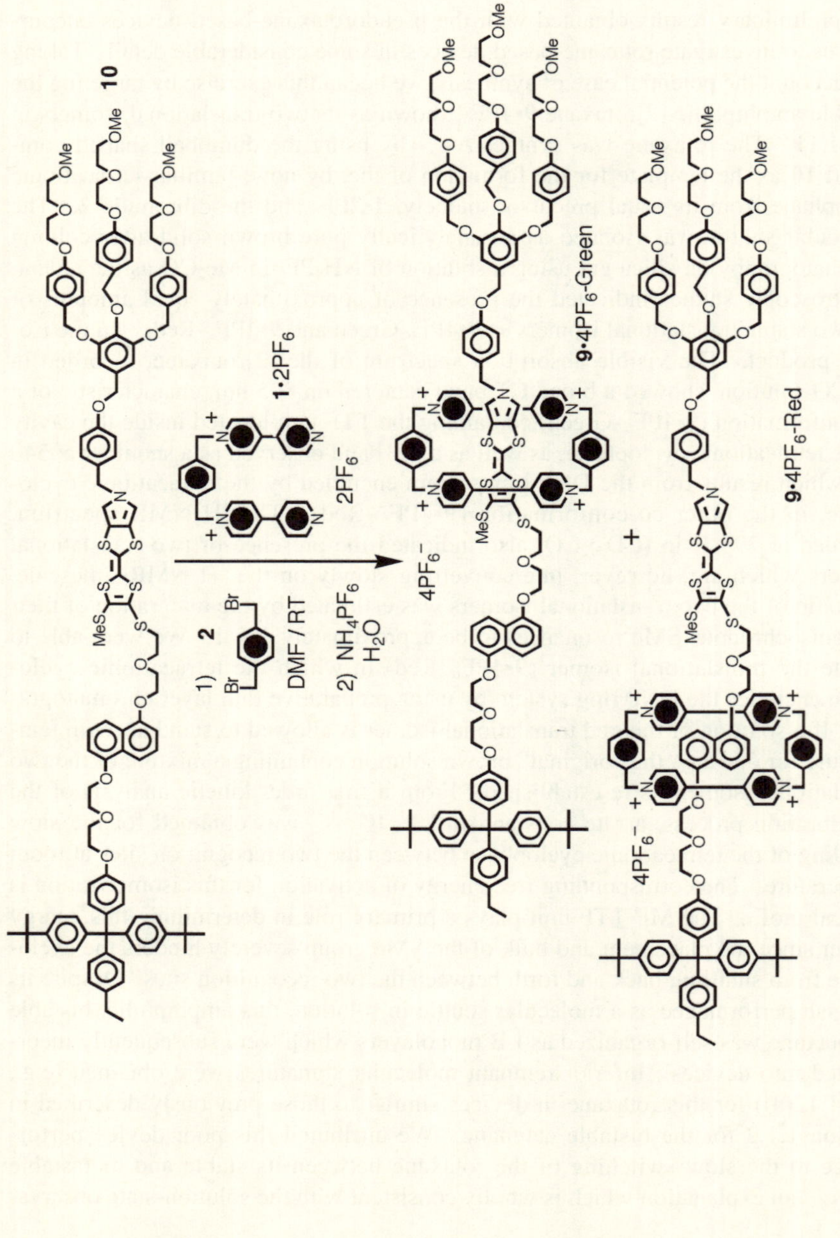


**Fig. 1.11** The template-directed synthesis of the [2]rotaxanes $9{\bullet}4PF_6$-Green and $9{\bullet}4PF_6$-Red.

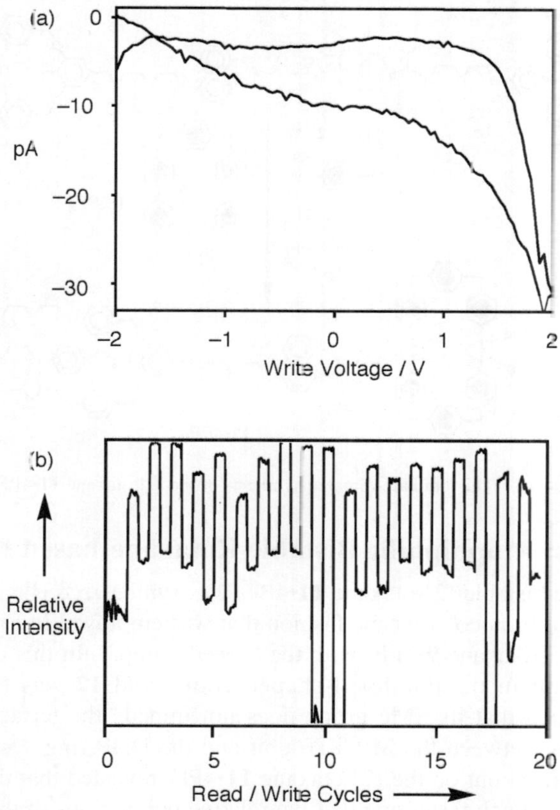


**Fig. 1.12** (a) The remnant molecular signature of the [2]rotaxane **9**•$4PF_6$ device which was monitored at + 0.1 V. (b) The switching cycles of this device.

tions. The very fact that the device performance was not up to par was not only to be anticipated but it also constituted yet another circumstantial demonstration of how the solution- and solid-state properties of a particular electromechanical switch go hand-in-hand in large measure. The far short of exemplary performance exhibited by the slow amphiphilic, bistable rotaxane-based switch in a device setting pointed us in the direction we were to take next. It seemed obvious that the energy barrier to the shuttling process had to be lowered so that the switching between the stable and metastable states of the switch would be relatively facile. The [2]rotaxane **11**•$4PF_6$, shown in Fig1.13, was designed to meet this requirement.

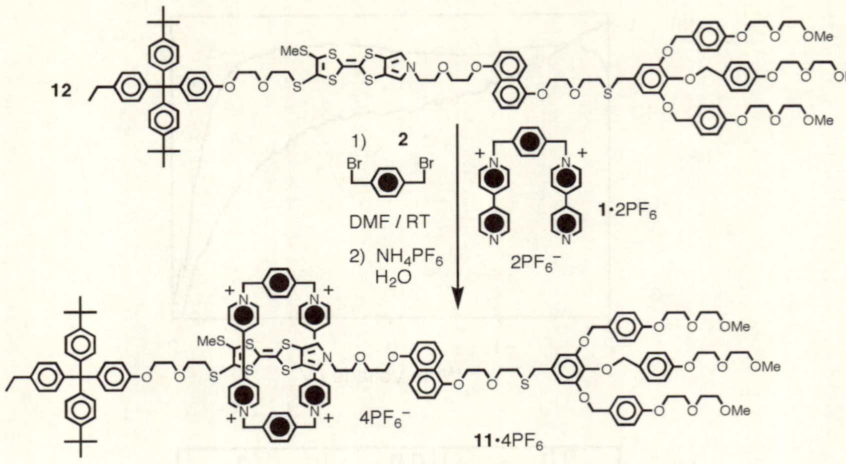


**Fig. 1.13** The template-directed synthesis of the [2]rotaxane **11**•4$PF_6$.

### 1.1.6 A Fast Amphiphilic Bistable Rotaxane-based Device

The amphiphilic bistable [2]rotaxane **11**•4$PF_6$ was synthesized (Fig. 1.13) using a template-directed protocol after the fashion that was employed in Fig. 1.12 for the preparation of [2]rotaxane **9**•4$PF_6$ with the "speed bump." In this case, however, the MP-TTF unit in the dumbbell-shaped compound **12** was turned around through 180° such that its SMe group does not impede the tetracationic cyclophane's shuttling between the MP-TTF unit and the DNP ring system. $^1$H NMR spectroscopy carried out on the [2]rotaxane **11**•4$PF_6$ revealed that the shuttling of the cyclophane is much faster and that the relative populations of the translational isomers are highly temperature dependent. At low temperatures (230 − 290 K), the **11**•4$PF_6$-Red isomer is the major co-conformation, while the **11**•4$PF_6$-Green isomer is the major co-conformation at high temperatures (320 − 390 K). These observations are also supported by UV/Vis spectroscopy: at room temperature in MeCN, absorption bands for both co-conformations centered around 540 and 795 nm were observed.

Just like the slow bistable [2]rotaxane **9**•4$PF_6$, this fast one forms stable Langmuir films. This [2]rotaxane molecule has a much larger footprint (140 Å$^2$ as compared to 40 Å$^2$ for the related [2]pseudorotaxane [**8**⊃**7**]$^{4+}$) as a result of incorporating a large dendron as the hydrophilic stopper in its dumbbell-shaped component. This difference means that there is more area per molecule to accommodate the molecular reorganizations that occur during switching. No domains were detected using either BAM or a range of scanning probe microscopies. This [2]rotaxane **11**•4$PF_6$ was self-organized by LB techniques and then transferred onto a $SiO_2$ substrate patterned with n-type polysilicon electrodes. The top electrodes were the same as before, namely, Ti supported on top by Al. MSTJs based on this amphiphilic, bistable rotaxane exhibited very similar device properties

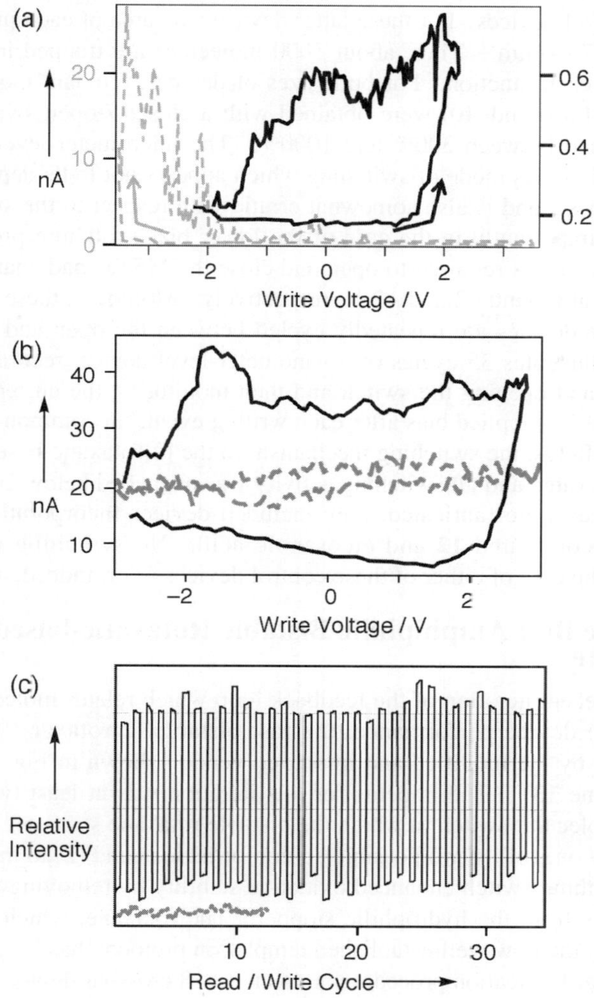


**Fig. 1.14** (a) The remnant molecular signature of a micrometer-scale [2]rotaxane **11**•4$PF_6$ device (solid line) and its dumbbell analogue device (dashed line), which were both monitored at +0.1 V. (b) The remnant molecular signature of a nanometer-scale [2]rotaxane **11**•4$PF_6$ device (solid line) and an eicosanoic acid control device (dashed line). (c) 35 Cycles of the nanometer-scale device.

(Fig. 1.14) as the devices were scaled down from micrometer- to nanometer-scale junctions. A switch-closed current level of approximately 1 nA (at +0.1 V) was observed in micrometer-level devices, but was reduced to 50 pA (at +0.1 V) in nanometer-level devices. For these latter devices, the area of each junction was in the range of 7000 $nm^2$. Thus, about 5000 molecules are trapped in between the electrodes at each junction. For both sizes of devices, remnant molecular signatures (Fig. 1.14(a) and (b)) were obtained with a closed / open switching amplitude that varied between 300% and 1000%. The micrometer-level devices also exhibited a high bias mode of switching which appears not to be dependent on the molecules per se, and is also somewhat erratic with respect to the switching voltages. More importantly in the case of the lower bias switching process was the fact that the voltages required to open and close the MSTJs had characteristic and stable values at around –2 and +2 V, respectively. Moreover, these values do not change as the devices are repeatedly cycled between the open and closed states. Fig 1.14(c) illustrates 35 cycles of a nanometer-level device, recorded by repeatedly closing and opening the switch and then monitoring the current through the junction at +0.1 V applied bias after each writing event. In common with the catenane-based MSTJs, the switching mechanism in the [2]rotaxane-based MSTJs was thermally activated and all switching activity was quenched below 260 K. Several control devices were fabricated: they included devices incorporating the dumbbell-shaped compound **12** and eicosanoic acid. No switching response was observed in the case of either of these control devices – or, indeed, any others!

### 1.1.7 The Best Amphiphilic Bistable Rotaxane-based Device to Date

In the most recent iteration of the feedback loop which relates molecular structure to solid-state device performance, the fast bistable [2]rotaxane $\mathbf{11}^{4+}$ has been superseded[24)] by the modified and improved version shown in Fig. 1.15, namely, the [2]rotaxane $\mathbf{13}^{4+}$. The impetus for this iteration was at least two-fold: (1) to produce a molecular switch in which the bistable rotaxane starts off exclusively in one, and only one, of its two possible co-conformations and (2) to incorporate into this all-or-nothing switch enhanced oxidative stability by removing all of the phenolic residues from the hydrophilic stopper. The rotaxane, which was obtained (Fig. 1.15) by the now well-established templation protocol, has been incorporated by the familiar fabrication procedures into an 8 × 8 crossbar device (Fig. 1.16(a)). Fig. 1.16(b) demonstrates one of these 64-bit devices in operation with the exception of one of the wires, i.e., 56 out of the 64 bits can be switched. This device has been used to store (Fig. 1.16(c)) the letters DARPA in ASCII code: the lifetime of the switch-closed state of this devices is about 10 minutes. The proposed electromechanical switching mechanism for the bistable [2]rotaxane $\mathbf{13}^{4+}$, which is outlined in Fig. 1.17, merits comparison with that already discussed in Fig. 1.16 for the bistable [2]rotaxane $\mathbf{4}^{4+}$. In the case of $\mathbf{13}^{4+}$, it is proposed that oxidation of the TTF unit to its radical cation ($TTF^{+\bullet}$) occurs when a +2 V bias is applied across the bistable [2]rotaxane and the tetracationic cyclophane moves to encircle the

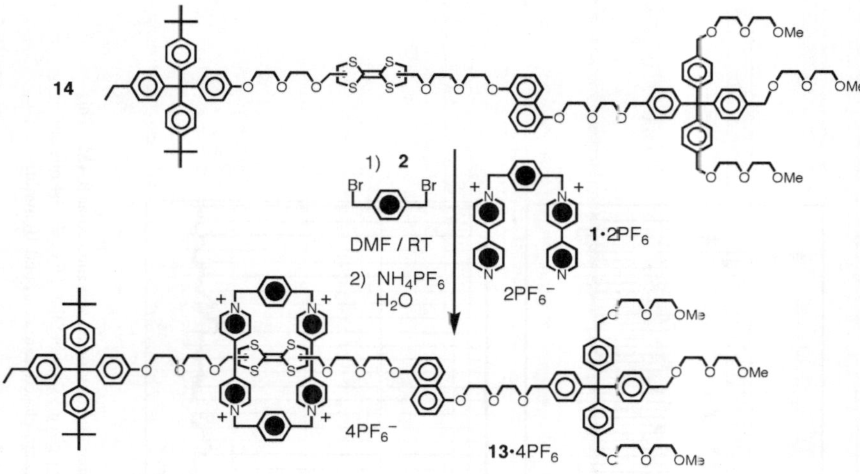


**Fig. 1.15** The template-directed synthesis of the [2]rotaxane **13•4PF$_6$**.

DNP ring system. When the bias is removed, however, and the TTF $^{+\bullet}$ is restored to its neutral state (TTF), the tetracationic cyclophane continues to reside on the DNP ring system. The cyclophane is subject to thermal activation and is also assisted in its return to the TTF unit by the reduction of bipyridinium units for a fleeting moment during the switching cycle. The 8 × 8 crossbar device is typically composed of bottom polysilicon electrodes (100 nm wide) and top Ti/Al electrodes (70 nm wide). Thus, each MSTJ has an area of 7000 nm$^2$ and so contains around 5000 molecules. A scanning electron micrograph (SEM) image of part of such a circuit is illustrated in Fig. 1.18.

## 1.1.8 Concluding Remarks

We have attempted to describe how an integrated systems-oriented approach to molecular electronics evolved quickly during no more than about one year and how it has continued to develop subsequently. It was essential to reach the stage when everything in relation to design of two-dimensional molecular electronic circuits was being done simultaneously, i.e., (1) the chemical design and synthesis of molecular switches, (2) their self-organization in devices, that (3) were being fabricated into molecular crossbar circuits which (4) can function as memory as well as demultiplexer/multiplexer logic in combination with memory. A feedback loop between device performance and molecular switching characteristics has already had a huge influence on the design and synthesis of amphiphilic bistable rotaxanes. In turn, their device performance has raised fundamental questions, such as just how the mechanically interlocked molecules switch in relation to their remnant molecular signatures and how important the interfaces between solid-state devices and the molecules they incorporate are. The electromechanical mech-

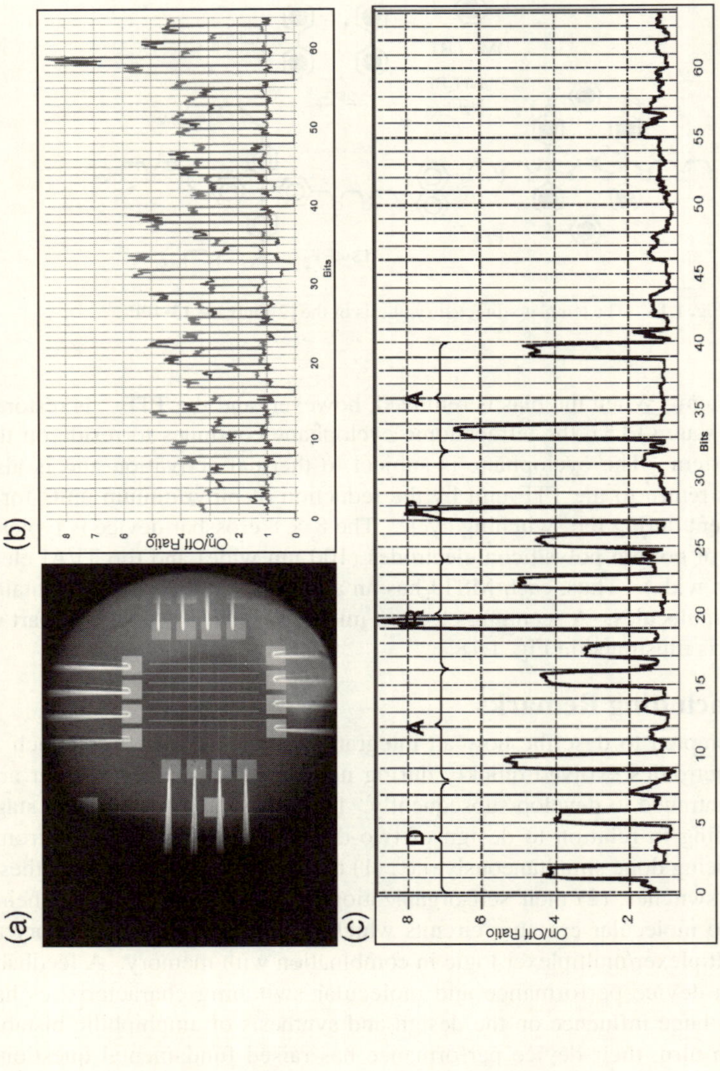


**Fig. 1.16** (a) The demonstration of point addressablity within a 64-bit molecular switch crossbar circuit utilized as a molecular RAM. This device was fabricated using the [2]rotaxane **13**•4$PF_6$. (b) An 8 × 8 crossbar memory circuit showing that 56 of the 64 bits are working well. (c) DARPA is written into the 64-bit memory. The half-life of the switch-closed state is around 10 minutes.

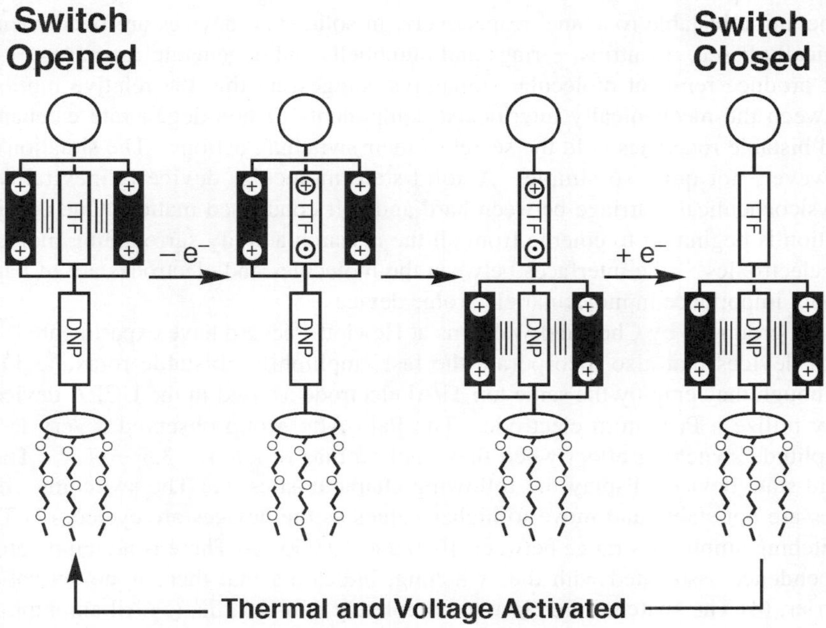


**Fig. 1.17** Proposed electromechanical mechanism for the operation of the device based on the amphiphilic, bistable [2]rotaxane **13**$^{4+}$.

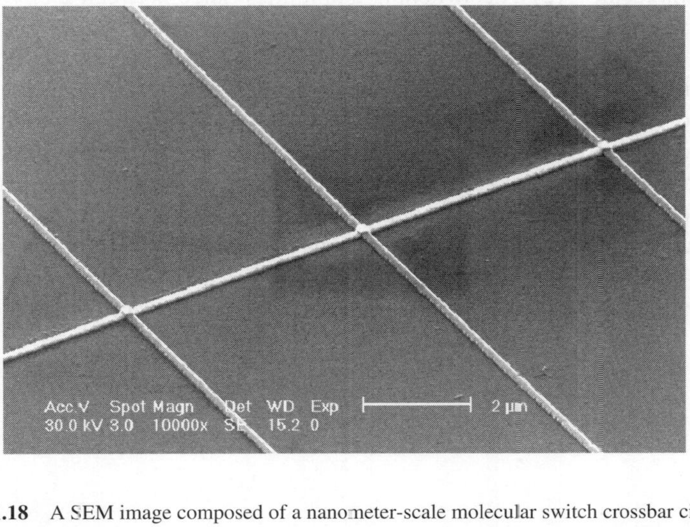


**Fig. 1.18** A SEM image composed of a nanometer-scale molecular switch crossbar circuit.

anisms proposed in Fig. 1.16 and 1.17 for the switching of a non-degenerate catenane and a bistable rotaxane, respectively, in solid-state devices are phenomenologically based. Controls – rings and dumbbells and degenerate catenanes – do not produce remnant molecular signatures, suggesting that the relative motions between the mechanically interlocked components of non-degenerate catenanes and bistable rotaxanes hold the secret of their switching actions. The situation is, however, not quite so simple. A solid-state molecular device is inevitably a physicochemical marriage between hard and soft condensed matter. One consideration is beginning to emerge from all the research activity surrounding molecular electronics – the interfaces between the molecules and electrodes are of paramount importance in molecular electronic devices.

A group led by Chen and Williams at Hewlett-Packard have experimented[25,26] with devices that also incorporate the fast amphiphilic, bistable rotaxane $\mathbf{11}^{4+}$. Although they employ the same top Ti/Al electrode as used in the UCLA devices, they utilize a Pt bottom electrode. The Palo Alto group observed a very large amplitude switching effect when they apply a bias of *ca.* +/– 3.5 – 7 V. Their solid-state devices display the following characteristics. (1) The switching voltages are not stable and move to higher values as the devices are cycled. (2) The switching amplitudes range between 1000 and 10,000. (3) There is no temperature dependence associated with the switching, indicating that there is no activation barrier. (4) The switches are close to completely non-volatile. (5) All amphiphilic molecules – including degenerate catenanes, dumbbell components of the rotaxanes and even simple amphiphiles – switch. By contrast, the devices fabricated by the UCLA group have the following characteristics. (1) The switching voltages, which are +/–2 V and are very stable, do not change from device to device as they are scaled down to small crossbars where the MSTJs contain only a few thousands of molecules. (2) The switching amplitudes range between 2 to 10. (3) The switches are thermally activated and no switch operates below 200 K. (4) The switches are quite volatile, relaxing from the closed to open state in a period of 10–60 minutes. (5) Only bistable, mechanically interlocked molecules switch. No controls switch. If higher bias voltages (+/– 3–4 V) are applied to the devices fabricated by the UCLA group, they observe a switching signature that is very similar to that seen by the Hewlett-Packard group – unstable write voltages, high amplitudes, no temperature dependences, non-volatile, etc. It is very unlikely that this signature at higher bias is molecular in origin. The difference between low and high bias switching in these various devices is a subject of intensive on-going research – as are the effects of the interfaces between the electrodes and the molecules.

## Acknowledgements

The research described in this article was supported primarily by the Defense Advanced Research Projects Agency (DARPA) with partial support from the National Science Foundation (NSF). P.C.C. thanks Pfizer for the Summer Undergraduate Research Fellowship in Synthetic Organic Chemistry. We are indebted to all our many collaborators, whose names appear along with ours in some of the cited literature, for their contributions to this research program during the past decade.

# 1.2 Supramolecular Polymers Containing Cyclodextrin

**Akira Harada**

## 1.2.1 Introduction

Recently, much attention has been focused on *nano*-science and *nano*-technology, construction of *nanometer*-scale architectures from bulk materials by top-down engineering. Advances in VLSI (very large-scale integrated systems) production techniques (lithography) have moved processing down from the micrometer to the sub-micrometer level. Ever smaller architectures have been required for the construction of new materials and devices. However, there is a limit to top-down approaches for many reasons including the interference of light waves for lithography. Researchers exploit new ways to fabricate smaller architectures. Hence, the bottom-up procedure from a molecule to create new architecture has become an important approach. This approach has already been utilized by nature. Most of the structures of biological systems have been made by self-assembly and self-organization of specific molecules.

Construction of supramolecular structures using specific intermolecular interactions is one of the most important procedures for the construction of *nano*-scale architectures by bottom-up procedures.[27)] Recently, with the advent of host-guest chemistry and chemistry of molecular recognition, supramolecular science has made its appearance.

Rotaxane (rotor and axle) and catenanes (chains) are important supramolecular structures.[28)] More recently, supramolecular polymer science has emerged. Although synthetic polymers are now ubiquitous, the rise of supramolecular chemistry has spurred investigations into alternative approaches for polymer synthesis. Whereas molecular polymerization relies upon the formation of covalent bonds between monomeric building blocks, the propagation step in a supramolecular polymerization proceeds via the formation of non-covalent bonds. When a guest part is attached to a cyclic host component, the conjugate may form intramolecular complexes or intermolecular complexes to give supramolecular polymers.

Recently, much attention has been focused on the design and synthesis of intrerlocked molecules, such as rotaxanes and catenanes, because of their unique structures and properties.[29)] Host-guest interactions are used for the efficient preparation of such interlocked molecules. Previously, we found and reported that cyclodextrins formed inclusion complexes to give crystalline compounds with high selectivity.[27] We also prepared polyrotaxanes in which many cyclic molecules are imprisoned in a single molecule by capping the complexes with bulky substituents.[49] More recently, we found that cyclodextrins formed complexes not only with non-ionic polymers but also with ionic polymers to give water-soluble complexes.[51]

We have prepared a molecular shuttle in which a cyclodextrin ring moves back and forth along a polymer chain by capping chain ends with bulky substituents. We also found that cationic groups in the main chain of the polymer serve as a potential surface. Using this result we prepared a rotaxane using cationic groups as stoppers.

When a guest group is covalently attached to a cyclic host, the molecule may form an intramolecular complex[30] or intermolecular complexes to give supramolecular polymers.[31] When supramolecular polymers are treated with bulky stopper groups, they may form poly[2]rotaxanes.[32] We chose cyclodextrin as a cyclic host and a phenyl group as a guest moiety because a phenyl group is suitable for fitting in a cyclodextrin cavity. However, benzoyl cyclodextrin did not form supramolecular polymers.[33] This suggests that some spacer groups are required for the efficient formation of intermolecular complexes. Therefore, we tested a hydrocinnamoyl group as a guest moiety. We have also tested cinnamoyl derivatives as a guest moiety. We found that an α-cyclodextrin derivative which has a cinnamoyl group as a guest moiety on the 6-position of cyclodextrin forms oligomeric supramolecular structures in aqueous solutions and that a β-cyclodextrin derivative which has a cinnamoyl group as a guest on the 6-position of cyclodextrin forms a supramolecular structure in the solid state.

## 1.2.2 Cyclodextrins

Cyclodextrins are cyclic compounds consisting of six to eight glucose units. They are called α-, β-, and γ-cyclodextrin, respectively (Fig. 1.19). They are known to form inclusion complexes with various low molecular weight compounds, ranging from non-polar aliphatic molecules to polar amines and acids.[34] Recently, they attracted renewed interest because they can be used as cyclic components for the construction of various supramolecular architectures.[35] We consider these cyclodextrin molecules as promising molecular components for the construction of molecular-based devices, because they are able to be threaded onto a long axle and to slide along a chain or to rotate around an axle, because they are basically photochemically and electrochemically inactive.

Cyclodextrins, for example, form inclusion complexes with metallocenes such as ferrocene, cobaltocene and nickelocene to form crystalline compounds.[36] The structures of the complexes are dependent on the size of the cyclodextrin. Although ferrocene and its derivatives are strongly bound in uncharged states, when they are oxidized, the complexes dissociate. Dendrimers containing ferrocene units at the ends of the molecule have been prepared The dendrimers form large supramolecular structures that can be broken apart or assembled upon

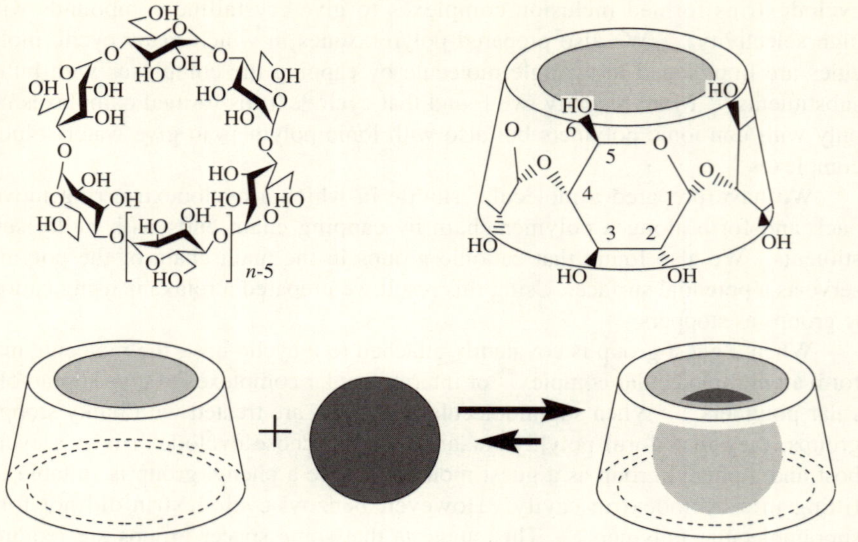

**Fig. 1.19** Cyclodextrins: α-cyclodextrin ($n$=6), β-cyclodextrin ($n$=7), and γ-cyclodextrin ($n$=8). Formation of the inclusion complex of cyclodextrin with a small guest molecule.

oxidation of the ferrocene units.[37] The opposite behavior has been found in the complex of cyclodextrins with cobaltocene. Cobaltocenium derivatives do not form inclusion complexes with cyclodextrins. However, they form complexes with β-cyclodextrin upon one-electron reduction to give the neutral cobaltocene.[38] The dendrimers containing cobaltocenium units on the ends of the molecule did not form complexes with cyclodextrins, but upon reduction of the cobaltocenium units, the dendrimers gave large complexes with cyclodextrins. Although compounds containing bipyridinium in the dicationic form are not included in the β-cyclodextrin cavity, they form weak complexes with β-cyclodextrin when they are reduced to their monocationic forms, and form stable pseudorotaxane with β-cyclodextrin when they are fully reduced to their uncharged forms.[39] They are examples of electrochemically driven molecular complexes.

## 1.2.3 Rotaxanes

Rotaxanes are considered to be one typical prototype of molecular devices machines, because they have a rooter and an axle in the molecule. α-Cyclodextrin was first used as a rotor of rotaxane by Ogino *et al.* in 1978.[40] They used metal complexes as stoppers. Since then some rotaxanes containing cyclodextrins have been reported (Fig. 1.20).[41-46] Most of them used metal complexes as stopper groups. However, the metal coordination bonds are so weak that cyclodextrin rings may escape from the bond. All the rotaxanes reported are ionic so as to solubilize the rotaxanes in water, except one example in which all the components are non-ionic and soluble in organic solvents.[47] In this case there are some inter-

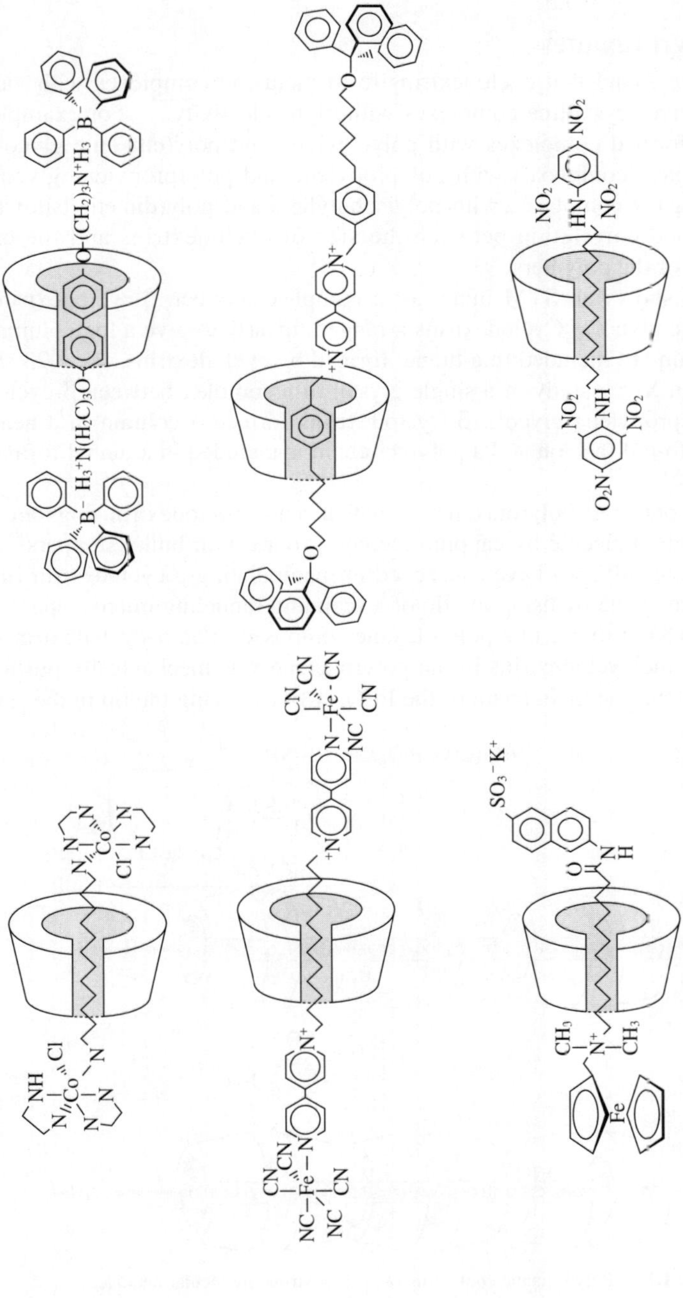

**Fig. 1.20** [2]Rotaxane containing cyclodextrin.[43–48]

actions between a cyclodextrin ring and stopper groups. In all cases, there is no clear evidence of a rotatory motion of cyclodextrins around an axle.

### 1.2.4 Polyrotaxanes

Previously, we found that cyclodextrins form inclusion complexes with various polymers to give crystalline complexes with high selectivity.[48] For example, α-cyclodextrin formed complexes with polyethylene and poly(ethylene glycol), β-cyclodextrin gave complexes with polypropylene and poly(propylene glycol). γ-Cyclodextrin gave complexes with polyisobutylene and poly(dimethylsiloxane)s. There is a good correlation between the size of cyclodextrins and the cross-sectional area of the polymers.

Fig. 1.22 shows an STM image of a complex between β-cyclodextrin and poly(propylene glycol). Cyclodextrins arranged linearly to give a log column and a polymer chain are included in a tunnel formed by cyclodextrins. Fig.1.23 shows the result of an X-ray study of a single crystal of a complex between β-cyclodextrin and poly(propylene glycol). β-Cyclodextrins formed a column in a head-to-head and tail-to-tail fashion and a polymer chain is included in a tunnel formed by cyclodextrins.

We have prepared polyrotaxanes in which many cyclodextrin rings are captured in a single molecule by capping the end groups with bulky stoppers[49] (Fig. 1.21). More recently, we have succeeded in manipulating a cyclodextrin ring or two in a polyrotaxane by using the tip of a scanning tunneling microscope.[50] Fig. 1.24 shows an STM image of a polyrotaxane composed of an α-cyclodextrin and a PEG. One of the cyclodextrins in the polyrotaxane was mechanically pushed by the STM tip along the main chain of the PEG. Upon moving the tip in the reverse

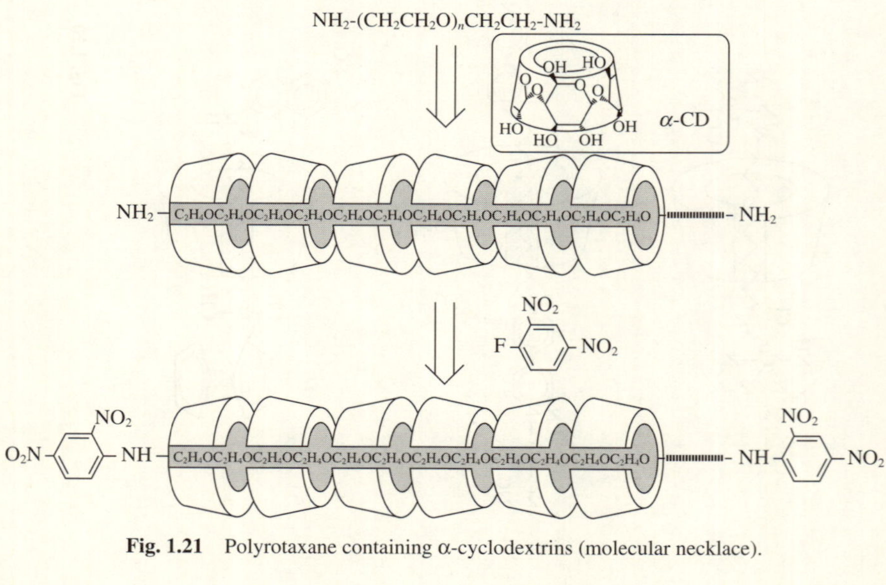


**Fig. 1.21** Polyrotaxane containing α-cyclodextrins (molecular necklace).

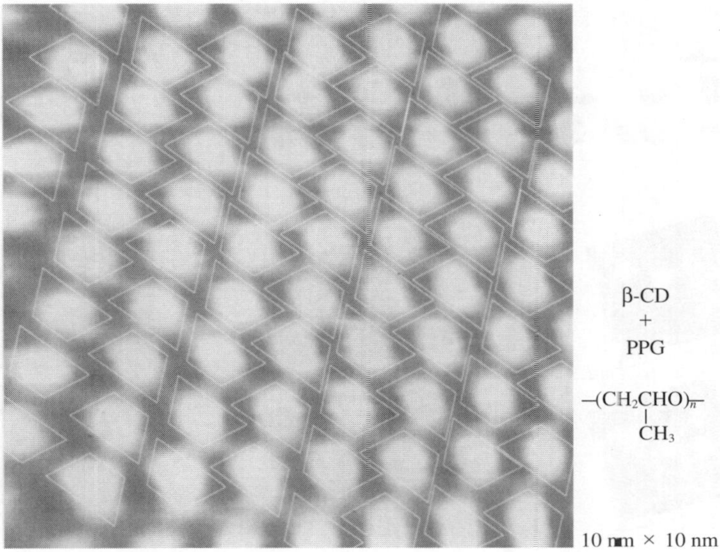


**Fig. 1.22** STM image of the complex between β-cyclodextrin and poly(propylene glycol).

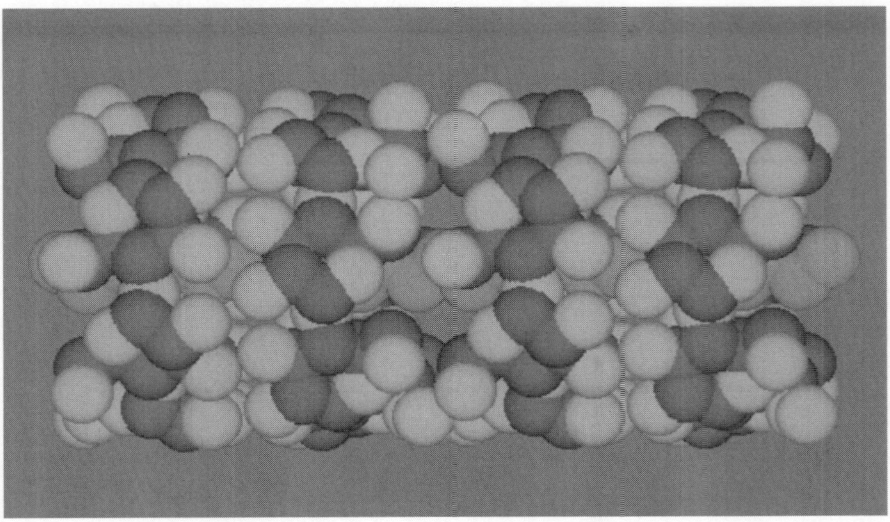

**Fig. 1.23** X-ray studies on the complex between β-cyclodextrin and poly(propylene glycol).

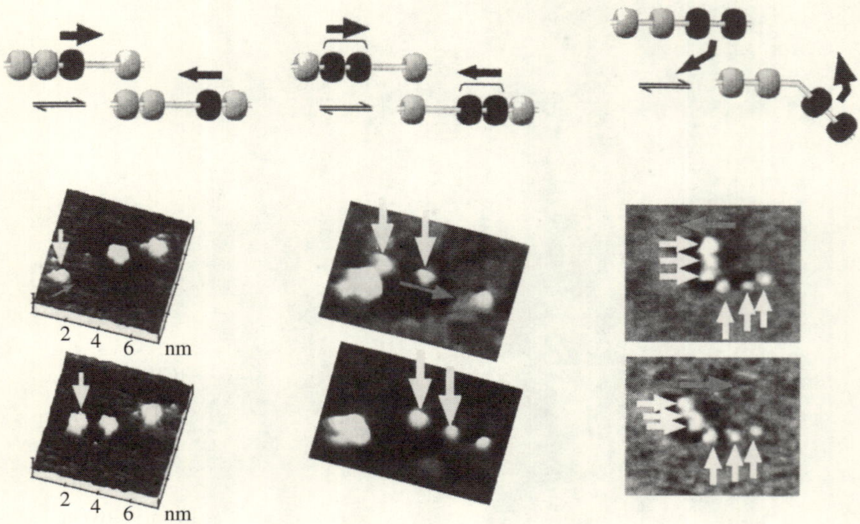


**Fig. 1.24** Molecular abacus, a cyclodextrin bead (or two) can be manipulated along a polymer chain by means of STM (scanning tunneling micrography).

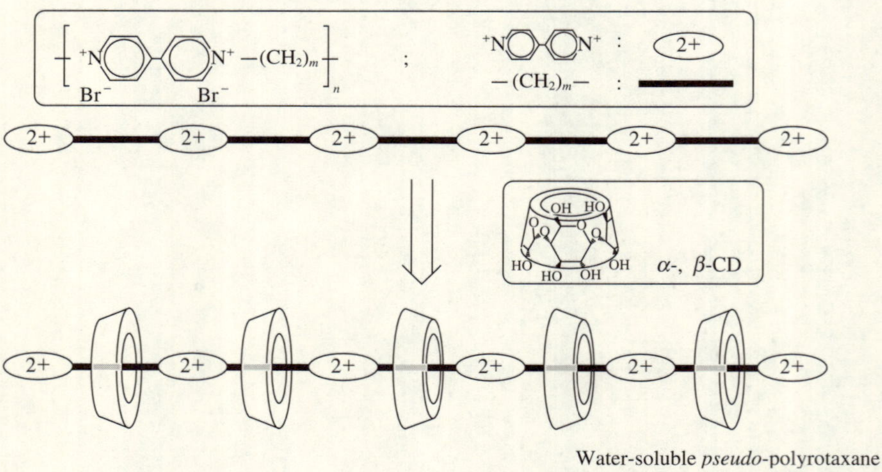


**Fig. 1.25** Complex formation of cyclodextrin with cationic polymers in an aqueous solution.

direction, the α-cyclodextrin retraced its path and returned to its original position. Thus, the shuttling of one α-cyclodextrin could be repeated. It was also possible to move a pair of cyclodextrins simultaneously.

More recently, we found that cyclodextrins form complexes not only with non-ionic polymers but also with ionic polymers such as linear polymers consisting of bipyridinium (viologen) bridged by polymethylene chains.[51)] Although these polymers do not give crystalline complexes with any cyclodextrins at all (Fig. 1.25), cyclodextrins form soluble complexes with these polymers in the aqueous phase.[51)] We found that the methylene peaks of the polymer in the $^1$H NMR spectra split in two upon the addition of α-cyclodextrin, although these peaks did not change upon the addition of γ-cyclodextrin. These peaks are broadened upon the addition of β-cyclodextrin. these results indicate that α-cyclodextrin remains at one of the methylene chains on the $^1$H NMR time scale, and β-cyclodextrin moves along the polymer chain faster than the $^1$H NMR time scale.

Based on these observations, we decided to make a molecular shuttle in which a cyclodextrin moves back and forth along a chain.

## 1.2.5 Molecular Shuttle

We have designed a molecular shuttle as follows: α-cyclodextrin as a ring, two polymethylene chains as stations, 4,4'-bipyridinum units as linkers, and dinitrophenyl groups as stoppers (Fig. 1.26). Shuttling behavior of the molecular shuttle is solvent- and temperature-sensitive and could be controlled by double interactions: hydrophobic interaction between a cyclodextrin ring and a station and a repulsive interaction between a cyclodextrin ring and a linker. A cyclodextrtin

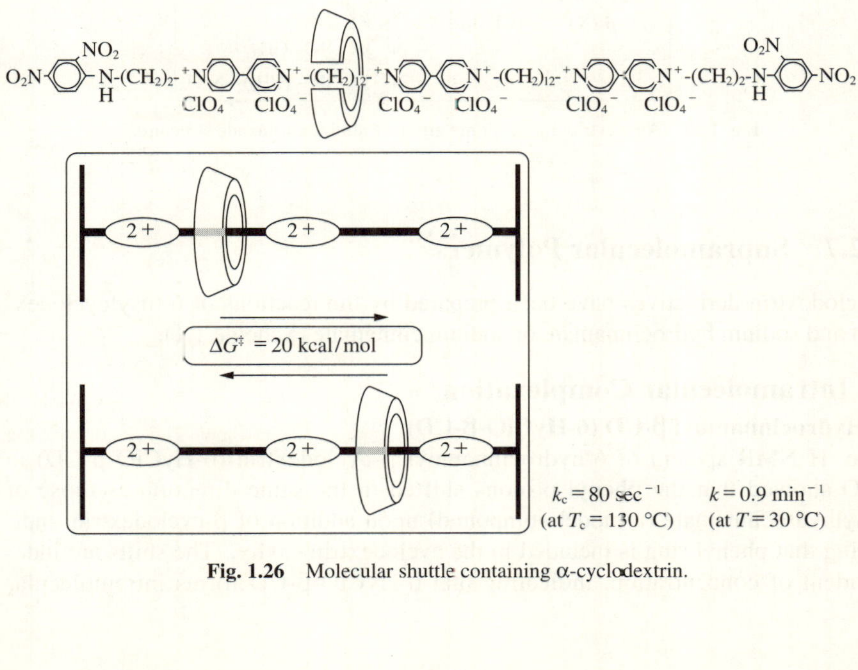


Fig. 1.26 Molecular shuttle containing α-cyclodextrin.

ring moves back and forth about 90 times/sec$^{-1}$ at 130 °C in DMSO and 0.8 times/sec$^{-1}$ at room temperature.[52] This is a new method to control the mobility of a bead in a molecular shuttle.

## 1.2.6 Electronic Trap

While we were preparing the molecular shuttle, we found that cationic groups serve as a potential surface for the translocation of a cyclodextrin ring on a polymer chain. Therefore, we studied the interactions between cyclodextrins and some dodecamethylene derivatives with different numbers of cationic groups at the chain end. It is difficult for a cyclodextrin ring to escape from the axle due to electrostatic repulsion between the cyclodextrin ring and the cationic groups (Fig.1.27). Multiple cationic groups stabilize the rotaxane structure by inhibiting the cyclodextrin ring from coming off.[53] Bulky stoppers are not necessarily required for the construction of rotaxanes.

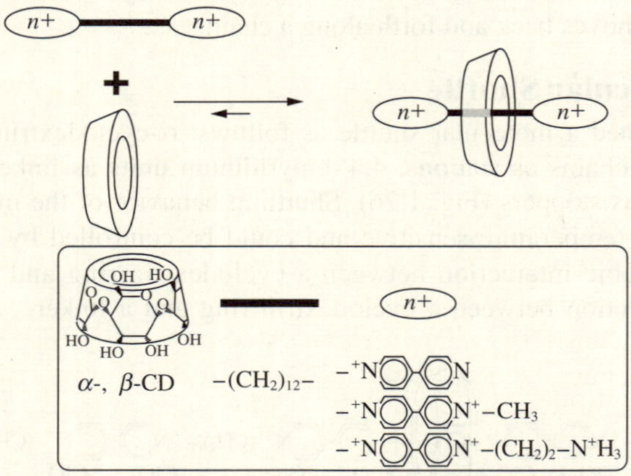


**Fig. 1.27** An electric trap: cationic groups stabilize a rotaxane structure.

## 1.2.7 Supramolecular Polymers[54]

Cyclodextrin derivatives have been prepared by the reactions of 6-tosylcyclodextrin and sodium hydrocinnamate or sodium cinnamate (Scheme 1.1).

## A. Intramolecular Complexation

### 6-Hydrocinnamoyl β-CD (6-HyCiO-β-CD)

The $^{1}$H NMR spectra of 6-hydrocinnamoyl β-cyclodextrin (6-HyCiO-β-CD) in $D_2O$ showed that the phenyl protons shifted in the same direction as those of ethylhydrocinnamate (a model compound) upon addition of β-cyclodextrin, indicating that phenyl ring is included in the cyclodextrin cavity. The shifts are independent of concentration, indicating that 6-HyCiO-β-CD forms intramolecular

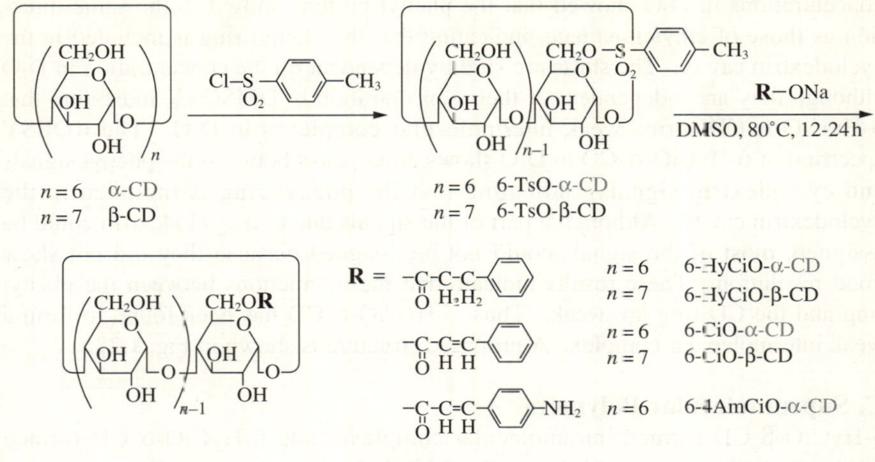

Scheme 1.1

complexes in $D_2O$. The ROESY spectrum of that 6-HyCiO-β-CD shows cross peaks between the phenyl signals and cyclodextrin signals, indicating that the phenyl ring is included in its own cyclodextrin cavity. All the signals due to β-cyclodextrin in the $^1$H NMR spectrum of 6-HyCiO-β-CD are assigned by measuring various 2D NMR (COSY, TOCSY, ROESY, HMQC). The C(3) and C(5) protons in the cavity shifted. When the glucose unit with a guest moiety is named ring A, the protons of the glucose rings D and E largely shifted, indicating that the phenyl ring is included in the cavity sandwiched by the ring A and rings D and E. A proposed structure of 6-HyCiO-β-CD in water is depicted in Fig. 1.28.

## B. Intermolecular Complexation

### 6-Hydrocinnamoyl α-CD (6-HyCiO-α-CD)

The $^1$H NMR spectra of 6-hydrocinnamamoyl α-CD(6-HyCiO-α-CD) in various

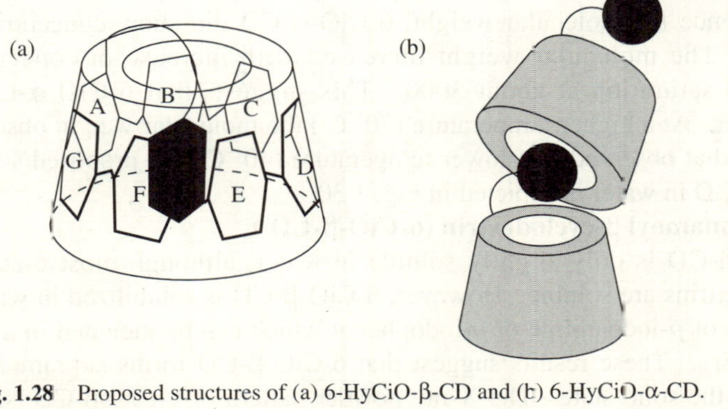


**Fig. 1.28** Proposed structures of (a) 6-HyCiO-β-CD and (b) 6-HyCiO-α-CD.

concentrations in $D_2O$ showed that the phenyl protons shifted in the same direction as those of ethylcinnamate, indicating that the phenyl ring is included in the cyclodextrin cavity. The shifts are slightly dependent on the concentration in $D_2O$ although they are independent of their concentration in DMSO-$d_6$, indicating that 6-HyCiO-α-CD forms weak intermolecular complexes in $D_2O$. The ROESY spectrum of 6-HyCiO-α-CD in $D_2O$ shows cross peaks between the phenyl signals and cyclodextrin signals, indicating that the phenyl ring is included in the cyclodextrin cavity. Although a part of the signals due to α-cyclodextrin could be assigned, most of the signals could not be assigned because they did not show good resolution. These results indicate that the interactions between the phenyl ring and the CD ring are weak. Thus, 6-HyCiO-α-CD has been found to form a weak intermolecular complex. A proposed structure is shown in Fig. 1.28.

## C. Supramolecular Polymers

6-HyCiO-β-CD formed intramolecular complexes and 6-HyCiO-α-CD formed weak intermolecular complexes, indicating that a hydrocinnamoyl group is too flexible to form intermolecular complexes. Therefore, we decided to use a more rigid spacer, such as a cinnamoyl group having a double bond.

**a. 6-Cinnamoyl α-cyclodextrin (6-CiO-α-CD)**

The $^1$H NMR spectra of 6-cinnamamoyl-α-cyclodextrin in various concentrations in $D_2O$ showed that the phenyl protons shifted in the same direction as those of methylcinnamate upon addition of α-cyclodextrin, indicating that the phenyl ring is included in the cyclodextrin cavity. The shifts are dependent on the concentration in $D_2O$, although they are independent of the concentration in DMSO-$d_6$, indicating that 6-CiO-α-CD forms intermolecular complexes in $D_2O$. The ROESY spectrum of 6-CiO-α-CD shows cross peaks between the phenyl signals and cyclodextrin signals, indicating that the phenyl ring is included in the other CD cavity.

The molecular weight of 6-CiO-α-CD was measured by vapor pressure osmometry (VPO) at various concentrations in water. Fig. 1.29 shows the results of the VPO measurements of the cyclodextrin and its derivatives at various concentrations at 40 °C. Although α-cyclodextrin showed no concentration dependence for molecular weight, 6-CiO-α-CD did show concentration dependence. The molecular weight increased with increase in concentration and reached saturation at about 3000. This suggests that 6-CiO-α-CD forms an oligomer. At a higher temperature (70 °C ) the molecular weight observed is lower than that observed at a lower temperature (40 °C). A proposed structure of 6-CiO-α-CD in water is depicted in Fig. 1.30.

**b. 6-Cinnamoyl β-cyclodextrin (6-CiO-β-CD )**

6-CiO-β-CD is only slightly soluble in water, although most 6-substituted β-cyclodextrins are soluble. However, 6-CiO-β-CD is solubilized in water upon the addition of *p*-iodoaniline or *p*-iodophenol which can be included in a β-cyclodextrin cavity. These results suggest that 6-CiO-β-CD forms supramolecular polymers in the solid state. The X-ray powder pattern of 6-CiO-β-CD is very similar

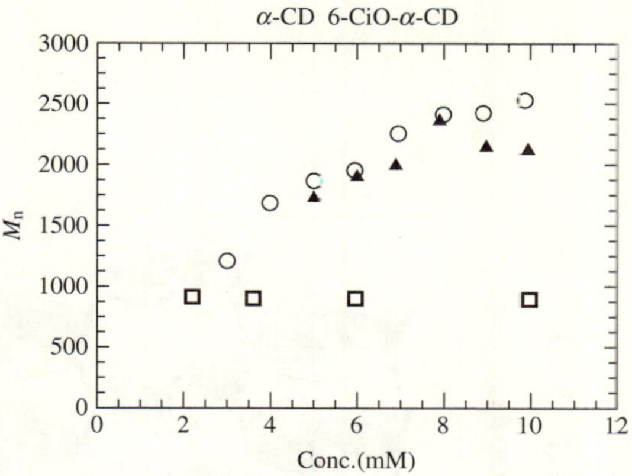


**Fig. 1.29** Concentration dependency of the number-average molecular weight of 6-CiO-α-CD in aqueous solutions measured by VPO.

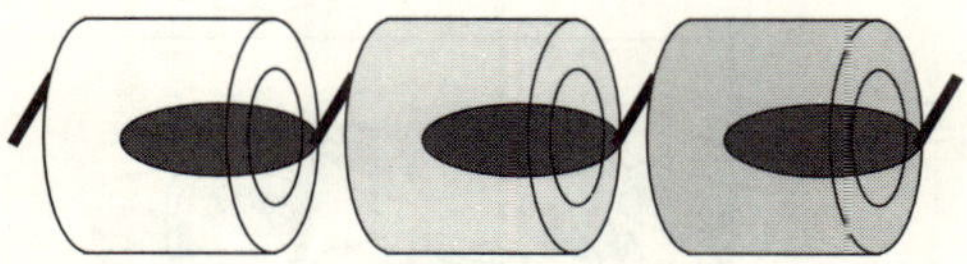

**Fig. 1.30** A proposed structure of 6-CiO-α-CD in aqueous solution.

to that of the complex between β-cyclodextrin and ethyl cinnamate [6], in which β-cyclodextrins form a layer structure. The pattern of 6-CiO-β-CD is different from those of β-cyclodextrin and β-cyclodextrin-*p*-acetamide, which have been reported to have a cage type structure and a channel type structure, respectively, by single crystal X-ray structural analysis. A proposed structure of 6-CiO-β-CD is shown in Fig. 1.31.

**c. Control of Supramolecular Structures by Competitive Guests**

6-CiO-α-CD and 6-CiO-β-CD form supramolecular oligomers and polymers. When guest molecules which bind strongly in a CD cavity are added, the modified cyclodextrins bind the competitive guest to give host-guest monomers.

The $^1$H NMR spectra of 6-CiO-α-CD in the absence and presence of *p*-iodoaniline ($K_c$=2800 $M^{-1}$) or *p*-iodophenol ($K_c$=2300 $M^{-1}$) showed that the protons of the guests shifted downfield with some broadening and the signals of 6-CiO-α-CD also shifted with some sharpening. These results indicate that the guest molecules added were included in the CD cavity and that the cinnamoyl group was exposed to water.

6-CiO-β-CD is insoluble in water. This is due to intermolecular complexation through host-guest interactions. When an excess amount of adamantane-1-carboxylic acid was added to the suspension of 6-CiO-α-CD, 6-CiO-β-CD was

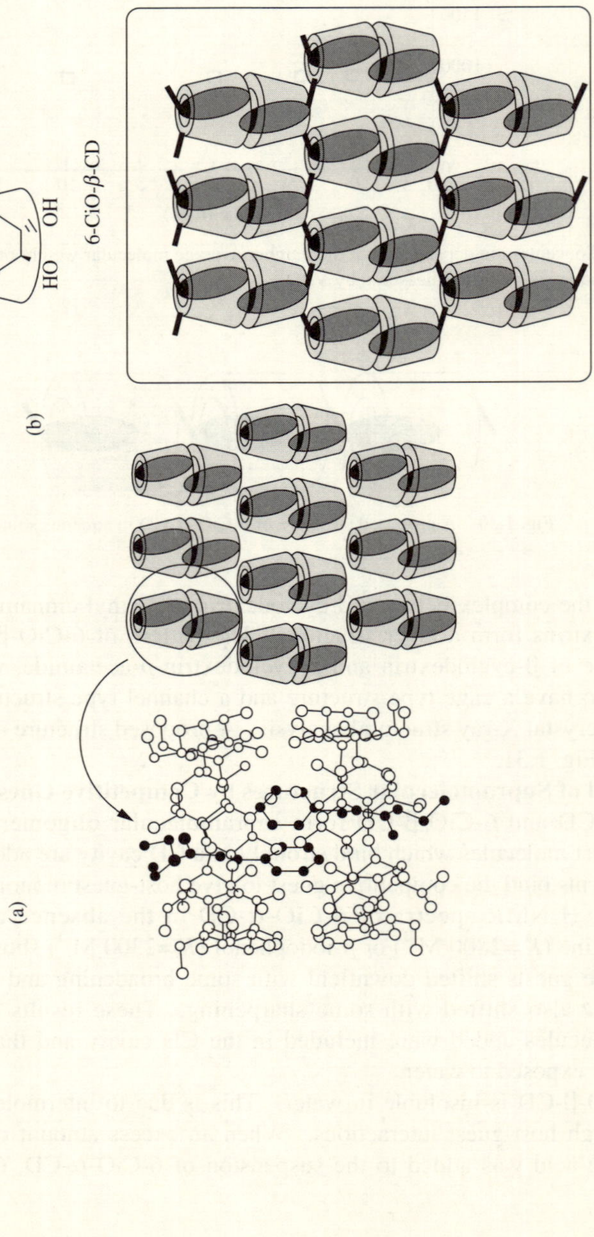


**Fig. 1.31** Crystal structure of β-CD-ethyl cinnamate complex (a) and a proposed structure of 6-CiO-β-CD in the solid state (b).

solubilized in water. The $^1$H NMR spectra showed clear 6-CiO-β-CD upon the addition of adamantane carboxylic acid, indicating that adamantane carboxylic acid solubilizes 6-CiO-β-CD in $D_2O$.

### 1.2.8 Daisy Chain Necklace

Cyclic tri[2]rotaxanes (daisy chain necklace) containing cyclodextrins have been prepared by closing a tri[2]rotaxane containing α-cyclodextrin and 6-(4-aminocinnamate) with 2,4,6-trinitrobenzene sulfonic acid sodium salt(Fig. 1.32)[55] If the molecule changes its conformation (or co-conformation), the ring may expand or shrink due to external conditions (temperature, solvent, photochemical, electrochemical). These compounds are important because the cycle can be used as a chemical valve, as can be seen in ion channels in biological membranes.

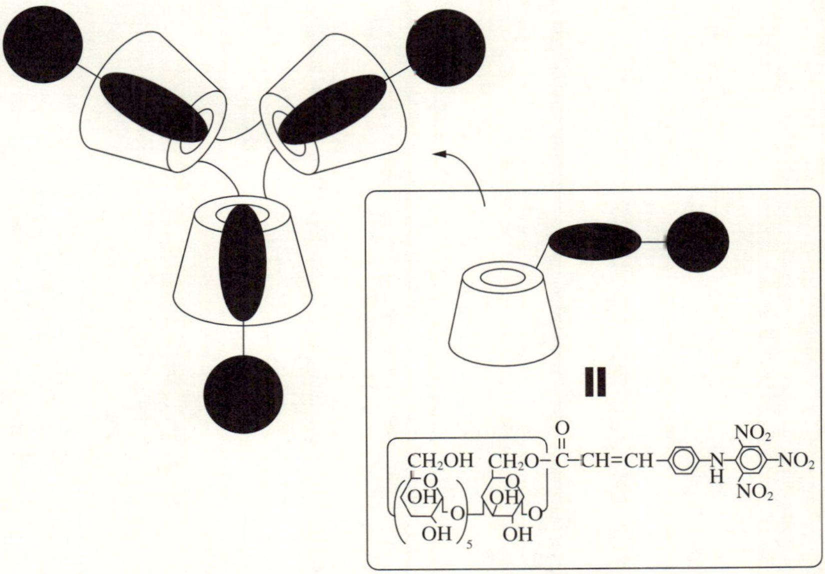


**Fig. 1.32** Daisy chain necklace, cyclic tri[2]rotaxane containing α-cyclodextrin.

### 1.2.9 Conclusion

Cyclodextrins were found to form inclusion complexes not only with small molecules but also macromolecules to give *pseudo*-polyrotaxanes. Polyrotaxanes in which many cyclodextrins were entrapped in a polymer chain were prepared by capping the chain ends with bulky stoppers. One or two cyclodextrins on a polymer chain could be moved along the polymer chain using an STM tip. Cyclodextrins were found to form complexes with ionic polymers as well. Ionic species serve as a potential surface for the shuttling of a cyclodextrin ring. Based on these observations, we prepared a molecular shuttle in which a cyclodextrin ring moves back and forth along the long axle. Moreover, we prepared rotaxanes using multi-cationic groups as stoppers: an electric trap.

Although 6-HyCiO-β-CD formed an intramolecular complex and 6-HyCiO-α-CD formed weak intermolecular complexes, 6-CiO-α-CD formed intermolecular complexes to give supramolecular oligomers and 6-CiO-β-CD formed insoluble supramolecular complexes in the solid state. These results indicate that flexible spacer groups allow the formation of intramolecular complexes and rigid spacer groups are required for the formation of supramolecular polymers. Supramolecular structures can be controlled by the addition of suitable guest molecules.

# 1.3 Cucurbituril : A Versatile "Bead" for Polyrotaxane Synthesis

**Theodora Krasia, Saghar Khodabakhsh, Dönüs Tuncel and Joachim H. G. Steinke**

## 1.3.1 Introduction

Rotaxanes as well as polyrotaxanes are supramolecular assemblies in which one or many cyclic units are threaded onto a linear segment. If these assemblies contain stopper groups that prevent the macrocyclic rings from dethreading then we refer to them as rotaxanes or polyrotaxanes, whereas the prefix "pseudo" is used to indicate the absence of bulky segments sufficient to prevent dethreading from occurring (Fig. 1.33).[56] Examples are known in which an increase in temperature or a change in solvent can blur this clear distinction and suddenly what was a rotaxane has turned into a pseudorotaxane.[57]

Cyclodextrins and crownethers are the most widely used macrocycles in the synthesis of rotaxanes and polyrotaxanes.[56] Their popularity benefits from their commercial availability and accessibility of a wide range of functional derivatives that can be obtained in a straightforward manner. Other macrocycles such as cyclophanes have been employed in some rare instances, although their synthesis is more involved and cumbersome.[58-60] In more recent years a long known macro-

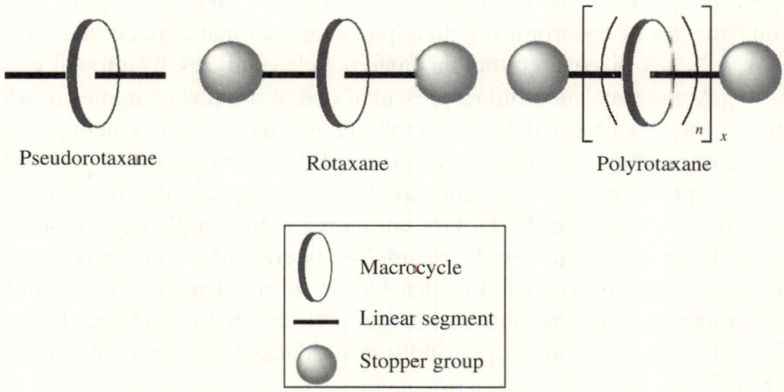


**Fig. 1.33** Illustration of a pseudorotaxane, rotaxane and polyrotaxane.

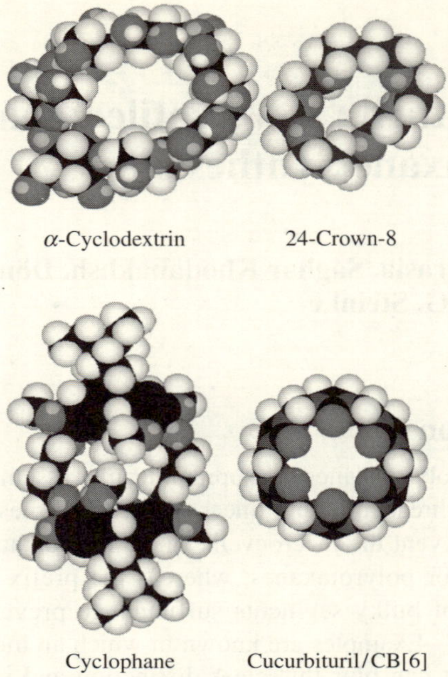


**Fig. 1.34** Macrocycles that have been used in the synthesis of polyrotaxanes.

cycle, cucurbituril, has been rediscovered for the synthesis of rotaxanes[61-68] and polyrotaxanes (Fig. 1.34).[69-73]

### 1.3.2 Cucurbituril

First synthesised by Behrend in 1905,[74] cucurbituril became the focus of detailed investigations by Mock and coworkers in the 1980s.[75] While studying the complexation behavior of cucurbituril with aliphatic and aromatic mono- and diammonium ions,[76-78] they discovered that this macrocycle catalyzes 1,3-dipolar cycloadditions.[79,80] At the time cucurbituril presented one of the few examples in which a synthetic receptor exhibits catalysis resembling enzyme reaction kinetics.[80] Mock *et al.* concluded that the rate acceleration of $10^5$ seen for the cycloaddition between an aliphatic azide and an aliphatic terminal alkyne is due to the release of the strain that is present in the ternary complex. Azide and alkyne moieties are slightly too large to simultaneously occupy the interior of cucurbituril strain-free. The ternary complex is "forced" together by the strong affinity of cucurbituril for ammonium ions, which generates an internal pressure that is released through the formation of the less strained cycloaddition product, a 1,3-disubstituted triazole (Fig. 1.35).[79]

Catalyzing 1,3-dipolar cycloadditions may be the most unusual feature of this supramolecular ligand, though by far not the only one. The size of cucurbituril

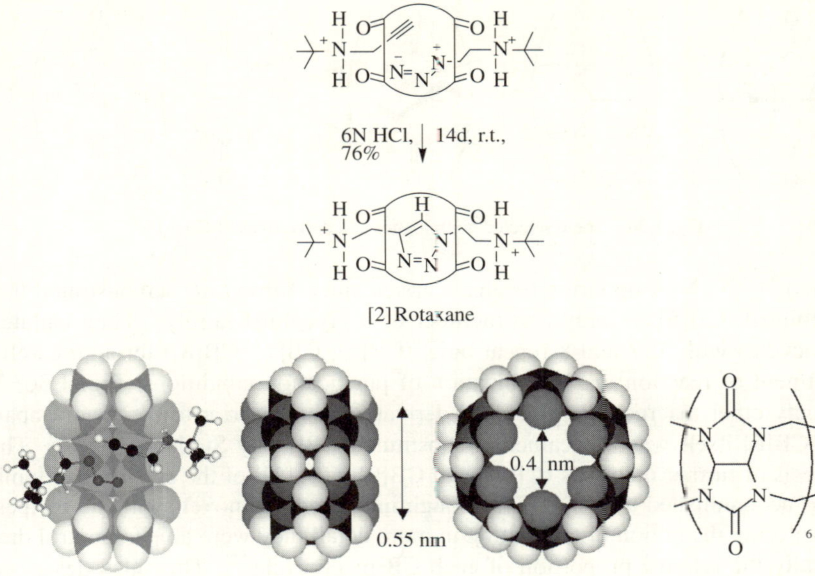


**Fig. 1.35** Mock's synthesis of a [2]rotaxane via a 1,3-dipolar cycloadditions catalyzed by cucurbituril.

and its cavity are quite similar to that of α-cyclodextrin. Access to the cavity in cucurbituril though is unusual as the entrance is decorated with a symmetrical sextet of carbonyl groups providing an opening of about 0.4 nm that must be negotiated in order to enter the significantly wider cavity, which is about 0.55 nm across (Fig. 1.35). In further contrast to α-cyclodextrin, the two fringes of the carbonyl groups in cucurbituril are preorganized not only to bind to ammonium ions via ion-dipole and to a lesser extent hydrogen bonding interactions, but also to mono- and dications in particular alkali and alkali earth metal cations.[75,81] Amino acids,[82] amino alcohols,[82] diols,[83] acids,[83,84] and diamines,[76,85] are all encapsulated by cucurbituril provided the distance between the functional groups is 4-10 atoms long. 1,4-Disubstituted aryl rings (not 1,3), and 5- and 6-membered alicyclic ammonium ions are illustrative of the limit in size and shape seemingly imposed by the carbonyl-fringed occuli of cucurbituril.[75,86]

The richness of the structural and electronic factors influencing host-guest interactions involving cucurbituril have inspired an ever increasing number of research groups to explore their potential in supramolecular chemistry. Pseudorotaxanes derived from cucurbituril have been investigated as nano-sized containers to protect dyes from oxidative degradation,[87] to remove dyes from waste water[88,89] and to reversibly encapsulate small guests.[88,90-92] Cucurbituril has been incorporated into pH-controlled molecular switches by exploiting differences in protonation level of amines (predominantly ion-dipole interactions) and the length of carbon spacer between the different nitrogen atoms (hydrophobic inter-

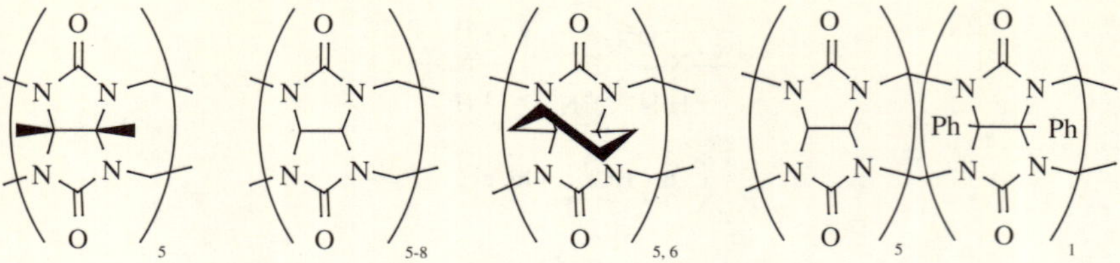

**Fig. 1.36** Some selected homologues and derivatives of CB[6].

actions).[66,93-95] New opportunities have arisen since Kim *et al.* demonstrated that cucurbituril (CB[6]) is only one member of a glycoluril family. They isolated macrocycles with 5, 7 and 8 repeat units (CB[5], CB[7], CB[8]) through careful adjustment of reaction and optimization of purification conditions (Fig. 1.36).[96] The only other macrocyclic glycoluril derivative known prior to their work apart from CB[6] itself was a decamethyl-substituted CB[5] by Stoddart *et al.*[97] The synthesis of further members (CB[4] and CB[9]- CB[11]) of the cucurbituril family was accomplished by Day *et al.* through improved and novel synthetic recipes. By exploring the notion of templating in more detail they were able to control dramatically the relative proportion of each CB macrocycle.[98] They also described the synthesis of CB derivatives by partially replacing glycoluril for substituted glycolurils in the cyclo-cooligomerization mixture.[99] Successful co-macrocyclization led to an isomeric product mixture, whereas the exclusive use of a 1,2-cyclohexyl-substituted glycoluril produced fully substituted CB[5] and CB[6] derivatives soluble in organic solvents.[65] Functionalization of only one of the six glycoluril repeat units was achieved by Nakamura *et al.*, who resorted to a combination of statistical monomer feed and preparative size exclusion chromatography to obtain pure unsymmetrically substituted CB[6].[64]

Kim *et al.* in particular have succeeded in incorporating cucurbituril within the ligand structure of a wide variety of 1-, 2- and 3-dimensional supramolecular coordination polymers (Fig. 1.37).[69,100] They have shown that the solid-state structure can be controlled through the geometry of the coordinating complex and thus through the coordination geometry determined by the selected transition metal.[69] Non-rotaxinated solid state structures in which cucurbituril is an integral part of a complex supramolecular structure are also being pursued.[101-103]

Buschmann *et al.* were the first to prepare a polyrotaxane containing cucurbituril by cleverly adapting interfacial polymerization conditions to the requirements of the water-soluble pseudorotaxane monomer.[104] Polymers in which cucurbituril has been threaded onto the side chain in a pseudorotaxane fashion have been reported by Kim *et al.*[69] Cucurbituril has also been integrated into a branched polymer structure. Through the functionalization of the surface of a dendrimer with a diamine motif, Kim *et al.* achieved the threading of CB[6] onto it, leading to a pseudorotaxinated surface architecture.[105]

Our interest in cucurbituril as a macrocyclic component ("bead") in polyrotaxane synthesis originates from our interest in the synthesis of well-defined poly-

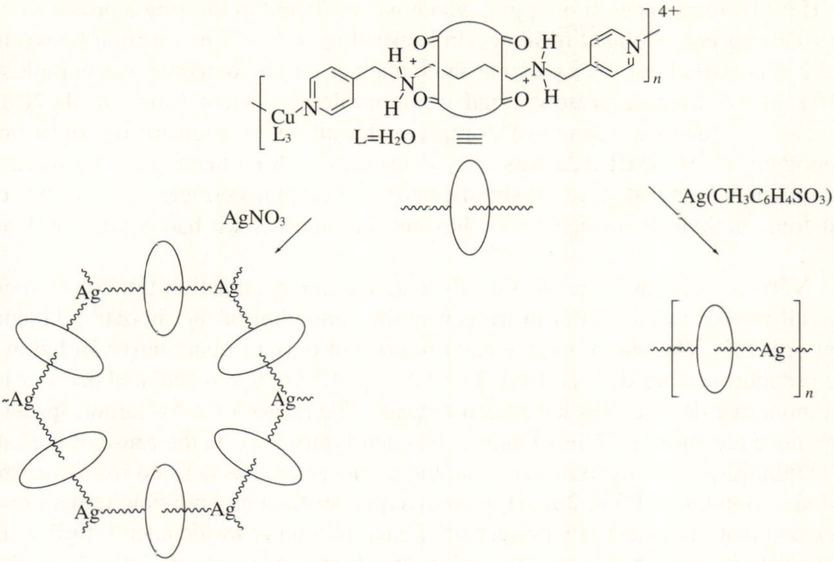


**Fig. 1.37** Examples of 1- and 2-dimensional supramolecular coordination polymers incorporating cucurbituril.

mer architectures. It remains a challenge to prepare polyrotaxanes (without resorting to rotaxane monomers) in which the number of incorporated macrocycles along the polymer backbone can be controlled precisely, even though many different and powerful synthetic approaches have been developed over the last decade.[56,71,106-108] A few years ago, we embarked on a research program with the aim to harness the catalytic activity of cucurbituril for the synthesis of main chain polyrotaxanes. We speculated that a monomer system could be designed where the polyaddition step would be controlled solely by the catalysis of cucurbituril, thereby linking polymerization chemistry inextricably with rotaxination. This should provide us with unprecedented ("perfect") control over the degree of threading.

The design of the initial diazide and dialkyne monomers **1** and **2** was based on the mono-functional derivatives used by Mock *et al.* (Fig. 1.35) in their seminal work on the catalytic activity of cucurbituril.[80] Thus the azide and ammonium groups in **1** were separated by a two-carbon spacer, whereas alkyne and ammonium groups in **2** were separated by just one methylene group (Fig. 1.38). The spacing between the two ammonium ions in each monomer was identical, six carbons long, which has been shown to be the optimal length for maximum complexation by cucurbituril.[76] Before we tried to polymerize monomers **1** and **2** we repeated the original experiment by Mock *et al.*, which involved dissolution of equimolar amounts of alkyne, azide and cucurbituril in acidified aqueous solution at room temperature. During the course of a few days we observed the appearance of $^1$H NMR resonances at 6.5 ppm, which we attributed to the single proton on the triazole ring encapsulated inside cucurbituril (Fig. 1.35). The reaction between **1** and **2** was carried out in exactly the same manner but the outcome was completely different. A triazole proton signal was completely absent from the $^1$H NMR. Instead we found evidence of complete threading of cucurbituril onto both monomers.[71] $^1$H NMR data was fully consistent with earlier reports by Mock *et al.*, in which they had assigned significant (0.7−1.0 ppm) upfield shifts to the central four methylene resonances of hexane diamine (as we had found for **1** and **2**).[76,77]

Variations in monomer and catalyst stoichiometry, reaction time or temperature all proved unsuccessful in triggering the formation of the dipolar cycloaddition product. The use of excess cucurbituril led only to quantitative isolation of the pseudorotaxanes derived from **1** and **2** (Fig. 1.38). We speculated that our initial monomer design must have been flawed. The choice of a six-carbon spacer to maximize recognition of the monomer by cucurbituril may be the cause of the catalytic inhibition. We hypothesized that the pseudorotaxanes derived from cucurbituril and monomers **1** and **2** must form much faster than any possible ternary complex and that the essentially perfect fit of each monomer inside cucurbituril would prevent the simultaneous encapsulation of alkyne and azide moieties by a third cucurbituril with the correct conformation required for catalysis.

In order to find evidence for this hypothesis we decided to prepare a polyrotaxane that would mimic the chain dimensions and therefore not only the electronic but also steric factors imposed by the growing polymer chain via a post-threading route. As the simplest model we selected a poly(hexamethylene amine) and its corresponding protonated form (Fig. 1.39).[70]

In this case we assumed the distance between two ammonium ions separated by a hexamethylene spacer to be very similar to the spacing of two ammonium ions separated by the triazole ring. The desired free polyamine was synthesized from Nylon 6/6 using borane as the reducing agent. When cucurbituril was threaded onto poly(hexamethylene iminium chloride) **3** under strongly acidic aqueous conditions we found the kinetics of the threading process to be very slow (Fig. 1.40).[70]

Only when the reaction mixture was heated at 90 °C for many days did we achieve significant levels of threading. The threading process could be followed conveniently by $^1$H NMR. Due to the symmetry of cucurbituril and that of the polymer repeat unit each methylene group resonance is split into two separate signals, those for methylene groups residing inside cucurbituril (Fig. 1.41(a)−(c)), those located next to a threaded repeat unit (Fig. 1.41(A)−(C)) and the remaining unthreaded hexamethylene segments.[70]

The slow threading kinetics are a result of the high complex stability between cucurbituril and protonated hexane diamine derivatives. Wenz *et al.* made a very similar observation while studying the threading of various cyclodextrins along a polycationic polymer chain.[109,110] Maximum levels of threading were established only after many weeks. They successfully modelled their post-threading system

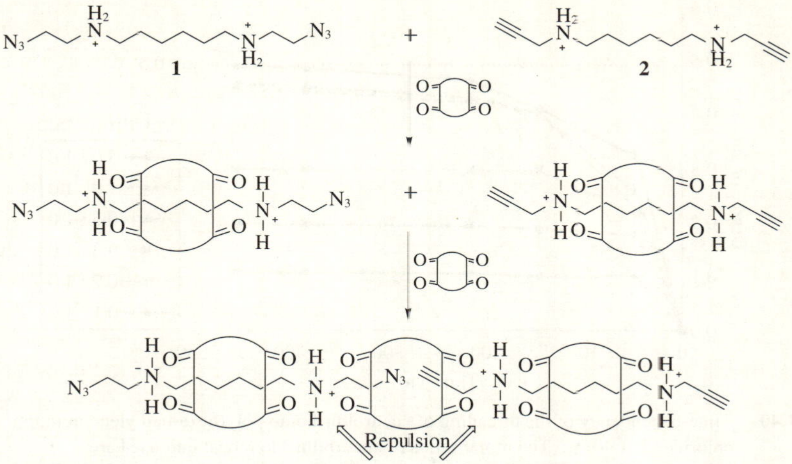

**Fig. 1.38** Attempted polymerization of monomers 1 and 2. Illustrated is the case using excess cucurbituril.

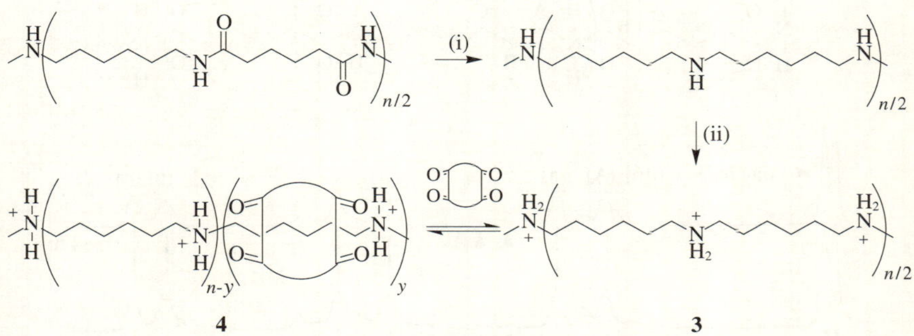

**Fig. 1.39** Synthesis of poly(hexamethylene amine) and post-threading of its protonated form, **3**, with cucurbituril. (i) $BH_3$-DMS, THF, 48 h, 66%, (ii) 20 wt% HCl

invoking a hopping mechanism with a high activation energy.[110] More recently Harada *et al.* interpreted their slow threading kinetics between a cyclodextrin and a bispyridinium dication containing a polymer chain in the same way.[111,112] Kim *et al.* also investigated a similar oligomeric bispyridinium motif with cucurbituril and they too inferred a high activation energy hopping mechanism to account for their slow rates of threading.[113] One structural feature is distinct to our post-threading system. Not only are the secondary ammonium ions positioned along the polymer chain available for "in-chain" complexation, it is also known that cucurbituril is capable of strong binding to secondary ammonium ions in a "side-on" mode.[75-77] We are currently looking for ways of separating out the side-on

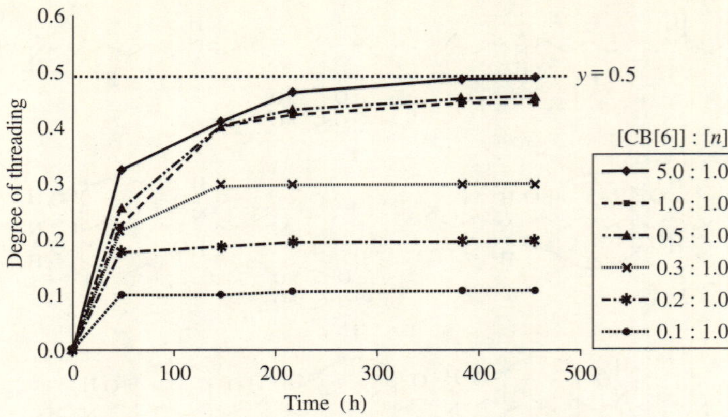


**Fig. 1.40** Time-dependency of the threading of cucurbituril onto poly(hexamethylene iminium chloride) **3** at 90 °C. The molar ratios of cucurbituril to repeat unit used are: 0.1 : 1.0; 0.2 : 1.0; 0.5 : 1.0; 1.0 : 1.0; and 5.0 : 1.0.

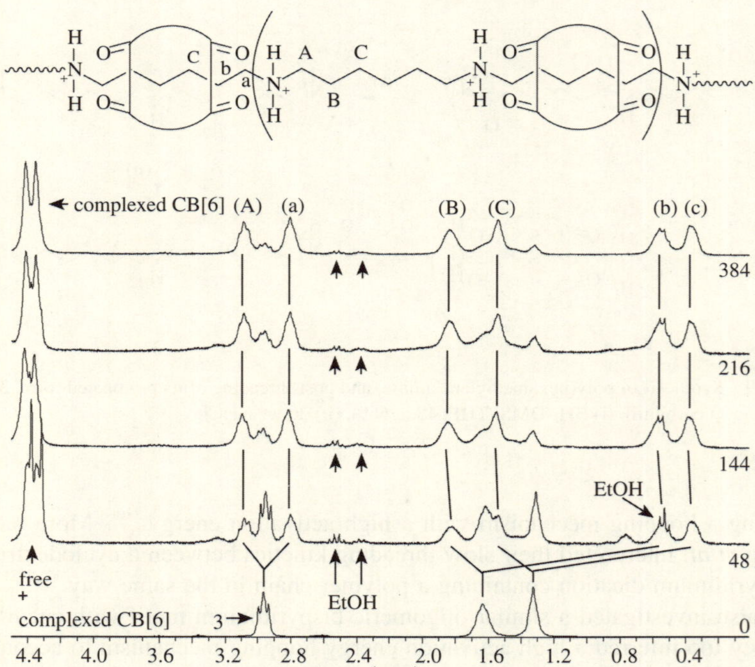


**Fig. 1.41** Time-dependence of post-threading followed by $^{1}$H NMR at 90 °C in $DCl/D_2O$ (20% w/w) with a molar ratio of cucurbituril to hexamethylene iminium repeat unit of 0.5 : 1.0.

and in-chain modes of complexation. Preliminary evidence suggests that the increase in the degree of threading at lower temperatures over time is part of a two-stage threading process possibly caused by an initial dominance of the side-on threading mode (Fig. 1.42).[114]

As can be gleaned from Fig. 1.40, even after extensive periods of heating and employing a 5-fold molar excess of cucurbituril the level of threading approaches 50% asymptotically. The most likely interpretation is that it is not possible for cucurbituril to occupy adjacent hexamethylene amine chain segments. This is probably due to several factors such as reduced ion-dipole interaction between the carbonyl groups and an ammonium ion that is already interacting with another carbonyl portal and repulsive dipole-dipole interactions between carbonyl groups belonging to adjacent portals. The X-ray crystal structure of the pseudorotaxane formed between the dihydrochloride salt of hexane diamine and cucurbituril is shown in Fig. 1.43.[115] This structural model of a threaded "repeat unit" found in pseudopolyrotaxane **4** shows that the protonated amine forms two hydrogen bonds to adjacent carbonyl oxygens and rises only slightly above the plane defined by the six carbonyl oxygen of each portal. This latter feature is consistent with our hypothesis that adjacent repeat units cannot be encapsulated because of unfavorable steric and electronic interactions.

Furthermore complexation studies with *N,N*-bis(hexylamine) serving as a two-repeat unit model compound showed evidence only for the formation of a 1 : 1 complex with excess CB[6]. Based on the information we had been accumulating it was proposed that the maximum level of threading had led to a structurally highly defined polymer in which threaded and non-threaded repeat units alternate.[70] It also proved that some features of our initial monomer structures were seriously flawed. In order to stay away from any complication caused by threading we designed a new set of dialkyne, **5**, and diazide, **6**, monomers (Fig. 1.44).[71]

In both monomers a 1,3-disubstituted phenyl ring separates the functional groups, which are too bulky to fit inside cucurbituril. The spacing between the ammonium groups and either alkyne or azide functional group have been left unchanged from our earlier design. This time polymerization took place as evidenced by the appearance of a proton resonance associated with the newly formed triazole ring (Fig. 1.45 and 1.46).

[2], [3] and [4]rotaxane and semirotaxane model compounds were also prepared in order to further ascertain that our initial NMR assignment was correct.[61] In all cases we observed the proton resonance of the triazole ring close to 6.5 ppm (indicating encapsulation by cucurbituril; 8.5 ppm free and protonated) and the proton resonance of the phenyl ring around 7.2 ppm. The ratio of the integral of both resonances provided us with a convenient method of calculating the degree of polymerization.[71] The repeat unit of this polyrotaxane is shown in Fig. 1.46.

An infinite polymer chain would require the integrals of the phenyl ring proton and the triazole proton to be the same. In the particular case shown in Fig. 1.46 the ratio is 1 : 0.9 which is converted using Carothers' equation to give a

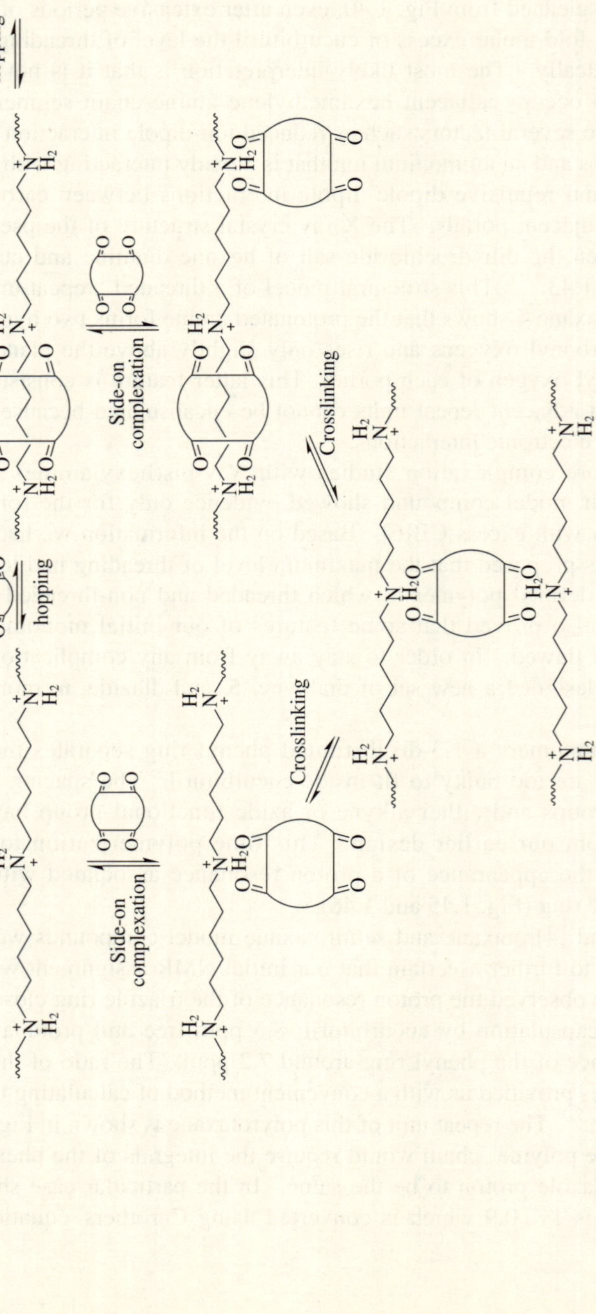


**Fig. 1.42** Possible modes of complexation during post-threading of 3.

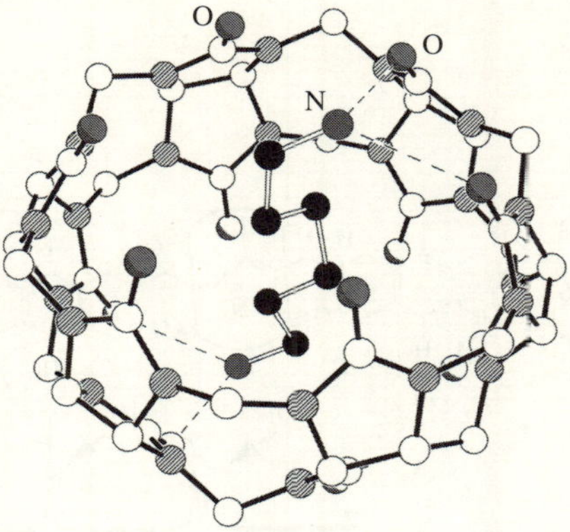


**Fig. 1.43** X-ray crystal structure of a threaded "repeat unit" of pseudopolyrotaxane **4**.

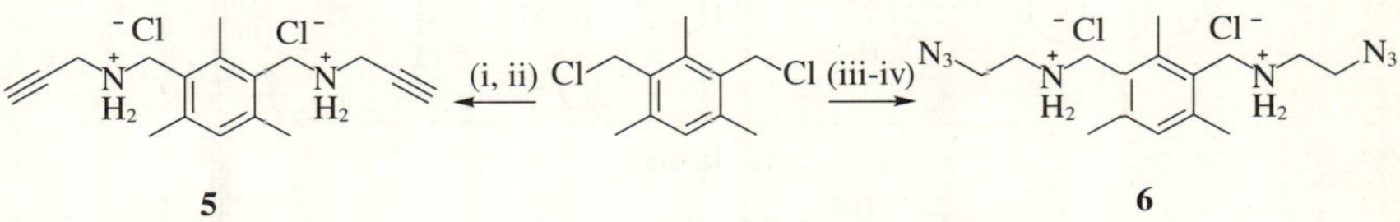

**Fig. 1.44** Synthesis of of monomers **5** and **6**. (i) $NH_2CH_2C{\equiv}CH$, neat, 0°C r.t., 16h, 100% (crude), (ii) 1N HCl in $Et_2O$, r.t., 86%, (iii) $NH_2CH_2CH_2OH$, neat, 5h, 150−160 °C, 66%, (iv) $SOCl_2$, $CHCl_3$, 5h, r.t., 70%, (v) $NaN_3$, $H_2O$, 75°C, 16h, 87%, (vi) 1N HCl in $Et_2O$, r.t., 96%.

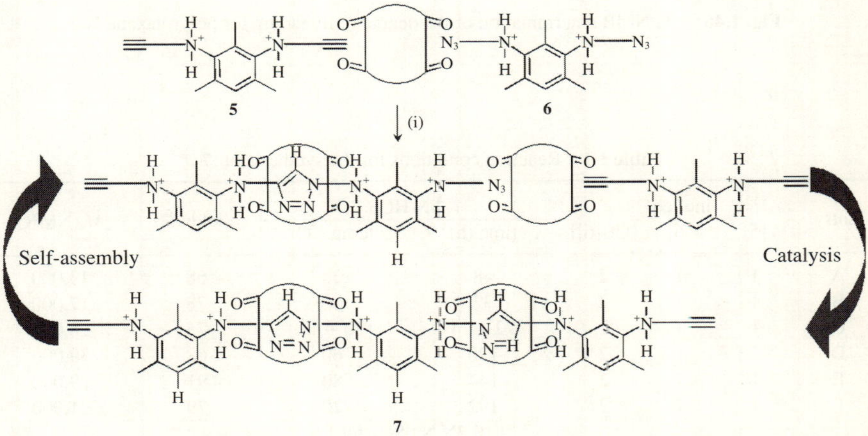


**Fig. 1.45** Catalytically self-threading synthesis of polyrotaxane **7**. (i) Please refer to Table 1.1 for details.

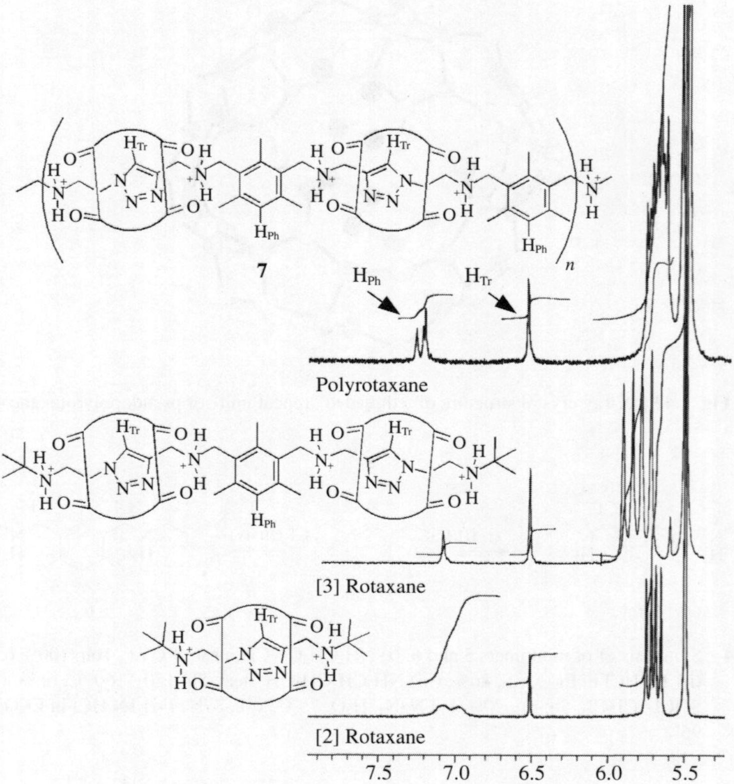


**Fig. 1.46** $^1$H NMR determination of the degree of threading for polyrotaxane **7**.

**Table 1.1** Reaction conditions for the synthesis of **7**

| Entry | mol eq. | | | 6N HCl | | Yield (%) | $M_n$(NMR) |
|---|---|---|---|---|---|---|---|
| | [5] | [6] | [CB[6]] | time (h) | temp.(°C) | | |
| A | 1 | 1 | 2 | 48 | 20 | 68 | 13,000 |
| B | 1 | 1 | 2 | 72 | 60 | 78 | 17,000 |
| C | 1 | 1 | 2 | 72 + 24 | 20 + 40 | 75 | 28,000 |
| D | 1 | 1 | 2 | 120 | 60 | 68 | 39,000 |
| E | 1 | 1 | 2 | 144 | 80 | 91 | 29,000 |
| F | 1 | 1 | 2 | 192 | 20 | 79 | 21,000 |
| | | | | 0.2N $Na_2SO_4$(for F) | | | |

degree of polymerization of 10 (Table 1.1, entry A). Since we are dealing with an $A_2 + B_2$ monomer system this number must be divided by two in order to correspond to the subscript in the bracket of the repeat unit (Fig. 1.46). We were able to improve the degree of polymerization significantly through the optimization of reaction conditions (DP = 15, $M_n$ = 39,000, Table 1.1, entry D). Polymerization attempts were also carried out without cucurbituril but in all other aspects identical to those summarized in Table 1.1.

In all instances only starting material was isolated with the $^1$H- and $^{13}$C-NMR confirming the absence of any triazole signals. This clearly illustrates that the dipolar cycloaddition reaction only takes place in the presence of cucurbituril, which has a dual function: to catalyze the cycloaddition reaction and in the process become inextricably threaded onto the polymer chain. The sequence of events involved in this catalytic self-threading process is presented in Fig. 1.45. The combination of using monomers with bulky in-chain stopper groups and the catalyst self-threading onto the polymer backbone limits this catalysis to a turnover of exactly 1.

One of the questions we asked ourselves arose from the contemporary synthesis of homologues of CB[6] by Kim *et al.*[96)] and Day *et al.*[98)] Would CB[5] or CB[7] also exhibit catalytic activity similar to CB[6]. In an exploratory fashion we used pure CB[5], a 7 : 3 mixture of CB[5] and CB[7] and pure CB[6] (after recrystallization) (as confirmed by MS and $^1$H- and $^{13}$C-NMR) obtained using the original procedure by Kim *et al.*[96)] Equimolar amounts of azide **8**, alkyne **9** and CB[5], CB[6] or CB[5]/CB[7] respectively were allowed to react at room temperature for 96 hours (Fig. 1.47). Only in the reaction in which CB[6] was present did we observe the appearance of the triazole proton signal over time. No reaction took place in the presence of any other macrocycle, including α-cyclodextrin (Fig. 1.48).[116)]

Heating of each sample after 96 hours for 48 hours at 75 °C was sufficient to thermally activate the cycloaddition reaction as evidenced by the appearance of the distinct proton resonance of the triazole. Although one cannot rule out that more sterically demanding azide and alkyne substrates than **8** and **9** may result in CB[7] to become catalytically active, space filling models show the interior of CB[5] to be too small to accommodate both azide and alkyne simultaneously, so that its catalytical inactivity is not surprising (Fig. 1.49). This set of experiments

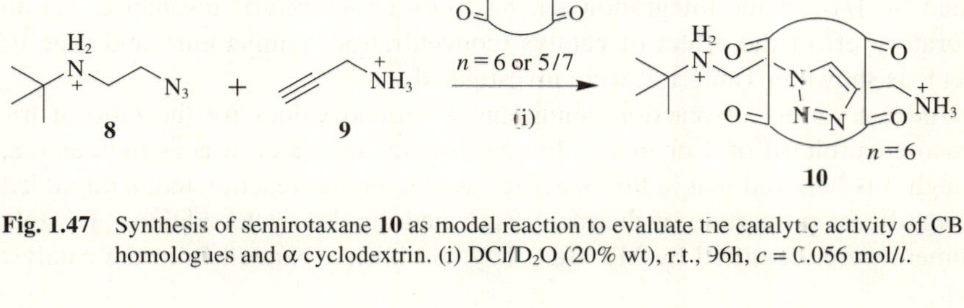

**8** **9** **10**

**Fig. 1.47** Synthesis of semirotaxane **10** as model reaction to evaluate the catalytic activity of CB homologues and α cyclodextrin. (i) $DCl/D_2O$ (20% wt), r.t., 96h, $c$ = 0.056 mol/$l$.

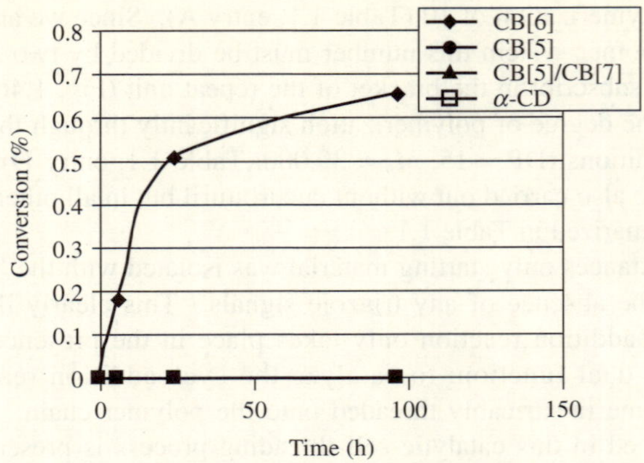


**Fig. 1.48** Plot of conversion versus time for the formation of semirotaxane **10** in the presence of CB[6], CB[5]/CB[7] (7 : 3 mixture (by MS and NMR)), CB[5] and α-CD.

can be interpreted as providing further evidence for the rationalization of the catalytic mechanism of cucurbituril as first proposed by Mock *et al.*[80] The formation of a ternary complex (not possible with CB[5]) and the subsequent productive release of strain in the form of the smaller cyclic product (not possible with CB[7]) (Fig. 1.49). Catalysis is not the result of effective molarity since otherwise some if even only small level of catalytic activity would have been detected from the CB[5]/CB[7] mixture.

The search for a way to trigger cucurbituril into behaving like a true catalyst exhibiting a turnover of greater than one led us to synthesize a set of monomers which we had designed in such a way so as to prevent cucurbituril from threading onto a monomer in a non-productive mode (encapsulation) and to prevent cucurbituril from complexing too strongly to the triazole product (product inhibition). This analysis produced a pair of "mini" monomers, **11** and **12**, in which the spacer length between two ammonium groups has been reduced to zero, which in turn should render it strongly unfavorable for two cucurbituril molecules to occupy adjacent locations along the anticipated growing polymer chain (Fig. 1.50).[73]

The progress of the reaction was again monitored by $^1$H NMR spectroscopy. Our criterion for catalytic turnover was that each cucurbituril molecule would lead to the formation of more than one triazole ring. This ratio was calculated from the integration of the sum of all triazole proton resonances (encapsulated and free) divided by 1/12 of the integration arising from a cucurbituril resonance. In an exploratory effort the effect of catalyst concentration, temperature and type of solvent, as shown in Table 1.2, were investigated.[73]

Under a variety of reaction conditions we found values for the ratio of triazole-to-cucurbituril of 2 or more. In one instance the value was as high as 6.8, although it is believed that in this specific case the higher reaction temperature led to a significant proportion of thermally activated dipolar cycloaddition. Triazole oligomers were identified by MS after they had been separated from the catalyst

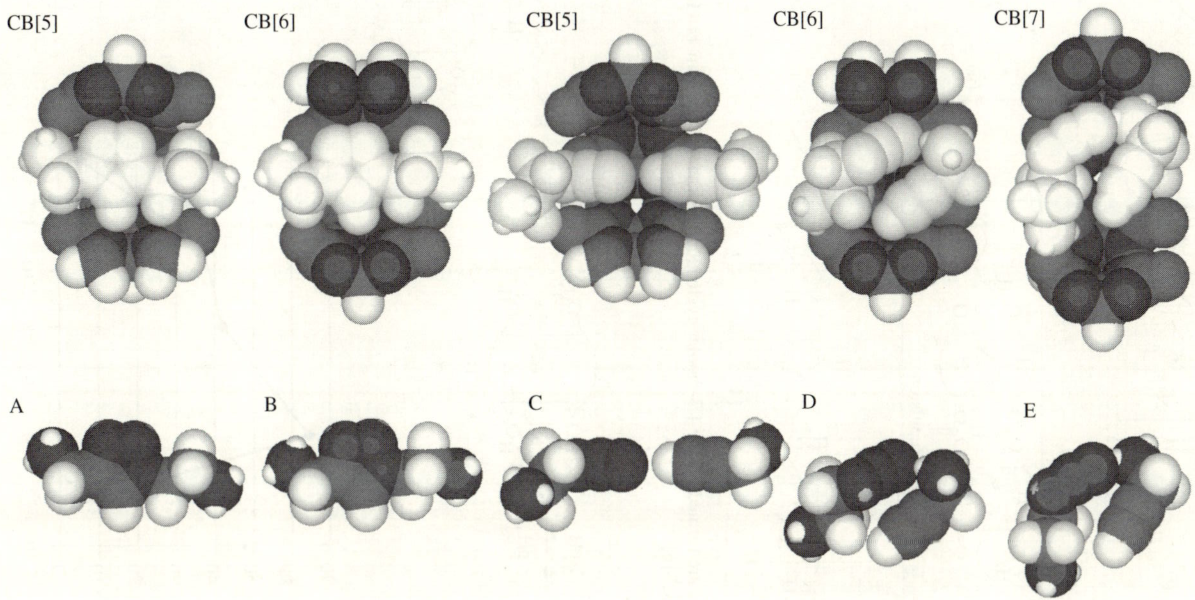


**Fig. 1.49** Illustrative space filling models showing the cross sections of CB[5], CB[6] and CB[7] onto which has been superimposed representations of the ternary complex (C–E) and the encapsulated triazole product (A and B, except for CB[7]) based on molecular mechanics calculations.

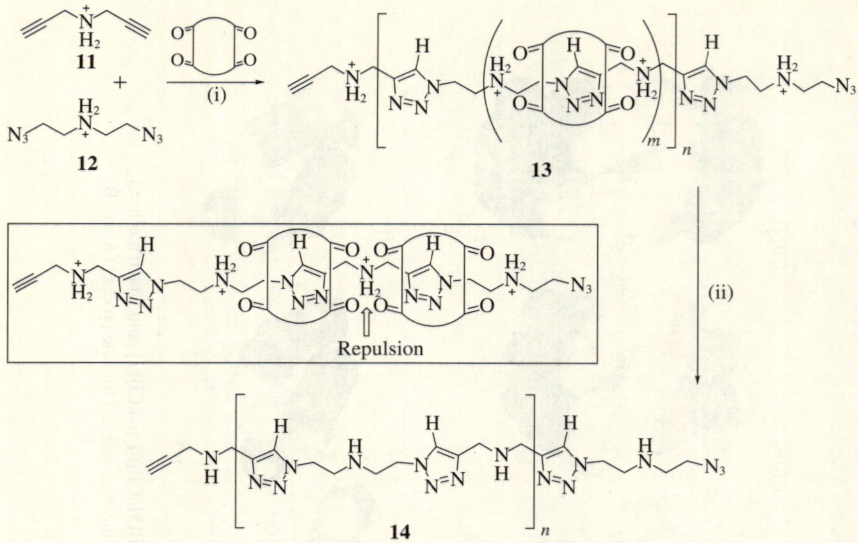

Fig. 1.50 Oligotriazoles are formed via the 1,3-dipolar cycloaddition of "mini" monomers **11** and **12** catalyzed by CB[6].

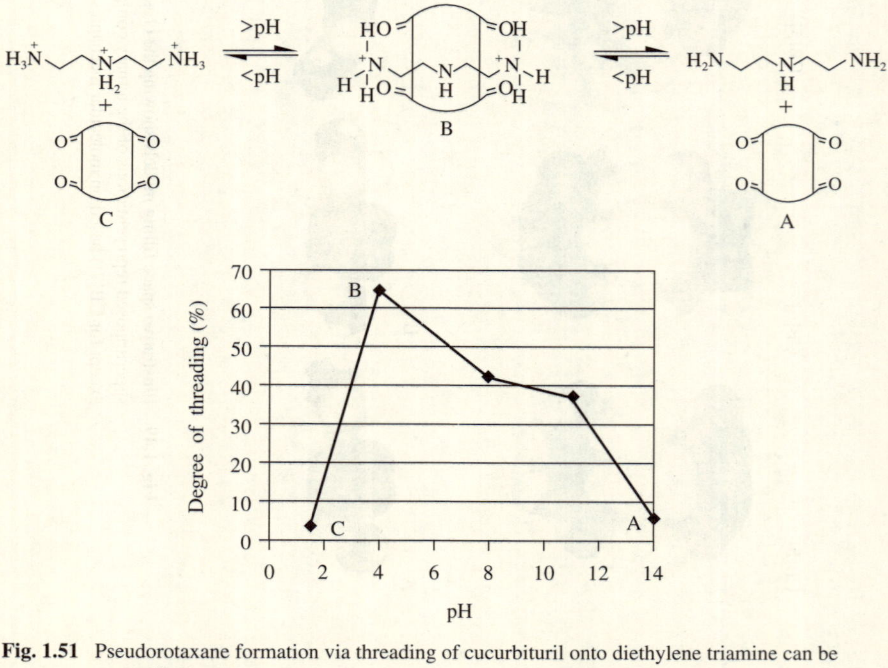


Fig. 1.51 Pseudorotaxane formation via threading of cucurbituril onto diethylene triamine can be controlled by pH.

through extraction of the free-based oligomers into an organic layer.[73] Although it is quite clear that considerable optimization efforts are required to establish reaction conditions for the formation of polytriazole chains, the results also demonstrated unambiguously and for the first time that cucurbituril can function as a true catalyst. This however demands quite specific reaction conditions. The acidity of the solvent medium (Table 2, PT1-A and PT1-B) and the concentration of the catalyst relative to the monomers appear crucial to maximize catalytic turnover (Table 1.2, PT1-B and PT2).

More recently we revisited the most straightforward synthesis of cucurbituril-containing polyrotaxanes, that of post-threaded organic polycations. Although we have made use of the selectivity of cucurbituril towards diammonium ions separated by spacer groups of different lengths, at no point have we, or any other group for that matter, considered the possibility of controlling the level of threading of cucurbituril in pseudopolyrotaxanes through the adjustment of pH. Similar to the idea behind the pH-triggered molecular switches containing cucurbituril, the pseudopolyrotaxane would be a stimuli-responsive supramolecular assembly in which cucurbituril serves as the triggerable recognition unit. Diethylene triamine served as the model compound to establish proof of this concept (Fig. 1.51).[117]

An aqueous solution set to pH 14 contained equimolar amounts of cucurbituril and the triamine. As shown in Fig. 1.51 the initial degree of threading of less than 5% (upper limit reflecting experimental error) increases with decrease in pH and reaches a maximum of 65% at pH 3.5 before the degree of threading drops again to a value below 5%. We repeated the experiment under identical conditions this time replacing the triamine with linear poly(ethylene imine) (Fig. 1.52). The ratio between encapsulated and free methylene protons was used to calculate the degree of threading. The maximum level of threading of 40% was reached at a pH of around 9.[117]

A change of pH to either lower or higher values led to a decrease in the degree of threading. Complete dethreading was found for a pH of 1.5 and 14. The kinetics of the threading process with poly(ethylene imine) are fast in compar-

**Table 1.2** Reaction conditions and $^1$H NMR data for the synthesis of **14**

| Entry | mol eq. [5] | [6] | [CB[6]] | $t$ (h)/$T$ (°C) | [triazole]/ [CB[6]] |
|---|---|---|---|---|---|
| PT1-A[a] | 1.0 | 1.0 | 2.0 | 168/25 | 0.4 |
| PT1-B | 1.0 | 1.0 | 2.0 | 168/25 | 0.7 |
| PT2 | 1.0 | 1.0 | 0.5 | 168/25 | 2.0 |
| PT3 | 1.0 | 1.0 | 0.5 | 336/25 | 2.1 |
| PT4 | 1.0 | 1.0 | 0.5 | 336/25 + 96/80 | 2.4 |
| PT5 | 1.0 | 1.0 | 0.1 | 168/25 | 1.4 |
| PT6 | 1.0 | 1.0 | 0.1 | 168/25 + 96/80 | 6.8 |
| PT7 | 1.0 | 1.0 | 0 | 336/25 | 0.0 |

[a] All reactions were carried out in 0.2 M $Na_2SO_4$ in $D_2O$ apart from **PT1-A** where DCl/$D_2O$ (20% wt) was used instead.

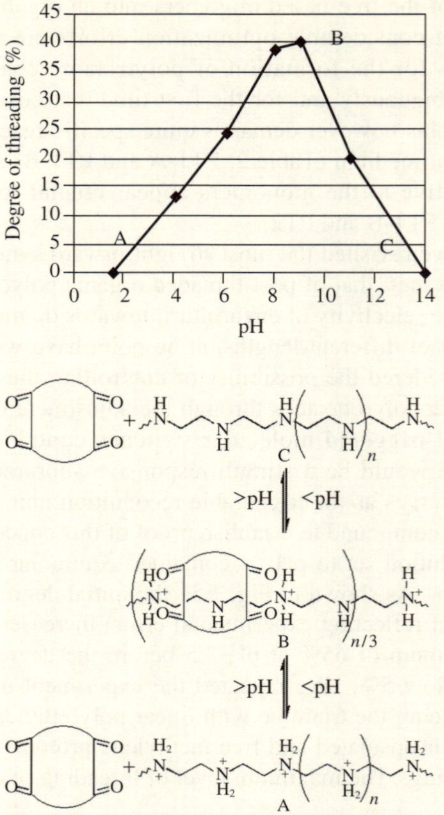


**Fig. 1.52** Stimuli-responsive pseudopolyrotaxane via threading of cucurbituril onto linear poly(ethylene imine). The degree of threading is controlled by pH.

ison to those found for poly(hexamethylene imine). This we believe is due to two distinctly different threading mechanisms. In the case of the fully protonated poly(hexamethylene imine) any movement of cucurbituril along the polymer chain must involve hopping from one chain segment to the next. Although hopping cannot be ruled out as a pathway for the movement of cucurbituril along a partially protonated poly(ethylene imine) chain, clearly there are alternative pathways with lower activation energy available. Being partially protonated poly(ethylene imine) acts as a buffer with the capacity to exchange protons as part of a dynamic equilibrium between individual polymer chains. Thus reversible deprotonation would allow cucurbituril to progress by one or perhaps two repeat units along the polymer chain before statistical re-protonation would temporarily tie it down again at one location.

### 1.3.3 Conclusions

We hope to have demonstrated the versatility of cucurbituril as a macrocycle for polyrotaxane synthesis. Its unique catalytic properties have been exploited for the synthesis of catalytically self-threading polyrotaxanes in which polymerization and rotaxination occur simultaneously thereby guaranteeing a precisely defined number of macrocycles along the polymer backbone. Cucurbituril has also been used as a catalyst for the formation of oligotriazoles exhibiting so far only very modest catalytic turnover. The high complexation constants of cucurbituril with protonated hexamethylene diamine subunits make it possible to form pseudorotaxanes via post-threading in which the degree of threading can be controlled by the simple choice of the desired stoichiometry. Finally we have presented our initial results on the threading behavior of cucurbituril onto linear poly(ethylene imine). Poly(ethylene imine) and cucurbituril form a stimuli-responsive pseudorotaxane in which the degree of threading is controlled directly by the pH.

### 1.3.4 Summary

Most commonly crownether and cyclodextrin derivatives have been employed in the synthesis of supramolecular polymers. In more recent years cucurbituril, a macrocycle that has been known for almost a hundred years, has been discovered to be a versatile and unique building block for the synthesis of polyrotaxanes.

### Acknowledgments

We would like to thank the EPSRC for providing an IPSI project studentship for D. T. Thanks are also due to Dr. Welham of the University of London ULIRS service for running MALDI-TOF spectra, Dr. Ball for ES and Mr. John Barton for all other mass spectrometry data and the EPSRC RAPRA service for GPC characterization data. We are grateful to Dr. A.J.P. White and Prof. D. J. Williams for the X-ray structure and analysis.

# References

< **Section 1.1** >

1. a) Aviram, A., Ratner, M. A., *Chem. Phys. Lett.*, 1974, **29**, 277.
   b) Goldhaber-Gordon, D., Montemerlo, M. S., Love, J. C., Opiteck, G. J., *Proc. IEEE*, 1997, **85**, 521.
   c) Joachim, C., Gimzewski, J. K., Aviram, A., *Nature*, 2000, **408**, 541.
   d) Carroll, R. L., Gorman,C. B., *Angew. Chem. Int. Ed.*, 2002, **41**, 4378.
2. a) Chen, J., Reed, M. A., Rawlett, A. M., Tour, J. M., *Science*, 1999, **286**, 1550.
   b) Reed, M. A., Tour, J. M., *Sci. Am.*, 2000, **282**, 86.
   c) Donhauser, Z. J., Mantooth, B. A., Kelly, K. F., Bumm, L. A., Monnell,J. D., Stepleton, J. J., Price, D. W., Rawlett, A. M., Allara, D. L., Tour, J. M., Weiss, P. A., *Science*, 2001, **292**, 2303.
3. a) Rueckes, T., Kim, K., Joselevich, E., Tseng, G. Y., Cheung, C.-L., Lieber C. M., *Science*, 2000, **289**, 94.
   b) Huang, Y., Duan, X., Cui, Y., Lauhon, L. J., Kim, K.-H., Lieber, C. M., *Science*, 2001, **294**, 1313.
   c) Bachtold, A., Hadley, P., Nakanishi, T., Dekker, C., *Science*, 2001, **294**, 1317.
   d) Avouris, P., *Acc. Chem. Res.*, 2002, **35**, 1026.
   e) Dai, H., *Acc. Chem. Res.*, 2002, **35**, 1035.
4. a) Collier, C. P., Wong, E. W., Belohradsky, M., Raymo, F. M., Stoddart, J. F., Kuekes, P. J. Williams, R. S., Heath, J. R., *Science*, 1999, **285**, 391.
   b) Wong, E. W., Collier, C. P., Belohradsky, M., Raymo, F. M., Stoddart, J. F., Heath, J. R., *J. Am. Chem. Soc.*, 2000, **122**, 5831.
   c) Perkins, J., Collier, C. P., Heath, J. R., Jeppesen, J. O., Luo, Y., Nielsen, K. A., Pease, A. R., Stoddart, J. F., Wong, E. W., *Molecular Electronics and Bioelectronics*, 2001, **12**, 69.
   d) Pease, A. R., Stoddart, J. F., *Struct. Bonding*, 2001, **99**, 189.
5. a) Schill, G., *Catenanes, Rotaxanes and Knots*, Academic Press, New York, 1971.
   b) Walba, D. M., *Tetrahedron*, 1985, **41**, 3161.
   c) Dietrich-Buchecker, C. O., Sauvage, J.-P., *Chem. Rev.*, 1987, **87**, 795.
   d) Philp, D., Stoddart, J. F., *Synlett*, **1991**, 445.
   e) Jäger, R., Vögtle, F., *Angew. Chem. Int. Ed. Engl.*, 1997, **36**, 930.
   f) Breault, G. A., Hunter, C. A., Mayers, P. C., *Tetrahedron*, 1999, **55**, 5262.
   g) Sauvage, J.-P., Dietrich-Buchecker, C. O. eds., *Molecular Catenanes, Rotaxanes, and Knots*, Wiley-VCH, Weinheim, 1999.
   h) Cantrill, S. J., Pease, A. R., Stoddart, J. F., *J. Chem. Soc., Dalton Trans.*, **2000**, 3715.
   i) Raehm, L., Hamilton, D. G., Sanders, J. K. M., *Synlett*, **2002**, 1743.
6. a) Amabilino, D. B., Stoddart, J. F., *Chem. Rev.*, 1995, **95**, 2725.
   b) Claessens, C. G. Stoddart,, J. F., *J. Phys. Org. Chem.*, 1997, **10**, 242.
   c) Raymo, F. M., Stoddart, J. F., *Chem. Eur. J.*, 1997, **3**, 1933.
   d) Raymo, F. M., Stoddart, J. F., *Chemtracts – Organic Chemistry*, 1998, **11**, 491.
   e) Raymo, F. M., Stoddart, J. F., *Chem. Rev.*, 1999, **99**, 1643.
   f) Stoddart, J. F., Tseng, H.-R., *Proc. Natl. Acad. Sci. U.S.A.*, 2002, **99**, 4797.
7. a) Belohradsky, M., Raymo, F. M., Stoddart, J. F., *Collect. Czech. Chem. Commun.*, 1996, **61**, 1.
   b) Belohradsky, M., Raymo, F. M., Stoddart, J. F., *Collect. Czech. Chem. Commun.*, 1997, **62**, 527.
8. a) Gellman, S. H. ed., Special Issue on Molecular Recognition, *Chem. Rev.*, 1997, **97**, 1231.
   b) Mammen, M., Chio, S.-K., Whitesides, G. M., *Angew. Chem. Int. Ed.*, 1998, **37**, 2755.
   c) Takeuchi, M., Ikeda, M., Sugasaki, A., Shinkai, S., *Acc. Chem. Res.*, 2001, **34**, 865.
   d) Hof, F., Craig, S. L., Nuckolls, C., Rebek, Jr. J., *Angew. Chem. Int. Ed.*, 2002, **41**, 1488.
9. a) Lindsey, J. S., *New J. Chem.*, 1991, **15**, 153.
   b) Whitesides, G. M., Mathias, J. P., Seto, C. T., *Science*, 1991, **254**, 1312.
   c) Philp, D., Stoddart, J. F., *Angew. Chem. Int. Ed. Engl.*, 1996, **35**, 1154.
   d) Lehn, J.-M., *Supramolecular Chemistry*, VCH, Weinheim, 1996.

e) Fujita, M., *Acc. Chem. Res.*, 1999, **32**, 53.
f) Rebek Jr., J., *Acc. Chem. Res.*, 1999, **32**, 278.
g) Bong, D. T., Clark, T. D., Granja, J. R., Ghadiri, M. R., *Angew. Chem. Int. Ed.*, 2001, **40**, 988.
h) Seidel, S. R., Stang, P. J., *Acc. Chem. Res.*, 2002, **35**, 972.

10. Fyfe, M. C. T., Stoddart, J. F., *Acc. Chem. Res.*, 1997, **30**, 393.
11. a) Ashton, P. R., Ballardini, R., Balzani, V., Gómez-López, M., Lawrence, S. E., Martínez-Díaz, M.-V., Montalti, M., Piersanti, A., Prodi, L. Stoddart, J. F., Williams, D. J., *J. Am. Chem. Soc.*, 1997, **119**, 10641.
b) Ashton, P. R., Ballardini, R., Balzani, V., Baxter, I., Credi, A., Fyfe, M C. T., Gandolfi, M. T., Gómez-López, M., Martínez-Díaz, M.-V., Piersanti, A., Spencer, N., Stoddart, J. F., Venturi, M., White, A. J. P., Williams, D. J., *J. Am. Chem. Soc.*, 1998, **120**, 11932.
c) Elizarov, A. M., Chiu, H.-S., Stoddart, J. F., *J. Org. Chem.*, 2002, **67**, 9175.
12. a) Bissell, R. A., Córdova, E., Kaifer A. E., Stoddart, J. F., *Nature*, 1994, **369**, 133.
b) Raehm, L., Kern, J.-M., Sauvage, J.-M., *Chem. Eur. J.*, 1999, **5**, 3310.
c) Ashton, P. R., Baldoni, V., Balzani, V., Credi, A., Hoffmann, H. D. A., Martínez-Díaz, M.-V., Raymo, F. M., Stoddart, J. F., Venturi, M., *Chem. Eur. J.*, 2001, **7**, 3482.
13. a) Ashton, P. R., Ballardini, R., Balzani, V., Credi, A., Dress, R., Ishow, E., Kocian, O., Preece, J. A., Spencer, N., Stoddart, J. F., Venturi, M., Wenger, S., *Chem. Eur. J.*, 2000, **6**, 3558.
b) Brouwer, A. M., Frochot, C., Gatti, F. G., Leigh, D. A., Mottier, L., Paolucci, F., Roffia, S., Wurpel, G. W. H., *Science*, 2001, **291**, 2124.
c) Balzani, V., Credi, A., Marchioni, F., Stoddart, J. F., *Chem. Commun.*, **2001**, 1860.
d) Ballardini, R., Balzani, V., Clemente-León, M., Credi, A., Gandolfi, M. T., Ishow, E., Perkins, J., Stoddart, J. F., Tseng, H.-R., Wenger, S., *J. Am. Chem. Soc.*, 2002, **124**, 12782.
e) Gatti, F. G., León, S., Wong, J. K. Y., Bottari, G., Altieri, A., Morales, M. A. F., Teat, S. J., Frochot, C., Leigh, D. A., Brouwer, A. M., Zerbetto, F., *Proc. Natl. Acad. Sci. U.S.A*, 2003, **100**, 10.
14. Balzani, V., Credi, A., Raymo, F. M., Stoddart. J. F., *Angew. Chem. Int. Ed.* 2000, **39**, 3348.
15. Collier, C. P., Mattersteig, G., Wong, E. W., Luo, Y., Beverly, K., Sampaio, J., Raymo, F. M., Stoddart, J. F., Heath, J. R., *Science*, 2000, **289**, 1172.
16. a) Asakawa, M., Ashton, P. R., Balzani, V., Credi, A., Hamers, C. Mattersteig, G., Montalti, M., Shipway, A. N., Spencer, N., Stoddart, J. F., Tolley, M. S., Venturi, M., White, A. J. P., Williams, D. J., *Angew. Chem. Int. Ed.*, 1998, **37**, 333.
b) Balzani, V., Credi, A., Mattersteig, G., Matthews, O. A., Raymo, F. M., Stoddart, J. F., Venturi, M., White, A. J. P., Williams, D. J, *J. Org. Chem.*, 2000, **65**, 1924.
17. Brown, C. L., Jonas, U., Preece, J. A., Ringsdorf, H., Seitz, M., Stoddart, J F., *Langmuir*, 2000, **16**, 1924.
18. Asakawa, M., Higuchi, M., Mattersteig G., Nakamura, T., Pease, A. R., Raymo, F. M., Shimizu, T., Stoddart J. F., *Adv. Mater.*, 2000, **12**, 1099.
19. Anelli, P.-L., Ashton, P. R., Ballardini, R., Balzani, V., Delgado, M., Gandolfi, M. T., Goodnow, T. T., Kaifer, A. E., Philp, D., Pietraszkiewicz, M., Prodi, L., Reddington, M. V., Slawin, A. M. Z., Spencer, N., Stoddart, J. F., Vicent, C., Williams, D. J., *J. Am. Chem. Soc.*, 1992, **114**, 193.
20. Jeppesen, J. O., Perkins, J., Becher, J., Stoddart, J. F., *Org. Lett.*, 2000, **2**, 3547.
21. Collier, C. P., Jeppesen, J. O., Luo, Y., Perkins, J., Wong, E. W., Heath, J. R., Stoddart, J. F., *J. Am. Chem. Soc.*, 2001, **123**, 12632.
22. Pease, A. R., Jeppesen, J. O., Stoddart, J. F., Luo, Y., Collier, C. P., Heath, J. R., *Acc. Chem. Res.*, 2001, **34**, 433.
23. Jeppesen, J. O., Perkins, J., Becher, J., Stoddart, J. F., *Angew. Chem. Int. Ed.*, 2001, **40**, 1216.
24. Luo, Y., Collier, C. P., Jeppesen, J. O., Nielsen, K. A., DeIonno, E., Ho, G., Perkins, J., Tseng, H.-R., Yamamoto, T., Stoddart, J. F., Heath, J. R., *Chem. Phys. Chem.*, 2002, **3**, 519.
25. Chen, Y., Ohlberg, D. A. A., Li, X., Stewart, D. R., Williams, R. S., Jeppesen, J. O., Nielsen, K. A., Stoddart, J. F., Olynick, D. L., Anderson, E., *Appl. Phys. Lett.*, 2003, **82**, 1610.
26. Chen,Y., Jung, G.-Y., Ohlberg, D. A. A., Li, X., Stewart, D. R., Jeppesen, J. O., Nielsen, K. A., Stoddart, J. F., Williams, R. S., *Nanotechnology*, 2003, **14**, 462.

**< Section 1.2 >**

27. Harada, A., *Modular Chemistry, NATO ASI Ser., Col. 499* (Michl, J. ed.), Kluwer, Dordrecht, 1997.
28. Harada, A., *Acc. Chem. Res.*, 2002, **34**, 456.
29. a) Amabiliino D. B., Stoddart, J. F., *Chem. Rev.*, 1955, **95**, 2725.
b) Harada, A., *Adv. Polym. Sci.*, 1997, **133**, 141.

c) Sauvage J. -P., Dietrich-Buchecker, C. O. eds., *Molecular Catenanes, Rotaxanes and Knots*, Wiley-VCH, Weinheim, 1999.

30. a) Hamasaki, K., Ikeda, H., Nakamura, A., Ueno, A., Toda, F., Suzuki, I., Osa, T., *J. Am. Chem. Soc.,* 1993, **115**, 5035.
b) Ueno, A., Minato, S., Suzuki, I., Fukushima, M., Okubo, M., Osa, T., Hamada, F., Mukai, K., *Chem. Lett.*, 1990, **605**.
c) Corradini, R., Dossena, A., Marchelli, R., Panagia, A., Sartor, G., Saviano, M., Lomvardi, A., Pavone,V., *Chem. Eur. J.*, 1996, **2**, 373.
d) Corradini, R., Dossena, A., Galaverna, G., Marchelli, R., Panagia, A., Sartor, *G., J. Org. Chem.*, 1997, **62**, 6283.
e) Wang, Y., Ikeda, T., Ueno, A., Toda, T., *Bull. Chem. Soc. Jpn.*, 1999, **67**, 1589.
31. a) Zanotti-Gerosa, A., Solari, E., Chiesi-Villa, L., Rissoli, C., *Chem. Commun.*, **1996**, 119.
b) Yamaguchi, N., Nagvekar, D. S., Gibson, H.W., *Angew. Chem. Int. Ed.*, 1998, **37**, 2361.
32. a) Ashton,P. R., Baxter, I., Cantrill, S., Fyfe, M. C. T., Glink, P. T., Stoddart, J. F., White, A. J. P., Willams, D. J., Wolf, R., *Angew. Chem. Int. Ed.*, 1998, **37**, 1294.
b) Ashton, P. R., Parsons, I. W., Raymo, F. M., Stoddart, J. F., White, A. J. P., Willams, D. J., Wolf, R., *Angew. Chem. Int. Ed.*, 1998, **37**, 1913.
c) Rowan, S. J., Cantrill, S. J., Stoddart, J. F., White, A. J. P., Willams, D. J., *Org. Lett.*, 2000, **2**, 759.
33. Tong, L. -H., Hou, Z. -J., Inoue, Y., Tai, A., *J. Chem. Soc., Perkin Trans. 2*, **1992**, 1253.
34. Harada, A., A. Cyclodextrins, *Large Ring Molecules* (Semlyen, J. A. ed.) , Wiley, Chichester, 1996.
35. Harada, A., Synthesis of Polyrotaxanes, *Synthesis of Polymers* (Schlüter, A. -D. ed.), Wiley-VCH, Weinheim, 1999.
36. a) Harada, A., Takahashi, S., *J. Chem. Soc., Chem. Commun.*, **1984**, 645.
b) Harada, A., Takahashi, S., *J. Chem. Soc., Dalton Trans.*, **1988**, 729.
c) Odagaki, Y., Hirotsu, K., Higuchi, T., Harada, A., Takahashi, S., *J. Chem. Soc., Perkin Trans. 1.*, **1990**, 1230.
37. Castro,R., Cuadrado, I., Alonso, B., Casado, C. M., Moran, M., Kaifer, A. E., *J. Am. Chem. Soc.*, 1997, **119**, 5760.
38. Gonzalez, B., Casado, C. M., Alonso, B., Cuadrado, I., Moran, M., Wang, Y., Kaifer, A. E., *J. Chem. Soc., Chem. Commun.*, **1998**, 2569.
39. Mirzoian, A., Kaifer, A. E., *Chem. Eur. J.*, 1997, **3**. 1052.
40. Ogino, H., Ohta, K., *Inorg. Chem.*, 1984, **23**, 3312.
41. Manka, J. S., Lawrence, D. S., *J. Am. Chem. Soc.*, 1990, **112**, 2440.
42. Rao, T. V. S., Lawrence, D. S., *J. Am. Chem. Soc.,* 1990, **112**, 3614.
43. Isnin, R., Kaifer, A. E., *J. Am. Chem. Soc.*, 1991, **113**, 8188.
44. Wylie, R. S., Macartney, D. H., *J. Am. Chem. Soc.*, 1992, **114**, 3136.
45. Wenz, G., Bey, E., Schmidt, L., *Angew. Chem. Int. Ed. Engl.*, 1992, **31**, 783.
46. Murakami, H., Kawabuchi, A., Kotoo, K., Kunitake, M., Nakashima, N., *J. Am. Chem. Soc.*, 1997, **119**, 7605.
47. Harada, A., Li, J., Kamachi, M., *J. Chem. Soc., Chem. Commun.*, **1997**,1413.
48. a) Harada, A., *Adv. Polym. Sci.*, 1997, **133**, 141.
b) Nepogodiev, S. A., Stoddart, J. F., *Chem. Rev.*, 1998, **98**, 1959.
c) Wenz, G., *Angew. Chem. Int. Ed. Engl.*, 1994, **33**, 803.
49. Harada, A., Li, J., Kamachi, M., *Nature*, 1992, **356**, 325.
50. Shigekawa, H., Miyake, K., Sumaoka, J., Harada, A., Komiyama, M., *J. Am. Chem. Soc.*, 2000, **122**, 5411.
51. Harada, A., Adachi, A., Kawaguchi, Y., Okada, M., Kamachi, M., *Polym. J.*, 1996, **28**, 159.
52. Kawaguchi, Y., Harada, A., *Org. Lett.*, 2000, **2**, 1353.
53. Kawaguchi, Y., Harada, A., *J. Am. Chem. Soc.*, 2000, **122**, 3797.
54. Harada, A., Kawaguchi, Y., Hoshino, T., *J. Incl. Phenom. Macrocycl. Chem.*, 2001, **41**, 115.
55. Hoshino, T. Miyauchi,, M., Kawaguchi, Y., Yamaguchi, H., Harada, A., *J. Am. Chem. Soc.*, 2000, **122**, 9876.

**< Section 1.3 >**

56. Gibson, H. W., Bheda, M. C., Engen, P. T., *Prog. Polym. Sci.*, 1994, **19**, 843.
57. Ashton, P. R., Baxter, I., Fyfe, M. C. T., Raymo, F. M., Spencer, N., Stoddart, J. F., White, A. J. P., Williams, D. J., *J. Am. Chem. Soc.*, 1998, **120**, 2297.
58. Owen, G. J., Hodge, P., *Chem. Commun.*, **1997**, 11.

59. Mason, P. E., Parsons, I. W., Tolley, M. S., *Polymer*, 1998, **39**, 3981.
60. Anderson, S., Anderson, H. L., *Angew. Chem. Int. Ed. Engl.*, 1996, **35**, 1956.
61. Tuncel, D., Steinke, J. H. G., *Chem. Commun.*, **2002**, 496.
62. Park, K. M., Kim, S. Y., Heo, J., Whang, D., Sakamoto, S., Yamaguchi, K., Kim, K., *J. Am. Chem. Soc.*, 2002, **124**, 2140.
63. Jansen, K., Buschmann, H. J., Zliobaite, E., Schollmeyer, E., *Thermochim. Acta*, 2002, **385**, 177.
64. Isobe, H., Sato, S., Nakamura, E., *Org. Lett.*, 2002, **4**, 1287.
65. Zhao, J. Z., Kim, H. J., Oh, J., Kim, S. Y., Lee, J. W., Sakamoto, S., Yamaguchi, K., Kim, K., *Angew. Chem. Int. Ed.*, 2001, **40**, 4233.
66. Lee, J. W., Kim, K. P., Kim, K., *Chem. Commun.*, **2001**, 1042.
67. Wagner, B. D., Fitzpatrick, S. J., Gill, M. A., MacRae, A. I., Stojanovic, N., *Can. J. Chem.*, 2001, **79**, 1101.
68. Day, A. I., Blanch, R. J., Arnold, A. P., Lorenzo, S., Lewis, G. R. Dance, I., *Angew. Chem., Int. Ed.*, 2001, **41**, 275.
69. Kim, K., *Chem. Soc. Rev.*, 2002, **31**, 96.
70. Tuncel, D., Steinke, J. H. G., *Chem. Commun.*, **2001**, 253.
71. Tuncel, D., Steinke, J. H. G., *Chem. Commun.*, **1999**, 1509.
72. Meschke, C., Buschmann, H. J., Schollmeyer, E., *Polymer*, 1999, **40**, 945.
73. Krasia, T. C., Steinke, J. H. G. *Chem. Commun.*, **2002**, 22.
74. Behrend, R., Meyer, E., Rusche, F., *Liebigs Ann. Chem.*, 1905, **339**, 1.
75. Mock, W. L., *Top. Curr. Chem.*, 1995, **175**, 1.
76. Mock, W. L., Shih, N. Y., *J. Org. Chem.*, 1986, **51**, 4440.
77. Mock, W. L., Shih, N. Y., *J. Am. Chem. Soc.*, 1988, **110**, 4706.
78. Mock, W. L., Shih, N. Y., *J. Am. Chem. Soc.*, 1989, **111**, 2697.
79. Mock, W. L., Irra, T. A., Wepsiec, J. P., Adhya, M., *J. Org. Chem.*, 1989, **54**, 5302.
80. Mock, W. L., Irra, T. A., Wepsiec, J. P., Manimaran, T. L., *J. Org. Chem.*, 1983, **48**, 3619.
81. Buschmann, H. J., Cleve, E., Jansen, K., Wego, A., Schollmeyer, E., *J. Incl. Phenom. Macrocycl. Chem.*, 2001, **40**, 117.
82. Buschmann, H. J., Jansen, K., Schollmeyer, E., *Thermochim. Acta*, 1998, **317**, 95.
83. Buschmann, H. J., Jansen, K., Schollmeyer, E., *Acta. Chim. Slov.*, 1999, **46**, 405.
84. Buschmann, H. J., Jansen, K., Schollmeyer, E., *Thermochim. Acta*, 2000, **346**, 33.
85. Meschke, C., Buschmann, H. J., Schollmeyer, E., *Thermochim. Acta*, 1997, **297**, 43.
86. Marquez, C., Nau, W. M., *Angew. Chem. Int. Ed.*, 2001, **40**, 3155.
87. Buschmann, H. J., Schollmeyer, E., *J. Incl. Phenom. Mol. Recognit. Chem.*, 1992, **14**, 91.
88. Kornmuller, A., Karcher, S., Jekel, M., *Water Res.*, 2001, **35**, 3317.
89. Karcher, S., Kornmuller, A., Jekel, M., *Water Res.*, 2001, **35**, 3309.
90. Whang, D., Heo, J., Park, J. H., Kim, K., *Angew. Chem., Int. Ed.*, 1998, **37**, 78.
91. Dantz, D. A., Meschke, C., Buschmann, H. J., Schollmeyer, E., *Supramol. Chem.*, 1998, **9**, 79.
92. El Haouaj, M., Ko, Y. H., Luhmer, M., Kim, K., Bartik, K., *J. Chem. Soc., Perkin Trans. 2*, **2001**, 2104.
93. Mock, W. L., Pierpont, J., *J. Chem. Soc., Chem. Commun.*, **1990**, 1509.
94. Jun, S. I., Lee, J. W., Sakamoto, S., Yamaguchi, K., Kim, K., *Tetrahedron Lett.*, 2000, **41**, 471.
95. Isobe, H., Tomita, N., Lee, J. W., Kim, H. J., Kim, K., Nakamura, E., *Angew. Chem. Int. Ed.*, 2000, **39**, 4257.
96. Kim, J., Jung, I. S., Kim, S. Y., Lee, E., Kang, J. K., Sakamoto, S., Yamaguchi, K., Kim, K., *J. Am. Chem. Soc.*, 2000, **122**, 540.
97. Flinn, A., Hough, G. C., Stoddart, J. F., Williams, D. J., *Angew. Chem. Int. Ed. Engl.*, 1992, **31**, 1475.
98. Day, A., Arnold, A. P., Blanch, R. J., Snushall, B., *J. Org. Chem.*, 2001, **66**, 8094.
99. Day, A. I., Arnold, A. P., *PCT Int. Appl.*, (Unisearch Limited, Australia ; Blanch, Rodney John)., WO0068232, 2000.
100. Whang, D., Jeon, Y. M., Heo, J., Kim, K., *J. Am. Chem. Soc.*, 1996, **118**, 11333.
101. Virovets, A. V., Samsonenko, D. G., Dybtsev, D. N., Fedin, V. P., Clegg, W., *J. Struct. Chem.*, 2001, **42**, 319.
102. Samsonenko, D. G., Sokolov, M. N., Virovets, A. V., Pervukhina, N. V., Fedin, V. P., *Eur. J. Inorg. Chem.*, **2001**, 167.
103. Fedin, V. P., Sokolov, M., Lamprecht, G. J., Hernandez-Molina, R., Seo, M. S., Virovets, A. V., Clegg, W., Sykes, A. G., *Inorg. Chem.*, 2001, **40**, 6598.
104. Meschke, C., Buschmann, H. J., Schollmeyer, E., *Macromol. Rapid Commun.*, 1998, **19**, 59.

105. Lee, J. W., Ko, Y. H., Park, S. H., Yamaguchi, K., Kim, K., *Angew. Chem. Int. Ed.*, 2001, **40**, 746.
106. Sohgawa, Y. H., Fujimori, H., Shoji, J., Furusho, Y., Kihara, N., Takata, T., *Chem. Lett.*, **2001**, 774.
107. Taylor, P. N., O'Connell, M. J., McNeill, L. A., Hall, M. j., Aplin, R. T., Anderson, H. L., *Angew. Chem. Int. Ed.*, 2000, **39**, 3456.
108. Raymo, F. M., Stoddart, J. F., *Pure Appl. Chem.*, 1997, **69**, 1987.
109. Wenz, G., Keller, B., *Angew. Chem. Int. Ed. Engl.*, 1992, **31**, 197-199.
110. Wenz, G., Keller, B., *Macromol. Symp.*, 1994, **87**, 11.
111. Kawaguchi, Y., Harada, A., *J. Am. Chem. Soc.*, 2000, **122**, 3797.
112. Harada, A., *Acc. Chem. Res.*, 2001, **34**, 456.
113. Choi, S., Lee, J. W., Ko, Y. H., Kim, K., *Macromolecules*, 2002, **35**, 3526.
114. Tuncel, D., Steinke, J. H. G., unpublished results.
115. Krasia, T., Khodabakhsh, S., Man, T. T. K., Tuncel, D., White, A. J. P., Williams, D. J., Steinke, J. H. G., submitted.
116. Krasia, T., Steinke, J. H. G., in preparation.
117. Khodabakhsh, S., Steinke, J. H. G., unpublished results.

# 2
# Macromolecular Organic Assemblies

# 2.1 Nanometer-size Multispin Macromolecules and Their Magnetic Force Microscopic Images

**Tsuyoshi Michinobu and Hiroyuki Nishide**

## 2.1.1 Introduction

There is considerable interest in constructing magnetically active nano-scale materials because the magnetic structure size is a key factor in determining the upper limit of the stored data density in magnetic materials.[1,2] The major approach has typically been top-down engineering methods of grains or etching bulk materials. On the other hand, bottom-up procedures from molecules open up the possibility of creating new nanostructured building blocks. In particular, organic macromolecules are useful candidates in developing micro-electronic and -magnetic devices because they possess a single-molecular size of nanometer order.

Recent advances in organic synthesis have allowed the preparation of well-defined macromolecules with active functions. For example, macromolecules bearing a large number of unpaired electrons which reveal fascinating magnetic properties in term of molecular-based magnets have been studied.[3,4] One of the advantages of such a chemical bottom-up approach is that the size and shape of the molecules are easily designed and constructed by conventional macromolecular chemistry to give macromolecules with two- or three-dimensional topologies such as dendritic, ladder and star structures. If such nanometer-size macromolecules themselves become magnetically detectable, they are potentially a new class of magnetic materials.

There is another interest in nanoscale science and technology, i.e., the observation of nanometer-size structures with scanning probe microscopies (SPMs). Among the SPMs, several kinds of methods to directly image magnetic structures have been developed.[5] The instrument most widely used now is magnetic force microscopy (MFM), which has a high spatial resolution on a 10-nm scale.

In this section, we briefly describe the recent progress made in magnetically responsive macromolecules bearing unpaired electrons and their first application to MFM.

## 2.1.2 Macromolecules Bearing Radical Groups

It was long assumed that organic molecules are magnetically inactive and only transition and lanthanide elements and their related compounds are applicable to magnetic materials. Recently, however, a variety of organic radical molecules have opened up a new field of magnetic materials, offering tremendous potential in combination with the fine-tunable characteristics of organic molecules. The study of macromolecules bearing unpaired electrons of radical groups, or organic polyradicals, is one example of this trend.

The first type of organic polyradicals is composed of an unconjugated backbone bearing highly stable radicals in its side chains. Such macromolecules can be synthesized from (i) the preparation and subsequent polymerization of a radical monomer containing a vinyl group, (ii) oxidation or reduction of a precursor macromolecule and (iii) introduction of a radical unit to a macromolecular backbone through a polymer reaction. For the example of (i), poly(4-methacryloyloxy-2,2,6,6-tetramethylpiperidine-1-oxyl) **1** was prepared by the anionic polymerization of the corresponding monomer with butyllithium as the initiator.[6] Polymer reactions of precursor macromolecules often gave the desired polyradicals with reasonable yields; the almost quantitative conversion of poly(4-methacryloyloxyl-2,2,6,6-tetramethylpiperidine) to **1** via its oxidation with *m*-chloroperbenzoic acid has been reported.[7] The studies concluded that they are chemically stable at ambient conditions and that an electron exchange interaction takes place between the pendant (here nitroxide) radicals. They also suggested a weak through-space antiferromagnetic interaction between the pendant radicals. These polyradicals have been investigated as redox resins and antioxidants.

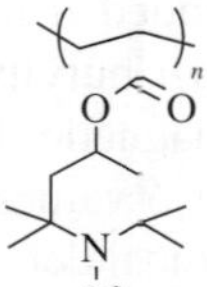

**1**

**Scheme 2.1**

In order to give a ferromagnetic character to these polyradicals, π-electrons play a vital role. Since the first report of an organic macromolecular ferromagnet in 1986 by A. A. Ovchinnikov *et al.*,[8] a π-conjugated backbone has been used to ferromagnetically connect radical species via the through-bond intramolecular exchange interaction. The design of a non-Kekulé and non-disjoint type molecule is a prerequisite for the ferromagnetic interaction in a π-conjugated polyradical.[9] Taking into account the non-Kekulé type biphenyl diradical **2**, the effective overlap of the two nonbonding molecular orbitals (NBMOs) on the *m*, *p*'-isomer favors the parallel orientation of the two unpaired electrons according to Hund's rule to stabilize the triplet ground state, which is called a non-disjoint system. On the other hand, the complete separation of the two NBMOs on the *m*,*m*'-isomer results in a smaller exchange interaction between the two unpaired electrons, and consequently, the almost total degeneration of the triplet and singlet states, which is called the disjoint system (Fig. 2.1).

m,p′-2
(non-disjoint)

m,m′-2
(disjoint)

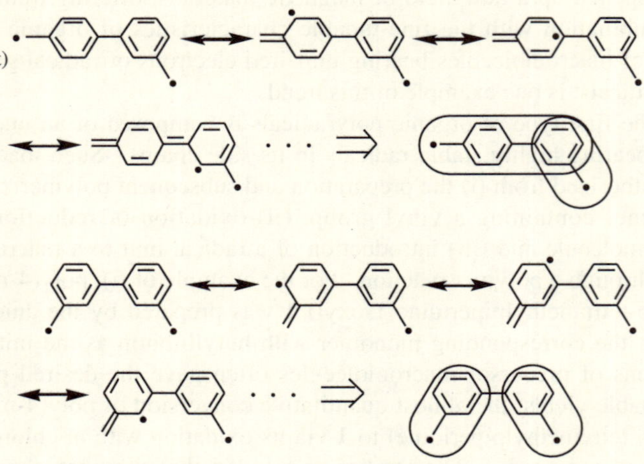

**Fig. 2.1** Non-disjoint and disjoint type connectivities of diradical **2**.

π-Conjugated polyradicals precisely prepared according to the non-Kekulé and non-disjoint rules are expected to have a very high spin quantum number ($S$) in proportion to the degree of polymerization. A. Rajca synthesized poly(1,3-phenylenephenylmethine)s and extended it to the star-shaped decaradical **3** and dendritic pentadecaradical **4**(Ar = *p*-*t*-butylphenyl).[10] These polyradicals displayed a strong through-bond ferromagnetic interaction between adjacent radicals at low temperatures but relatively low $S$ values, e.g., 3.5 for **4** with 15 radical sites. This is because a small number of radical defects or a failure in the generation of a radical interrupts the π-conjugated pathway and consequently suppresses an increase of $S$ value. In order to overcome this disadvantage, a calix[4]arene ring was newly selected as a macrocyclic structure with two pathways for the exchange interaction. The defect-insensitive macrocyclic polyradical **5** showed a high $S$ value of 3.8 as expected.[11] Polycondensation of the two tetrafunctionalized calix[4]arene subparts using the Negishi coupling reaction gave a polymacrocyclic network macromolecule. The corresponding polyradical **6** is designed with an alternating connectivity of the macrocyclic subpart ($S$ = 2) and linker moiety ($S$ = 1/2) in a non-Kekulé and non-disjoint fashion, inducing a high $S$ value irrespective of either the ferromagnetic or antiferromagnetic coupling between the subpart and the linker moiety. For the soluble part of **6** with $M_n > 10^5$, a striking increase of $S$ > 40 was realized.[12] Remarkably, the insoluble part of **6** exhibited an average $S$ value of 5000 and behaved magnetically as insulating spin glasses and blocked superparamagnets at low temperatures.[13] This is the first example of a purely organic-derived magnetic macromolecule with magnetic anisotropy.

Another approach to obtain a very high-spin polyradical is a π-conjugated

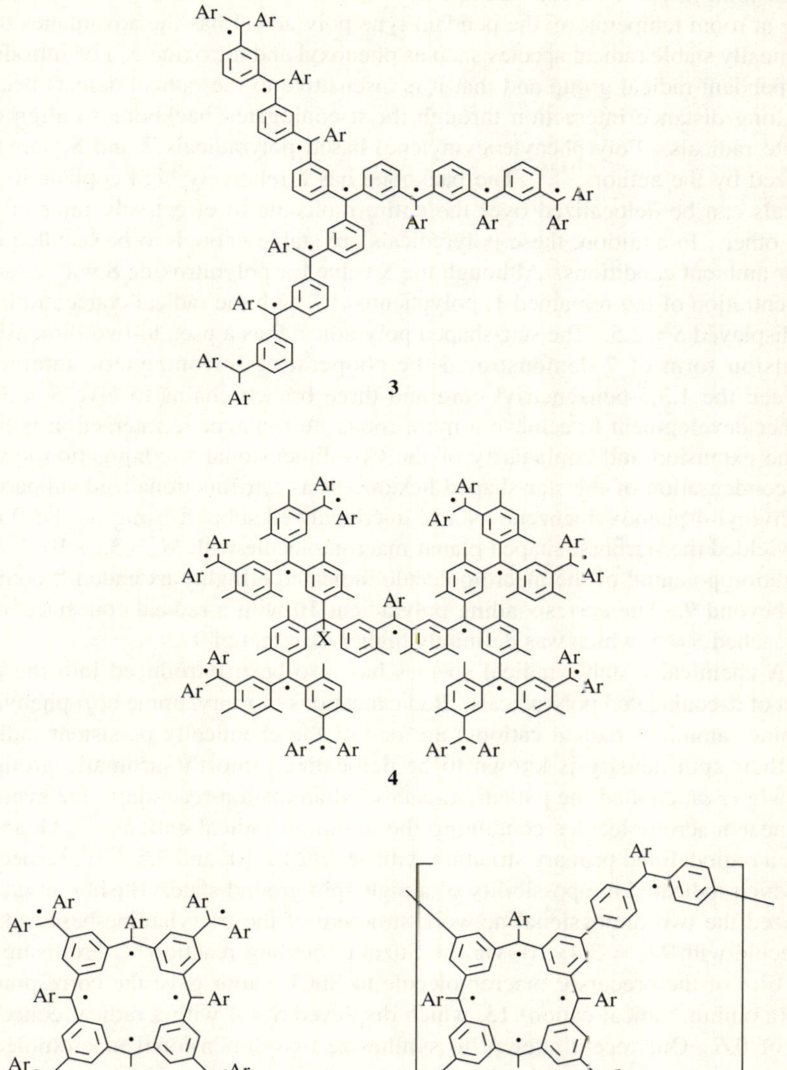


**Scheme 2.2**

macromolecule pendantly bearing multiple radical groups in a non-Kekulé and non-disjoint fashion. In contrast to Rajca's polyradicals, which lack chemical stability at room temperature, the pendant-type polyradical has the advantages that a chemically stable radical species such as phenoxyl and nitroxide can be introduced as a pendant radical group and that it is insensitive to the radical defects because of a long distance interaction through the π-conjugated backbone to align even remote radicals. Poly(phenylenevinylene)-based polyradicals **7** and **8** were synthesized by the authors.[14,15] The backbone has a relatively high coplanarity and radicals can be delocalized over the entire molecule to effectively interact with each other. In addition, these polyradicals are stable enough to be handled even under ambient conditions. Although the $S$ value for polynitroxide **8** with a radical concentration of 0.9 remained 1, polyphenoxyl **7** with the radical concentration of 0.7 displayed $S = 2.5$. The star-shaped polyradical **9** as a pseudo-two-dimensional extension form of **7** demonstrated the cooperative ferromagnetic interaction between the 1,3,5-benzenetriyl core and three branch chains to give $S = 3.5$.[16] Further development to achieve a more robust ferromagnetic interaction is based on the extension and coplanarity of the two-dimensional π-conjugation system. Polycondensation of the star-shaped hexamer **9** as a trifunctionalized subpart and 1,3-divinyl-4-phenoxylbenzene as a bifunctionalized subpart using the Pd/P catalyst yielded the starburst-shaped planar macromolecule with $M_n = 3.2\times10^4$.[17] The ionization potential of the macromolecule indicated a highly extended π-conjugation beyond **9**. The corresponding polyradical **10** with a radical concentration of 0.4 reached $S = 5$, which was definitely higher than that of **9**.

A chemically stable radical species has also been introduced into the main chain of π-conjugated polyradicals. Radical cations of triarylamine or *p*-phenylenediamine (aminium radical cations) are one of the chemically persistent radicals, and their spin density is known to be delocalized into the aromatic groups.[18] Hartwig *et al.* applied the palladium-catalyzed amination reaction to the synthesis of linear macromolecules containing the aminium radical cations.[19] **11** and **12** have a regiodefined primary structure with $M_n = 2.2\times10^4$ and $3.5\times10^4$, respectively, which indicated the possibility of a high-spin ground state. Bushby *et al.* synthesized the two-dimensional network structure of the triarylamine-based macromolecule with $M_w = 3.4\times10^4$ via the Suzuki coupling reaction.[20] Exposure of a thin film of the precursor macromolecule to $SbCl_5$ vapor gave the corresponding poly(aminium radical cation) **13**, which displayed $S = 4$ with a radical concentration of 0.6. Our recent attempt to synthesize two-dimensional macromolecules composed of *p*-phenylenediamine-based aminium radical cations was successful using the palladium-catalyzed amination reaction.[21] Polyradical **14** obtained by the oxidation with $NOPF_6$ exhibited $S = 4$ with a radical concentration of 0.7. All of the polyradicals mentioned here have more or less difficulty in optimizing radical generation to form the largest spin cluster. Nevertheless, the dimensionally extended network polyradical is certainly becoming the next trend for an organic macromolecule that is magnetic at room temperature.

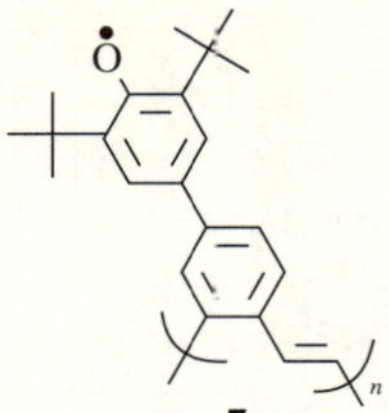

**7**

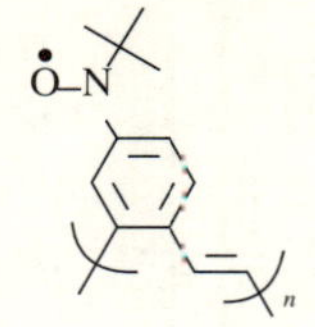

**8**

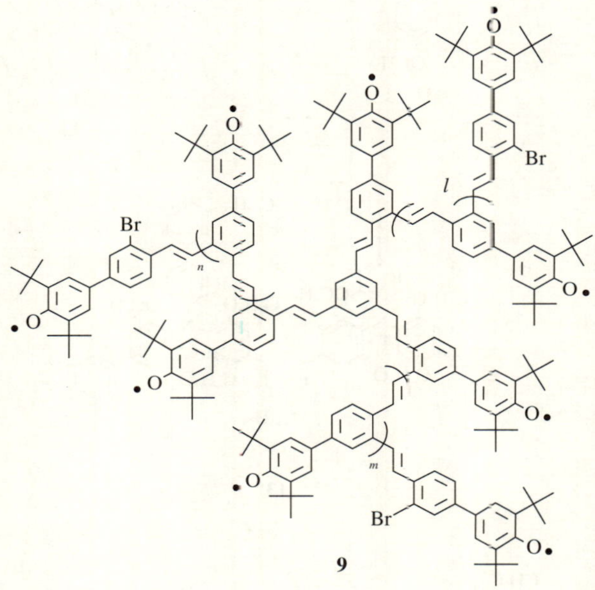

**9**

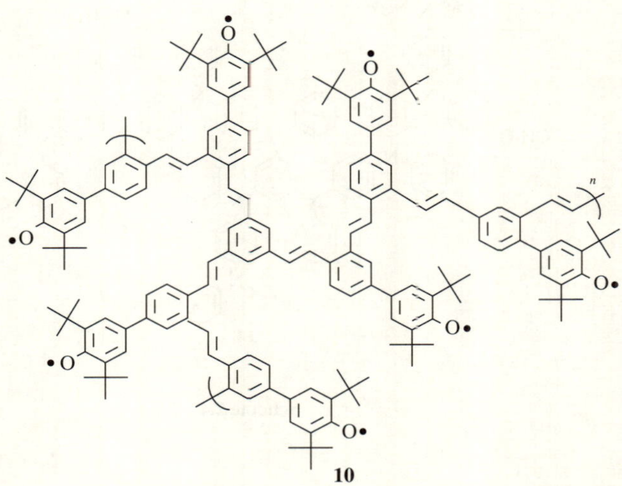

**10**

**Scheme 2.3**

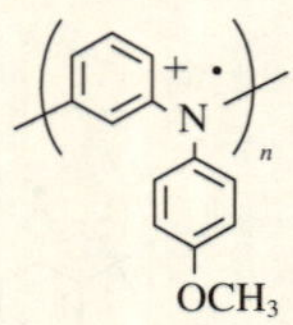

**11**

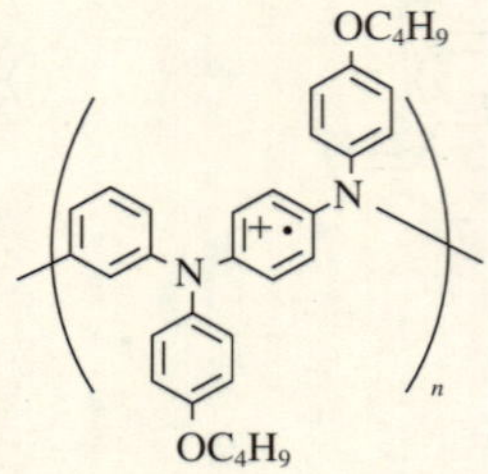

**12**

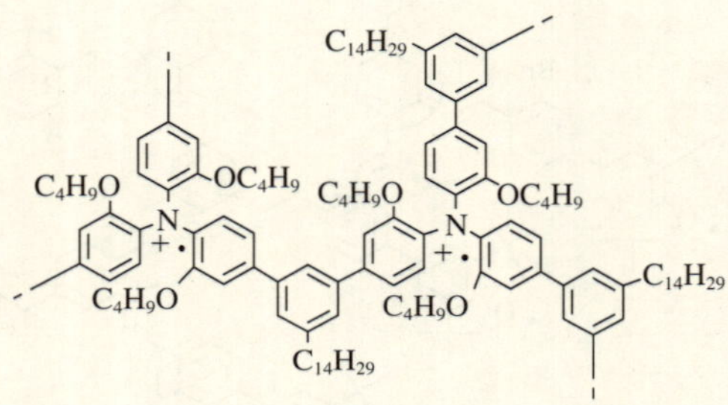

**13**

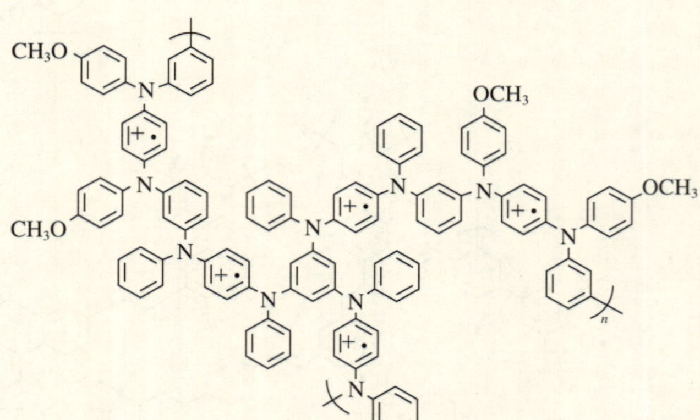

**14**

**Scheme 2.4**

### 2.1.3 Magnetic Force Microscopy

Magnetic force microscopy (MFM) has emerged as a type of non-contact operation mode of scanning probe microscopy.[22,23] The advantages for its magnetic imaging include a fairly high resolution on a 10-nm scale and simple operation under ambient conditions in most cases without any special sample preparation, which made MFM a standard analytical tool and has led to rapid progress in MFM applications to material science and data storage. On the other hand, quantitative interpretation of MFM images still remains a challenging topic and many active studies have been undertaken in this area.

MFM is, from an instrumental point of view, very closely related to atomic force microscopy (AFM) except that it is equipped with a ferromagnetic probe. The microscope's cantilever equipped with the ferromagnetic probe achieves a local magnetostatic interaction between the probe and the stray micromagnetic field from a sample on the substrate. The probe is assumed to effectively behave as a point dipole, which provides the basic idea of the magnetic force acting on the probe: If the effective magnetic moment of the probe is oriented along the vertical direction without inversion induced by the magnetic field of a sample and the magnetic distribution of a sample is not perturbed by the probe's dipole moment, the MFM response is proportional to the second differential with respect to the vertical component of the stray field originating from a sample at the probe location.

Separation of the surface topography and magnetic contrast is a practically important issue especially when scanning a rough surface. The Lift Mode technique developed by Digital Instruments, Inc. is efficient for the complete image acquisition. The first scan is in the tapping mode to recognize the surface or the shape of the sample, that is, AFM with the ferromagnetic probe. After lifting the probe to a certain height, which can be selected as desired, the second scan along the known trajectory above the same area uses the non-contact mode, and the magnetic gradient of the sample is detected by measuring the deflection of the cantilever equipped with the ferromagnetic probe in the static mode of the MFM

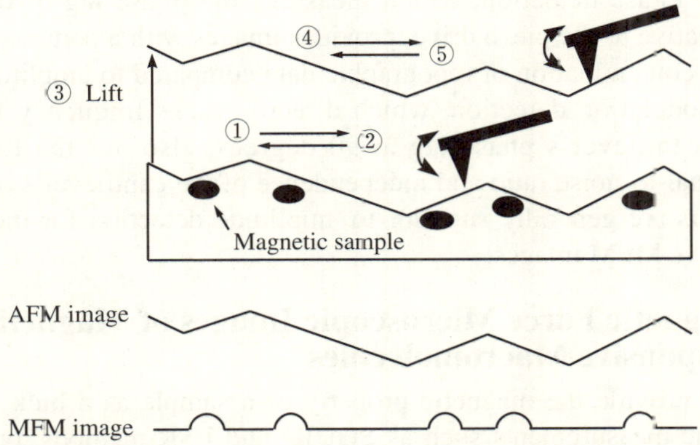


**Fig. 2.2** Schematics of magnetic force microscopy operation.

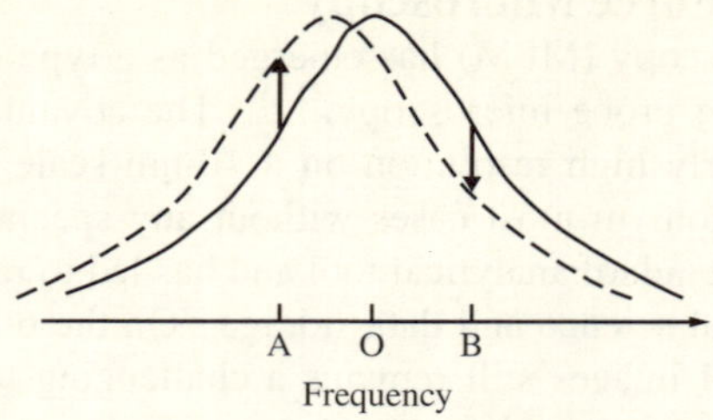


**Fig. 2.3** Shift of resonance curve in amplitude detection. The initial curve (solid line) is shifted to a lower frequency (dotted line) by the attraction between the ferromagnetic probe and the sample. If the drive frequency is set at point A, which is lower than the resonance frequency of point O, an amplitude shift is positive. On the contrary, if the drive frequency is set at point B, the amplitude shift is negative.

operation (Fig. 2.2). During the second scan, the van der Waals force between the probe and the sample almost disappears and the probe mainly responds to magnetic forces. The lateral resolution of the MFM image largely depends on the probe-sample distance and the probe geometry. Close proximity of the probe to the sample surface and sharpening of the probe apex improve resolution.

Actual MFM contrasts are recorded as amplitude, phase, or frequency shifts of the cantilever oscillation using a Nanoscope IIIa with a MultiMode AFM/MFM microscope (Digital Instruments, Inc.). For amplitude detection, the drive frequency of the cantilever is set slightly higher or lower than its resonance frequency and variations in the oscillation amplitude are tracked while raster-scanning across the sample surface. If an attractive force between the probe and the sample is detected, the resonance curve of the cantilever shifts toward a lower frequency. Consequently, the amplitude at a fixed drive frequency lower than the resonance frequency increases, but decreases at a higher drive frequency (Fig. 2.3). Since topographic information never provides inversion of the image, this method can be useful in verifying that the MFM image obtained is derived from the magnetic information. Phase detection, which measures the phase lag of the cantilever oscillation relative to the piezo drive, provides images with a better signal-to-noise ratio and less contamination of topographic data compared to amplitude detection. Frequency modulation detection, which directly tracks frequency variations by keeping the cantilever's phase lag at 90 degrees, also has the features of an improved signal-to-noise ratio and independence of the cantilever's quality factor. These methods are generally superior to amplitude detection for more extensive and quantitative MFM images.

### 2.1.4 Magnetic Force Microscopic Images of Magnetically Responsive Macromolecules

MFM cannot provide the magnetic property of a sample as a bulk evaluated by other magnetic measurements such as SQUID and ESR methods, but it can provide information on an individual magnetic macromolecule of nanometer size.

Magnetically responsive macromolecules of nano-size and well-defined shape that we have synthesized are very suitable for this purpose. Hence, three kinds of magnetic macromolecules were subjected to MFM. For the MFM measurement, a Nanoscope IIIa with a MultiMode AFM/MFM microscope was used. A commercially available MFM probe (type MESP) coated with a ferromagnetic CoCr alloy possessing a magnetic moment of $1 \times 10^{-13}$ emu was magnetized in the direction perpendicular to the sample surface before use. We note here that our target sample was paramagnetic with a weak magnetic moment relative to ferromagnetic materials at which most MFM studies have generally aimed and been utilized. In such cases, the application of an external magnetic field is considered to be a good approach. Since the external magnetic field homogeneously applied over the probe and sample is not detected as an MFM response to some extent, it is often used for investigating the dynamics of a magnetic sample. We set the sample holder on a powerful magnet to enhance the induction magnetization of the sample. The magnet consists of a NdFeB alloy with which a MultiMode SPM scanner AS-130 (J) for the Nanoscope IIIa is equipped. The effective external magnetic field through a stainless sample holder was *ca.* 80 G, which seemed to be sufficient to affect the paramagnetic samples in this series of studies. The magnetic macromolecule would be affected by the external magnetic field, and the ferromagnetic MFM probe would respond to a change in the magnetic permeability at the position of the magnetic macromolecule.

First, we prepared a nanometer-size macromolecular particle bearing 2,2,6,6-tetramethylpiperidine-1-oxyl (TEMPO) radicals, which has been described in section 2.1.2 as the first type of organic polyradical.[24)] In order to obtain a monodispersed particle, the emulsifier-free emulsion polymerization of methacrylic acid, methyl methacrylate and methylenebisacrylamide was used. Because there were no emulsifiers on the particle surface, the introduction of the TEMPO moiety proceeded smoothly to yield polyradical nanoparticles of different radical concentrations. The radical concentration and persistency of the polyradical particles were estimated by SQUID measurements. The polyradical particles were stable under ambient conditions for more than six months, as was the TEMPO radical. A drop of a dilute dichloromethane dispersion of the polyradical particles was transferred onto a highly orientated pyrolytic graphite (HOPG), and the solvent was carefully blotted off by air-drying. The AFM gave *ca.* 100 nm-size dot images for the single particle. The following MFM clearly indicated magnetic gradient responses at exactly the same particle positions. The three-dimensional representation of the MFM image given as an amplitude shift demonstrates holes ascribed to the polyradical particles. By switching the drive frequency of the cantilever from higher to lower against the resonance frequency, the MFM image was reversed, supporting the magnetic response of the observed images and negating any noise image caused by the rough surface information in the AFM. In order to allow a comparison of the particles of different radical concentrations, the phase detection in MFM was used. The same apparatus and the same cantilever were used for all measurements, and the scan lift height or the distance between the probe and the sample surface was maintained constant in the range of 20–30 nm, controlled by the scanning parameters based on the assumption that the sensitivity of the cantilever was 0.03 V $nm^{-1}$. The intensity of the MFM image, i.e., the magnitude of the magnetic gradient, decreased with the radical concentration of the particle (Fig. 2.4). The MFM vertical scale could be represented, to a first approximation, by the magnetic moment of the probe, the quality factor of the cantilever's oscillation, its spring constant and the second differential with respect to the vertical component of the magnetic field, but it was given as an arbitrary unit in this case.

Next, a gadolinium(III) ($Gd^{3+}$) ion in place of the TEMPO radical was tested as the magnetic origin for the MFM image. A $Gd^{3+}$ ion has seven unpaired electrons on the 4f orbital and a large magnetic moment ($S$ = 3.5) in each unpaired electron. A macromolecular particle was prepared by emulsion polymerization of methacrylic acid, ethyl acrylate and allyl methacrylate. By adding a $Gd^{3+}$ ion salt to the particle aqueous suspension at pH 5, the complex of the $Gd^{3+}$ ion with a poly(methacrylic acid) particle was obtained without interference from the emulsifier. Potentiometric titration indicated that the overall complex formation constant is *ca.* $10^{14}$ $M^{-3}$ and that the $Gd^{3+}$ ion can form a very stable tris-coordinate complex with the methacrylic acid residues in the particle. This information enabled us to adjust the $Gd^{3+}$ ion concentration in the particle. Dot images of around 100 nm were observed in the AFM and the corresponding magnetic gradient responses on the particle position in the MFM, similar to the TEMPO particle. The MFM images of the $Gd^{3+}$ particle showed good sensitivity even at a longer probe-sample distance of 20 – 40 nm, probably because of the high-spin alignment in one atom. The MFM of the particle with a low $Gd^{3+}$ ion concentration gave an interesting hollow dot image (Fig. 2.5). On this particle sample, $Gd^{3+}$ ions readily complexed due to the large complex formation constant, and their deficiency brought about the homogeneous distribution only over the particle surface and not inside. A strong magnetic gradient is observed on the perimeter of the particle but very weakly in the interior side, probably giving rise to the hollow dot image.

Based on the above results, π-conjugated but non-Kekulé and non-disjoint type organic polyradicals were examined. For example, the high-spin polyradical **10** is stable even at room temperature and is a two-dimensionally extended macromolecule with a high molecular weight, which will become an appropriate sample for MFM imaging.[17)] The dilute dichloromethane solution of **10** was transferred to the HOPG surface and subjected to AFM. The planar disk-like shape, that is, the horizontal and vertical distances of *ca.* 35 nm and *ca.* 0.6 nm, respectively, were estimated for the single macromolecule **10**, which is much smaller than the TEMPO or $Gd^{3+}$ macromolecules. In the following MFM, the scan lift height was kept at 20 – 25 nm since the limit of the MFM lateral resolution is anticipated to be roughly identical to the lift height. The phase detection technique gave the distinct MFM topography of the hole images, which are ascribed to the high-spin polyradicals or the attraction of the magnetized MFM probe to the polyradicals. This MFM image was reversed when the probe was magnetized in the opposite direction (Fig. 2.6). The MFM images of the polyradical disappeared after about one

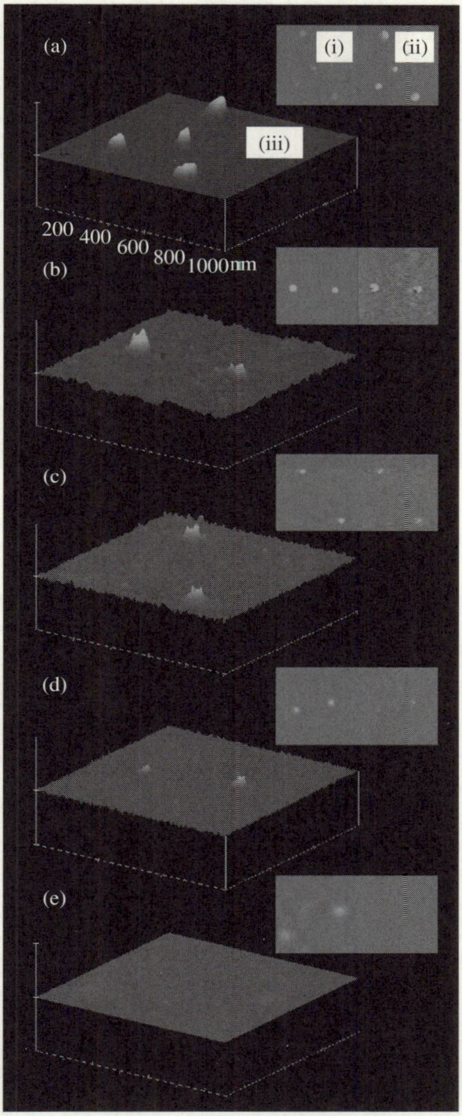


**Fig. 2.4** Effect of radical concentration in 100 nm-size polyradical particles. (i) AFM, (ii) MFM images obtained by phase shift detection, and (iii) three-dimensional representations of MFM images of the polyradical particles with various radical concentrations (mmol/g). (a) 1.1, (b) 0.85, (c) 0.60, (d) 0.17, (e) 0.

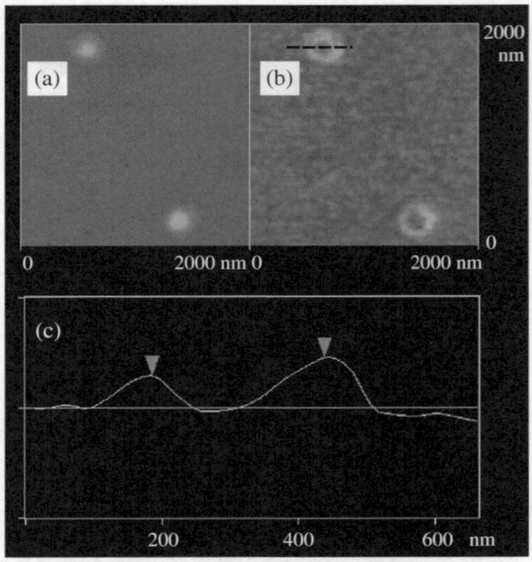


Fig. 2.5 MFM hollow dot image of the particle with low $Gd^{3+}$ ion concentration. (a) AFM, (b) MFM and (c) cross sectional view of the MFM image.

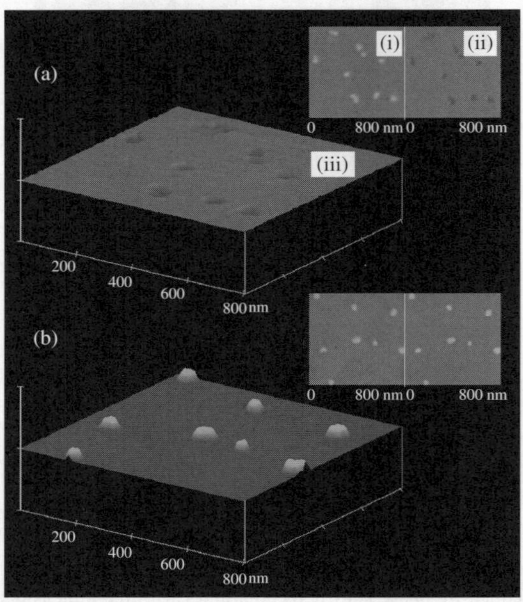


Fig. 2.6 Force microscopic images of the polyradical **10** with $M_n = 3.2 \times 10^4$. (a) (i) AFM, (ii) MFM and (iii) three-dimensional representation of MFM image using the MFM probe magnetized in one direction perpendicular to the sample surface and (b) force microscopic images of **10** under the same conditions except for the use of a probe magnetized in the opposite direction.

day due to inactivation of the radical under ambient conditions, suggesting that the radical generation is essential for the production of the MFM image. A dichloromethane solution of the polyradical **10** and diamagnetic polystyrene was spin-coated on a substrate. MFM produced dots in the image, the size of which almost coincided with the molecular size of **10**. The MFM image was maintained for one week, a fact which is ascribed to the prolonged lifetime of the radical embedded in a glassy polystyrene coating.

It is concluded that 1) MFM is an effective tool for detecting the very weak magnetic signals of magnetically responsive macromolecules, 2) the nanometer-size magnetic macromolecule is a potentially new material of molecular-based magnetic dots and 3) high-spin polyradicals are particularly promising for increasing the data density in magnetic materials because of their smaller size, designable molecular shapes, different radical concentrations and different spin quantum numbers.

## 2.2 Synthesis and Characterization of New Highly Soluble Organic Polyimides

Der-Jang Liaw

### 2.2.1 Introduction

Polyimides have been widely used in the advanced microelectronics industry such as passivation or stress-relief layers for high-density electronic packaging, interlayer dielectric layers for wafer-level semiconductor fabrication, or alignment layers for liquid crystals in advanced liquid crystal display devices (LCDs) owing to their outstanding thermal, mechanical and good insulation properties with low dielectric constant, good adhesion to common substrates and superior chemical stability.[25, 26] However, their applications were limited in many fields because they were insoluble and intractable. Therefore, considerable research has been undertaken to identify new ways to circumvent these restrictions. The introduction of flexible groups and/or bulky units in the polymer backbone has been a general approach to change the chemical structure of polyimides.[27-31] An important example is Ultem 1000® developed and commercialized by General Electric Co.,[32] exhibiting reasonable thermal stability and good mechanical properties together with good moldability.

The incorporation of 2,2'-disubstituted biphenylene in a para-linked polymer chain reduced the interactions of polymer chains.[33, 34] The phenyl rings are forced by the 2,2'-disubstitution into a noncoplanar conformation, decreasing the intermolecular forces between the polymer chains.[33, 34] The crystallization tendency is markedly lowered and the solubilities are significantly enhanced.[33-37] On the other hand, another effective approach to obtain organosoluble polyimides is the incorporation of substituted methylene linkages, such as isopropylidene,[38-40] hexafluoroisopropylidene[41-44] and diphenylmethylene[45, 46] units which provide kinks between the rigid phenyl rings in the backbone and lead to enhanced solubility of the polymers. The incorporation of these flexible linkages such as isopropylidene into the polymer backbone is expected to reduce the crystallinity and enhance the solubility and melt-moldability of the poly(ether imide)s.[38-40] The improved solubility of the polymers is ascribed to the presence of the kink unit in the polymer backbone that lowered the chain rigidity. It was observed that the polymers with diphenylmethylene unit showed better thermal stability than those containing isopropylidene and hexafluoroisopropylidene.[46] Therefore, the incorporation of noncoplanar 2,2'-dimethyl-4,4'-biphenylene and kink diphenylmethylene unit in the poly(ether imide) backbone is expected to provide organosoluble poly(ether imide)s with good thermal stability. Introduction of cardo (Latin meaning loop) groups into the backbone of polymers is another approach for improving solubility and thereby processability. A combination of functional groups offers increased thermal stability with increased solubility in organic solvents because of the specific contribution of the cardo groups[47-52] such as cyclododecylidene,[49] adamantane,[50] norbornyl[51] and tricyclo[5.2.1.0$^{2.6}$]decane[52] groups in the polymer backbone. A cardo group such as *t*-butylcyclohexylidene could be considered as a bulkier pendant group compared with the other pendant groups that are mentioned above (such as cyclododecylidene, adamantane, norbornyl and tricyclo [5.2.1.0$^{2.6}$]decane groups).[49-59] Bulkier pendant groups possibly may contribute to an enhanced solubility of the polymers.[53]

Organic charge-transporting materials are important for electrical and optical applications such as organic photoconductors, electroluminescent, and electrochromic devices.[60-62] To obtain charge transporting polymers, many investigators have recently prepared polymers containing triphenylamine units in the main chain.[61-64] Triphenylamines, which are electron-rich aromatics and can readily oxidize to form radical cation conducting salts, act as electroactive sites.[62] Son *et al.*[60, 61] have successfully prepared triphenylamine-containing polymers, which had hole-transporting ability. Recently, Wu *et al.*[64] reported that the charge injection and electroluminescent efficiency were improved remarkably by the incorporation of the hole-transporting polyimide containing a triphenylamine group in the backbone. Therefore, the introduction of triphenylamine units into polymer backbone is expected to be a potential structural modification to the rigid polymers. In addition, the incorporation of triphenylamine groups in the polyimide backbone would allow them to be used as potential hole-transporting materials.[65]

### 2.2.2 Characterization of the Novel Polyimides

#### A. Monomer Synthesis

The first step in the attempt to prepare organosoluble polyimides was to synthesize diols,[29-31, 54-59] **15A** ~ **15E**, which were carried out in accordance with Scheme 2.5. The diamines were synthesized through the aromatic nucleophilic substitution of corresponding diols with *p*-chloronitrobenzene or *p*-fluoronitrobenzene in the presence of potassium carbonate, followed by catalytic reduction with hydrazine monohydrate and Pd/C (such as synthesis of DBAPB and DBTFAPB in Scheme 2.6). As illustrated in Scheme 2.7, bis(ether anhydride)s were prepared by a three-stage synthetic procedure starting from the nucleophilic nitrodisplacement reaction of 4-nitrophthalonitrile with diols (**15A** ~ **15F**, Scheme 2.7) in dry DMF in the presence of potassium carbonate at room temperature. Use of high temperature (in excess of 100 °C) was avoided as it led to dark colored products in the case of **16A** ~ **16F**. The nitrodisplacement reactions led to a series of new bis(ether nitrile)s (**16A** ~ **16F**, Scheme 2.7). The bis(ether dinitrile)s were then hydrolyzed in an alkaline solution in the presence of hydrogen peroxide to obtain the corre-

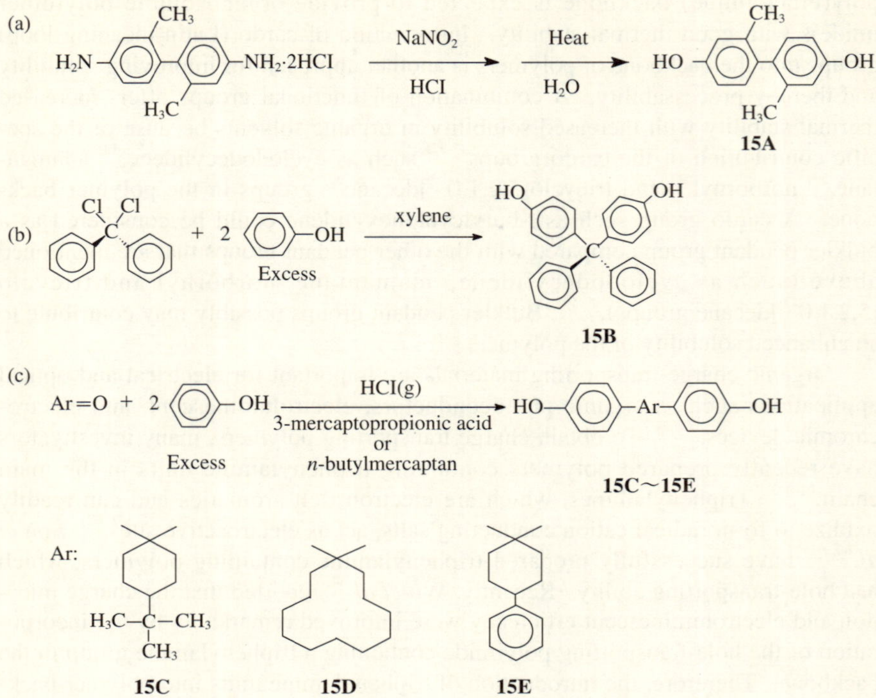


Synthesis of various diols (15A ~ 15E)[54-59].

**Scheme 2.5**

sponding bis(ether diacid)s (**17A** ~ **17F**, Scheme 2.7). It is worth noting that nitriles can be hydrolyzed to give either amides or carboxylic acids. Although the amide was formed initially, carboxylic acid was the most common product since amides were hydrolyzed with acid or base. When carboxylic acid was desired, the reagent of choice was aqueous KOH containing about 6% to 12% hydrogen peroxide, although acid-catalyzed hydrolysis was also carried frequently out. The hydrolysis procedure of **16A** ~ **16F** (except **16B**) was performed for two days. However, the hydrolysis procedure of bis(ether dinitrile) **16B** required more time due to the lower solubility of **16B** than **16A**[54]. After complete hydrolysis, the solution became clear. Before acidification by aqueous HCl, the removal of the residual ethanol is necessary. The presence of the residual ethanol in the aqueous solution always resulted in a viscous product during acidification. The bis(ether diacid)s were then cyclodehydrated to the bis(ether anhydride)s (**18A** ~ **18F**, Scheme 2.7) in a dehydrating agent such as a mixture of acetic anhydride and glacial acetic acid. The structures of these compounds were confirmed by elemental analysis, IR and NMR. For instance, the cyano group (C≡N) of compound

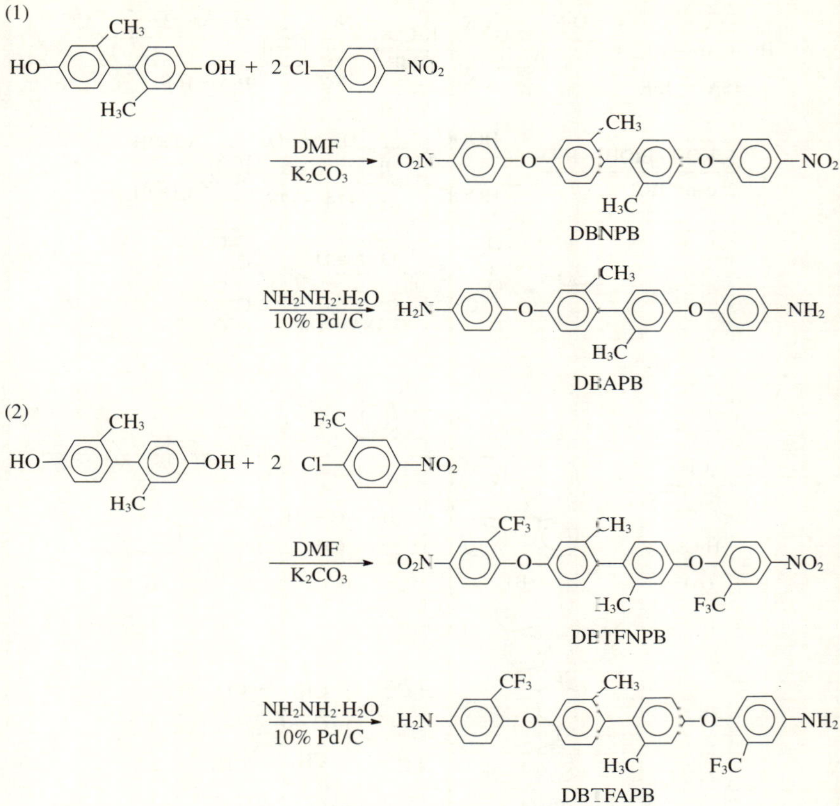


Synthesis of 2,2'-dimethyl-4,4'-bis(4-aminophenoxy)biphenyl (DBAPB) and 2,2'-dimethyl-4,4'-bis(2-trifluoromethyl-4-aminophenoxy)biphenyl (DBTFAPB)[48, 67, 68].

**Scheme 2.6**

**16A** was evident from the peak at 2222 $cm^{-1}$ in the IR spectrum. However, in the IR spectrum of **17A**, the cyano stretching vibration was absent, although a broad C(O)OH absorption appeared in the region of 2500~3600 $cm^{-1}$ and a C=O stretching absorption appeared at 1690 $cm^{-1}$. Furthermore, the IR spectrum of the bis(ether anhydride) **18A** showed characteristic cyclic anhydride absorption at 1837 and 1767 $cm^{-1}$ attributed to the asymmetric and symmetric stretching vibrations of C=O of the anhydride group. NMR spectra data provided clear evidence for the proposed structure of the bis(ether anhydride) monomers.

## B. Preparation of Novel Polyimides

The poly(ether imide)s derived from ether-containing bis(ether anhydride)s (Scheme 2.8)[54-59] or polyimides (Scheme 2.9) were prepared *via* a conventional two-stage procedure involving ring-opening polyaddition of the diamines (Tables

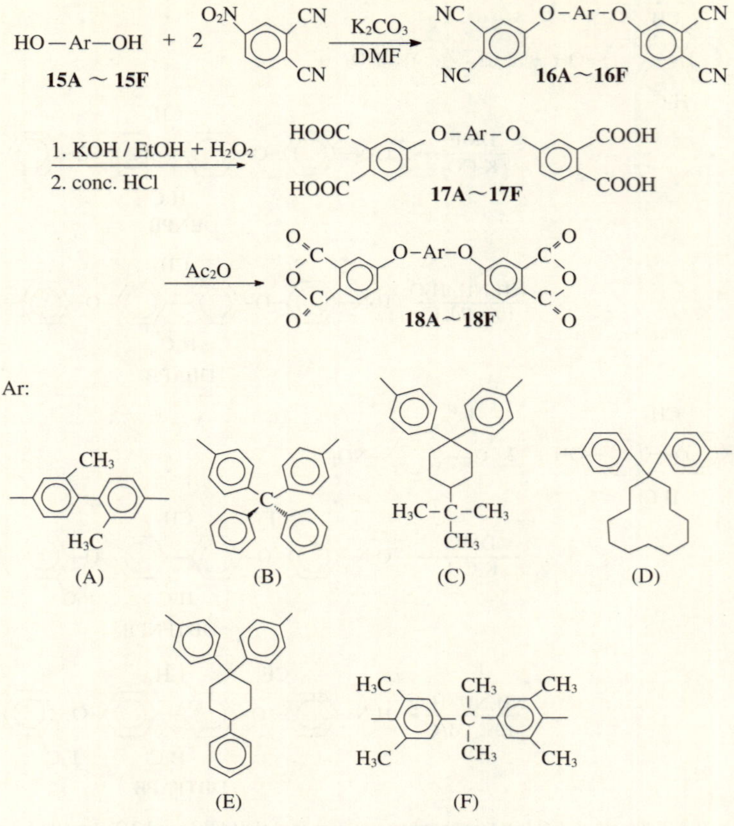


Synthesis of the new bis(ether anhydride)s. (18A ~ 18F)[54-59].

**Scheme 2.7**

2.1, 2.2 and 2.4) with the dianhydrides (Tables 2.3 and 2.5) to form poly(ether amic acid)s (Scheme 2.8) or poly(amic acid)s (Scheme 2.9), followed by thermal or chemical cyclodehydration to poly(ether imide)s or polyimides.[49-53] The polyimide or poly(ether imide) films were obtained by thermal cyclodehydration of the poly(amic acid)s or poly(ether amic acid)s. Most of the polymer films obtained were flexible, transparent and tough.

The IR spectra of poly(amic acid) BAPDM/PMDA(PAA) and the corresponding polyimide BAPDM/PMDA(PI) are shown in Fig. 2.7. The IR spectrum of poly(amic acid) BAPDM/PMDA(PAA) [Fig. 2.7(A)] showed major absorption peaks at 1666, 1537, and 1715 $cm^{-1}$, assigned to amide I, amide II, and carbonyl group (C=O), respectively, and characteristic absorption around 3250 $cm^{-1}$ for N-H and O-H groups. In the spectrum of the polyimide BAPDM/PMDA(PI). [Fig. 2.7(B)] obtained by thermal imidization, the characteristic bands at 1760 and

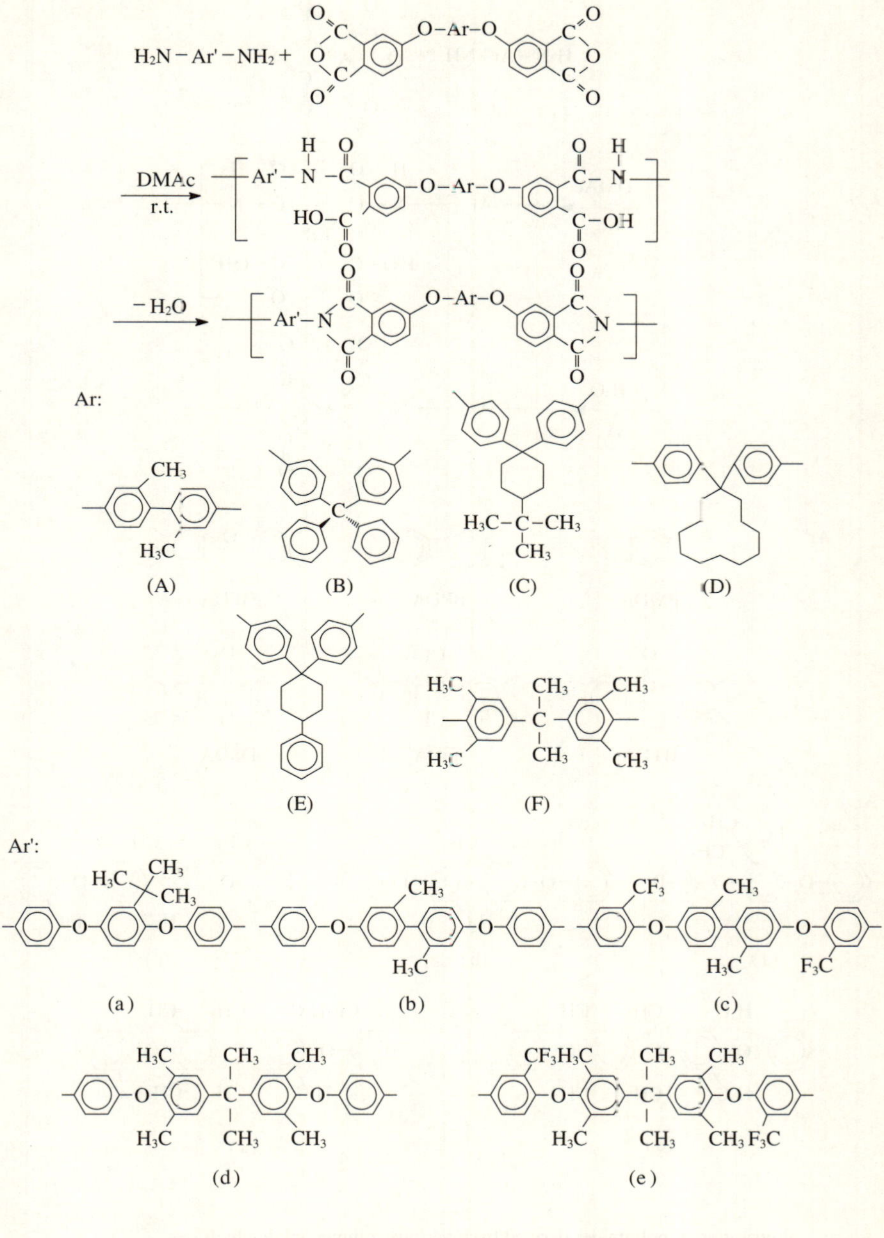

Preparation of various poly(ether imide)s[54-59, 57, 68].

**Scheme 2.8**

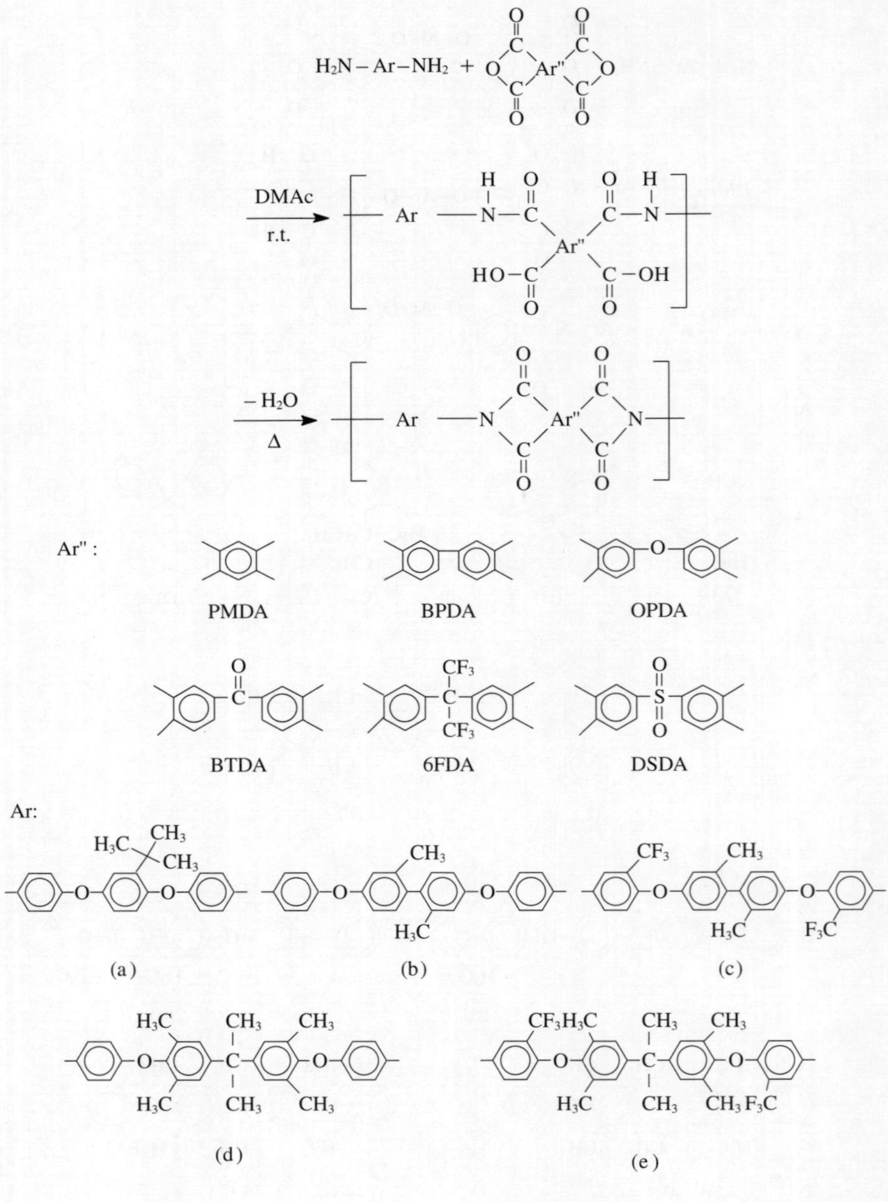

Preparation of polyimides derived from various commercial dianhydrides[33, 39, 68, 73].

**Scheme 2.9**

**Table 2.1** Names, abbreviations and chemical structures of synthesized diamines

| Names and abbreviations | Melting point (°C) | Chemical structures |
|---|---|---|
| 1,4-bis(4-aminophenoxy)-2-*t* -butylbenzene (BATB) | 134−135 | 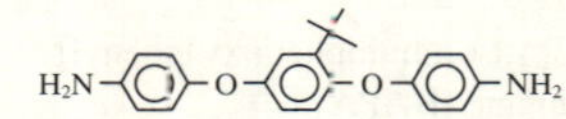 |
| 1,4-bis(4-aminophenoxy)-2,5-di-*t* -butylbenzene (BADTB) | 242−243 | 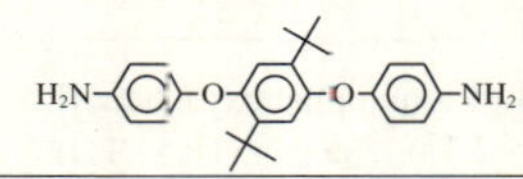 |
| 3,3'5,5'-tetramethyl-bis[4-(4-amino phenoxy)phenyl]sulfone (TBAPS) | 199 | 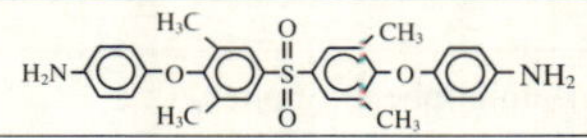 |
| 3,3'5,5'-tetramethyl-2,2-bis[4-(4-aminophenoxy)phenyl]propane (TBAPP) | 164−165 | 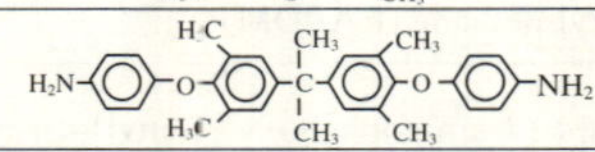 |
| 2,2-bis[4-[2-(4-aminophenoxy) ethoxy]phenyl]propane (BAEPP) | 117−118 | 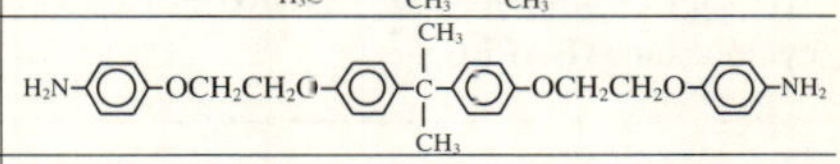 |
| 2,2-bis[4-[2-(4-aminophenoxy) ethoxy]phenyl]sulfone (BAEPS) | 169−171 | 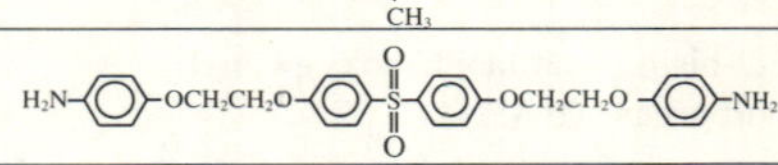 |
| 2,2'-dimethyl-4,4'-bis(4-aminophenoxy) biphenyl (DBAPB) | 138−139 | 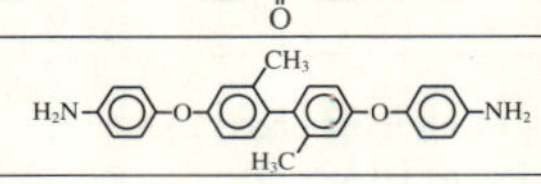 |
| α,α'-bis[3,5-dimethyl-4-(4-aminophenoxy)phenyl] -1,4-diisopropylbenzene (BDAPD) | 224−226 | 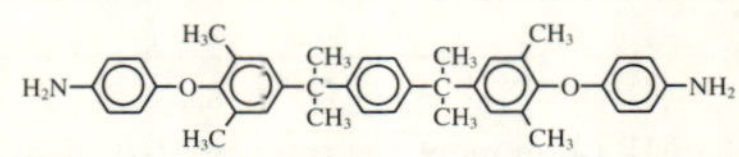 |
| 4,4'-diamino-3",4"-dimethyl-triphenylamine (DADT) | 184−185 | 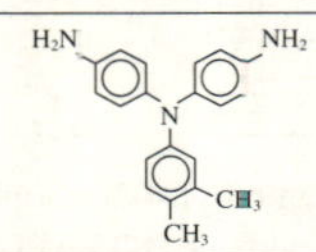 |
| 2,2'-dimethyl-4,4'-bis(2-trifluoromethyl--4-aminophenoxy)biphenyl (DBTFAPB) | 147−148 | 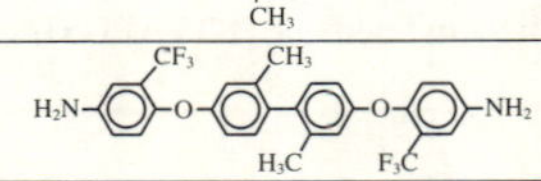 |
| 3,3',5,5'-tetramethyl-2,2-bis[4-(2-trifluoromethyl-4-aminophenoxy) phenyl]propane (TBTFAPP) | 212 | 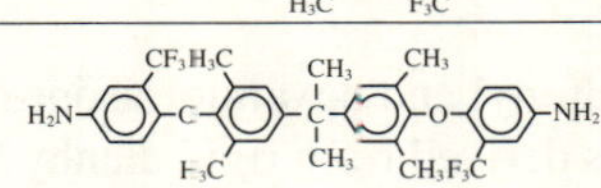 |

1721 $cm^{-1}$ are commonly attributed to the asymmetric and symmetric stretches of the carbonyl groups of imide, respectively. The C−N band at 1373 $cm^{-1}$ verified the formation of the imide structure. Furthermore, the bands at 1666 (amide I) and 1537 $cm^{-1}$ (amide II) disappeared, indicating good imidization while heating at temperatures up to 300 °C.

## C. Solubility

All the polyimides derived from synthesized diamines with common dianhydrides such as DSDA and 6FDA exhibited excellent solubility.[49-53] The high solubility was possibly due to the structural modification through the incorporation of flexi-

**Table 2.2** Names, abbreviations and chemical structures of diamines containing cardo group

| Names and abbreviations | Melting point (°C) | Chemical structures |
|---|---|---|
| 2,2-bis[4-(4-aminophenoxy)phenyl] adamantane (BAPA) | 241–243 | 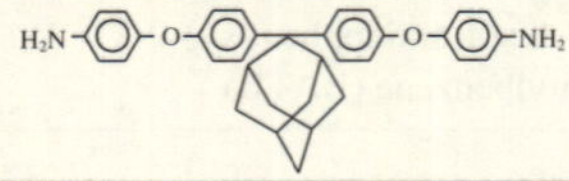 |
| 8,8-bis[4-(4-aminophenoxy)phenyl] tricyclo[5.2.1.0$^{2,6}$]decane (BAPTD) | 184–185 | 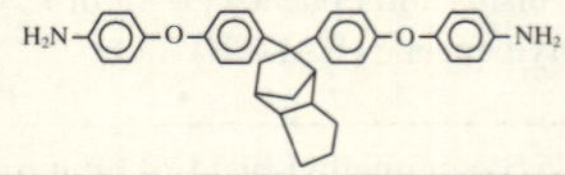 |
| bis[4-(4-aminophenoxy)phenyl] diphenylmethane (BAPDM) | 180–181 | 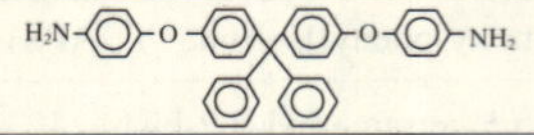 |
| 1,1-bis[4-(4-aminophenoxy)phenyl]-4-*t*-butyl cyclohexane (BAPBC) | 171–172 | 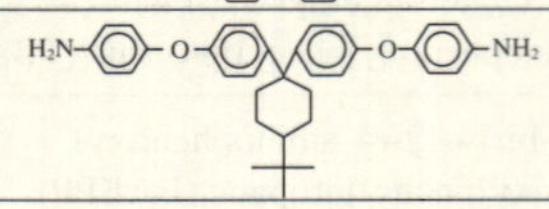 |
| 2,2-bis[4-(4-aminophenoxy)phenyl] norbornane (BAPN) | 149–150 | 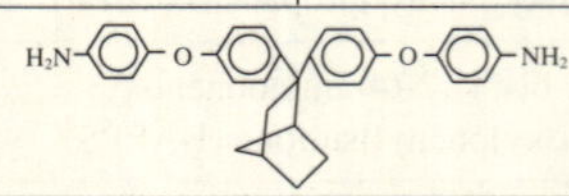 |
| 1,1-bis[4-(4-aminophenoxy)-phenyl] cyclododecane (BAPCD) | 222–224 | 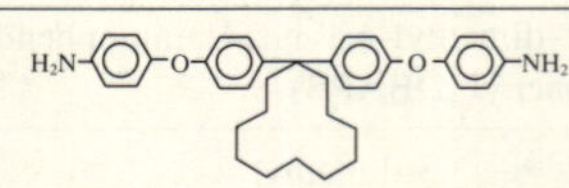 |
| 1,1-bis[4-(4-aminophenoxy)phenyl]-4-phenyl cyclohexane (BAPPC) | 157–159 | 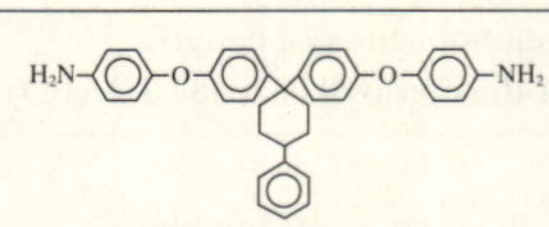 |
| bis[4-(2-trifluoromethyl-4-aminophenoxy) phenyl]diphenylmethane (BTFAPDM) | 199–200 | 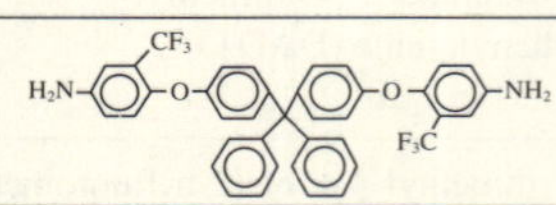 |

ble sulfonyl and hexafluoroisopropylidene groups, respectively. Most of the polyimides derived from rigid dianhydrides such as PMDA, BPDA and BTDA showed less solubility in organic solvents. In general, polyimides derived from such rigid dianhydrides have poor solubility. Among the novel polyimides in our studies, those derived from diamine TBAPP or BAPTD showed excellent solubility.[52, 66] It was observed that the TBAPP-based polyimides derived from dianhydrides BPDA and BTDA were soluble in polar aprotic solvents. The high solubility of TBAPP-based polyimides was possibly governed by the structural modification through the incorporation of the flexible isopropylidene, increasing the overall chain flexibility, and methyl substituted arylene ether groups into the polyimide structure.[35, 66] It has been verified that polyimide derived from diamine DAM3 (Table 2.4) had less solubility than that derived from diamine TBAPP.[66] This result clearly demonstrated that the tetramethyl substituents on the phenylene unit effectively improved the polymer solubility. It has been reported that methyl

**Table 2.3** Names, abbreviations and chemical structures of synthesized bis(ether anhydride)s

| Names and abbreviations | Melting point (°C) | Chemical structures |
|---|---|---|
| 2,2'-dimethyl-4,4'-bis[4-(3,4-dicarboxy phenoxy)]biphenyl dianhydride (DBDPB) | 217–218 | 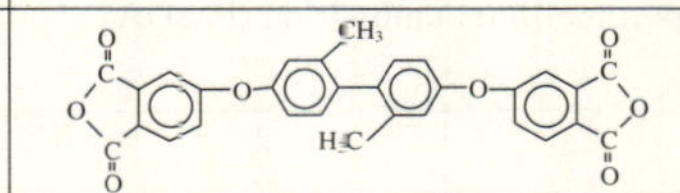 |
| bis[4-(3,4-dicarboxyphenoxy)phenyl] diphenylmethane dianhydride (BDPDM) | 262 | 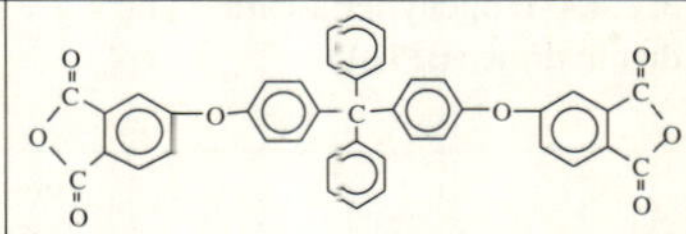 |
| 1,1-bis[4-(3,4-dicarboxyphenoxy)phenyl]-4-*t*-butylcyclohexane dianhydride (BDPTBC) | 211–212 | 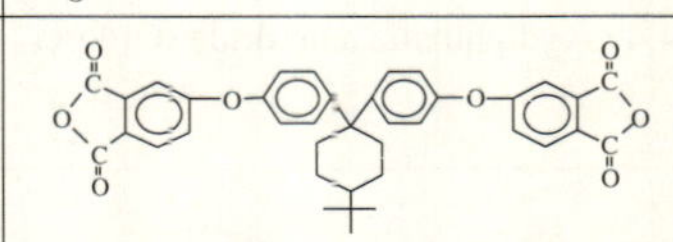 |
| 3,3',5,5'-tetramethyl-2,2-bis[4-(3,4-dicarboxy phenoxy)phenyl]propane dianhydride (TBDPP) | 164–165 | 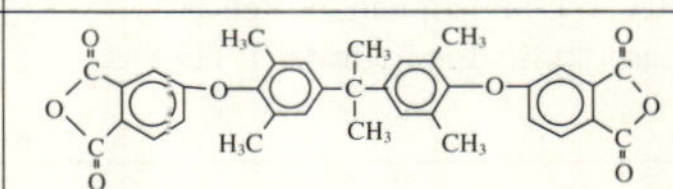 |
| 1,1-bis[4-(3,4-dicarboxyphenoxy)phenyl]cyclo dodecane dianhydride (BDPCD) | 212–213 | 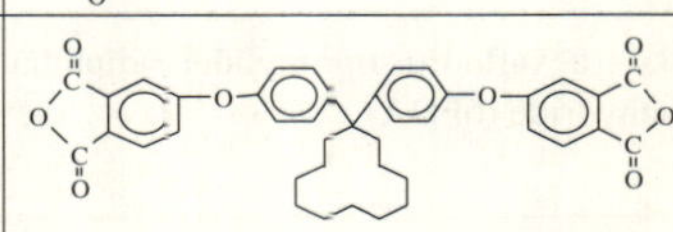 |
| 1,1-bis[4-(3,4-dicarboxyphenoxy)phenyl]-4-phenylcyclohexane dianhydride (BDPPC) | 217–218 | 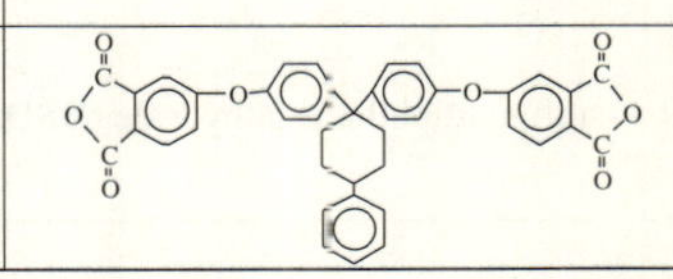 |

**Table 2.4** Names, abbreviations and chemical structures of common diamines

| Names and abbreviations | Melting point (°C) | Chemical structures |
|---|---|---|
| 1,4-bis(4-aminophenoxy) benzene (DAM1) | 172 | 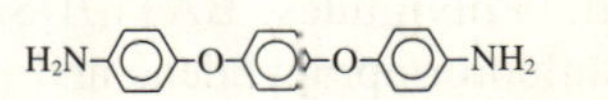 |
| 4,4'-bis(4-aminophenoxy) biphenyl (DAM2) | 200 |  |
| 2,2-bis[4-(4-aminophenoxy) phenyl]propane (DAM3) | 126 | 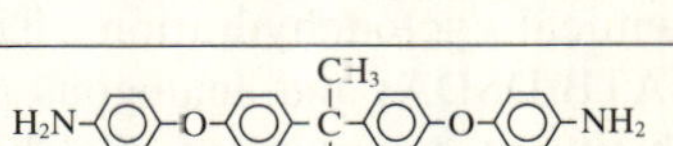 |
| 2,2-bis[4-(4-aminophenoxy)phenyl] hexafluoropropane (DAM4) | 200–203 | 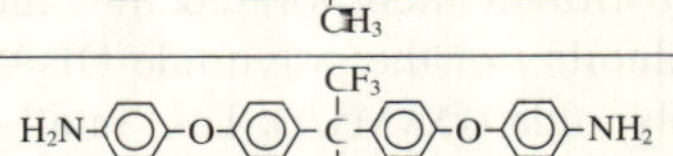 |

**Table 2.5** Names, abbreviations and chemical structures of common dianhydrides

| Names and abbreviations | Melting point (°C) | Chemical structures |
|---|---|---|
| pyromellitic dianhydride (PMDA) | 285 | 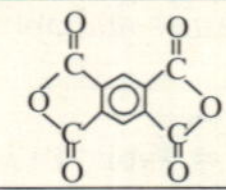 |
| 3,3',4,4'-biphenyltetra-carboxylic dianhydride (BPDA) | 300 | 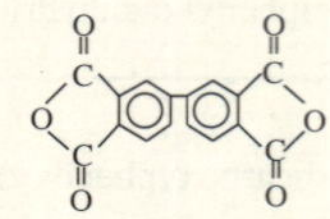 |
| 4,4'-oxydiphthalic anhydride (OPDA) | 226–227 | 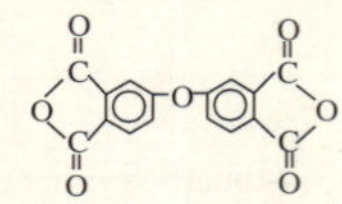 |
| 3,3',4,4'-benzophenone-tetra carboxylic dianhydride (BTDA) | 225 | 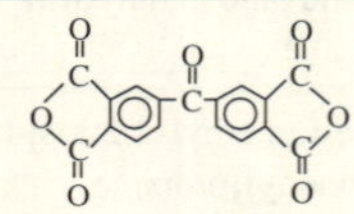 |
| 4,4'-(hexafluoroisopropylidene)diphthalic anhydride (6FDA) | 247 | 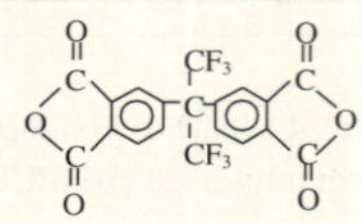 |
| 4,4'-sulfonyldiphthalic anhydride (DSDA) | 287–288 | 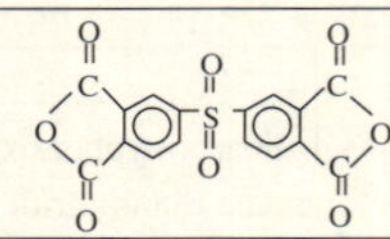 |

groups on the phenylene not only inhibit close packing of the polymer chain but also make the polymer less polar in nature.[48, 53] Solubility of polyimides derived from various commercial dianhydrides is listed in Table 2.6. Polyimides, BATB/PMDA, BATB/BPDA, BATB/OPDA and BATB/BTDA (Scheme 2.9), were almost insoluble in organic solvents but dissolved in concentrated sulfuric acid. Polyimides, BATB/DSDA and BATB/6FDA (Scheme 2.9), containing hexafluoroisopropylidene and sulfone linkages exhibit excellent solubility toward test solvents. For comparison, polyimide (DAM1/DSDA) was prepared by the condensation of 1,4-bis(4-aminophenoxy)benzene (DAM1) with DSDA through chemical cyclodehydration. The results from *t*-butyl-substituted polyimide (BATB/DSDA) and analogous polyimide (DAM1/DSDA) showed that *t*-butyl substituent incorporated into the polyimide did indeed increase its solubility. Solubility of the polyimide DBAPB/DSDA was superior to that of its analogous polyimide (DAM1/DSDA) in all organic solvents, indicating that the solubility of polyimide was improved by introducing the noncoplanar conformation of 2,2'-dimethyl-substituted biphenylene unit in the polymer backbone. It should be noted that polyimides derived from dianhydride BHTDA, containing hydroxyl

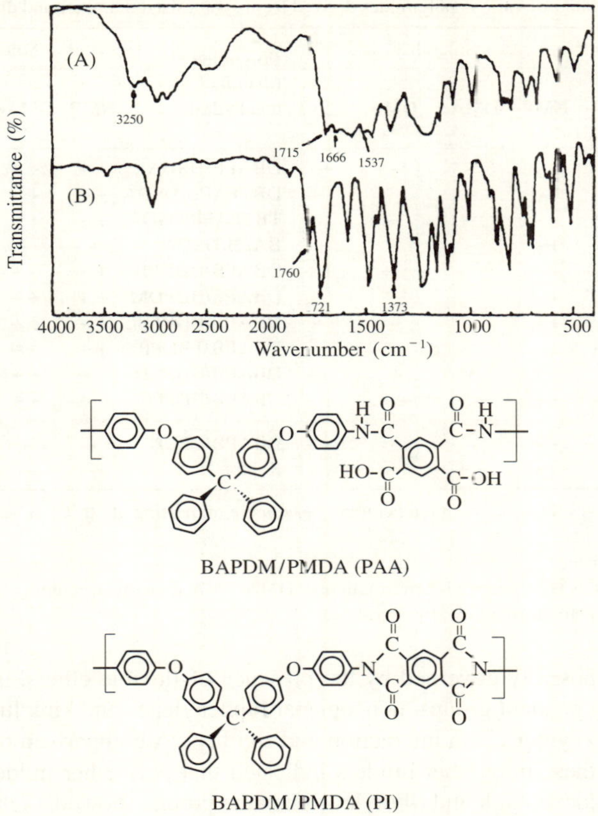


**Fig. 2.7** IR spectra of (A) poly(amic acid) BAPDM/PMDA (PAA) and (B) the corresponding polyimide BAPDM/PMDA (PI)[45, 46].

groups, showed as good solubility as polyimides derived from dianhydride 6FDA, containing hexafluoroisopropylidene units. Introduction of hydroxyl units into the polyimide backbone enhanced the solubility of the polymer due to the strong interaction of the hydroxyl group with polar solvents. Trifluoromethyl-containing polyimide DBTFAPB/DSDA derived from DBTFAPB, and TBTFAPP/DSDA derived from TBTFAPP containing trifluoromethyl groups revealed excellent solubility in contrast to the corresponding polyimide DBAPB/DSDA.[67, 68]

New bis(ether anhydride)s containing flexible ether, isopropylidene groups, bulky pendant groups, noncoplanar biphenylene or kink linkages were synthesized. Almost all of the poly(ether imide)s derived from such novel bis(ether anhydride)s were soluble in the test solvents, including *N*-methyl-2-pyrrolidinone, *N,N*-dimethylacetamide (DMAc), pyridine, cyclohexanone, tetrahydrofuran and even in chloroform at room temperature. The high solubility of these poly(ether

**Table 2.6** Solubility of various polyimides derived from synthesized diamines and dianhydrides[48-59, 68)]

| Polyimide (diamine/ dianhydride) | Solubility[a] | | | | Polyimide (diamine/ dianhydride) | Solubility[a] | | | |
|---|---|---|---|---|---|---|---|---|---|
| | NMP | DMAc | DMF | Py | | NMP | DMAc | DMF | Py |
| BATB/OPDA | +− | − | − | +− | DBAPB/DSDA | ++ | ++ | ++ | +− |
| BADTB/OPDA | − | − | − | − | DBTFAPB/DSDA | ++ | ++ | ++ | ++ |
| DBAPB/OPDA | +− | − | − | − | TBTFAPP/DSDA | ++ | ++ | ++ | ++ |
| TBAPS/OPDA | +− | − | − | − | BAPB/DSDA | − | − | − | − |
| TBAPP/OPDA | ++ | ++ | ++ | ++ | DBAPB/DBDPB | ++ | ++ | ++ | ++ |
| BAEPP/OPDA | +− | − | − | − | DBAPB/BDPDM | ++ | ++ | ++ | ++ |
| BAEPS/OPDA | +− | − | − | − | DBAPB/BDPTBC | ++ | ++ | ++ | ++ |
| BAPCD/OPDA | +− | − | − | − | DBAPB/TBDPP | ++ | ++ | ++ | ++ |
| BAPA/OPDA | ++ | + | + | ++ | DBAPB/BDPCD | ++ | ++ | + | − |
| BAPTD/OPDA | ++ | ++ | ++ | ++ | DBAPB/BDPPC | ++ | ++ | ++ | ++ |
| BATB/OPDA | +− | − | − | +− | DBAPB/BPDA | + | + | +− | − |
| BATB/DSDA | ++ | ++ | ++ | ++ | Kapton H | − | − | − | N/A[b] |

[a] Solubility: ++, soluble at room temperature; +, soluble on heating at 70 °C; + −, partially soluble; −, insoluble.

[b] N/A: not available

[c] Abbreviations: NMP: *N*-methyl-2-pyrrolidinone; DMAc: *N,N*-dimethylacetamide; DMF: *N,N*-dimethylformamide; Py: Pyridine.

imide)s was possibly governed by the presence of flexible ethers, isopropylidene groups, bulky pendant groups, noncoplanar biphenylenes and kink linkages, which reduced the polymer chain interaction and rigidity. A comparison of the polymer solubility of these poly(ether imide)s indicated that poly(ether imide)s containing the isopropylidene, kink and *t*-butylcyclohexane groups showed slightly better solubility than the others. An analogous polyimide DBAPB/BPDA containing biphenylene was nearly insoluble in the test solvents and showed less solubility than these poly(ether imide)s.

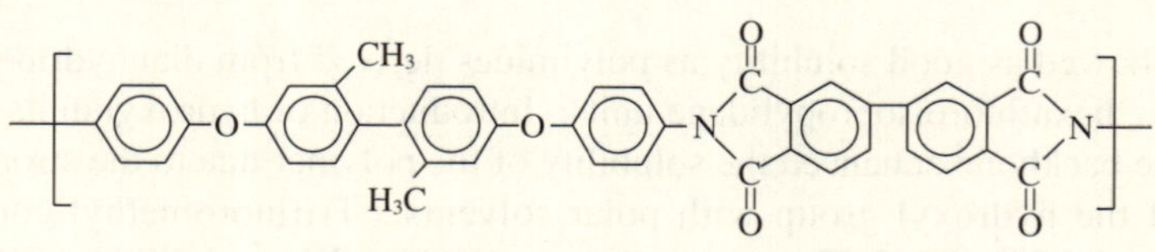

DBAPB/BPDA

The improvement in the solubility of the polyimides showed a similar tendency to that of the corresponding polyamides. For example, 2,5-di-*t*-butyl substituent in polymer backbone was too symmetrical for the solubility to be superior to that of *t*-butyl-substituted polymers.[48-53)] From the solubility behaviors of the polyimides, therefore, it could be concluded that the incorporation of bulky pendant groups along the polymer chain gave the most effective enhancement in the solubility of polymers. (Fig. 2.8)

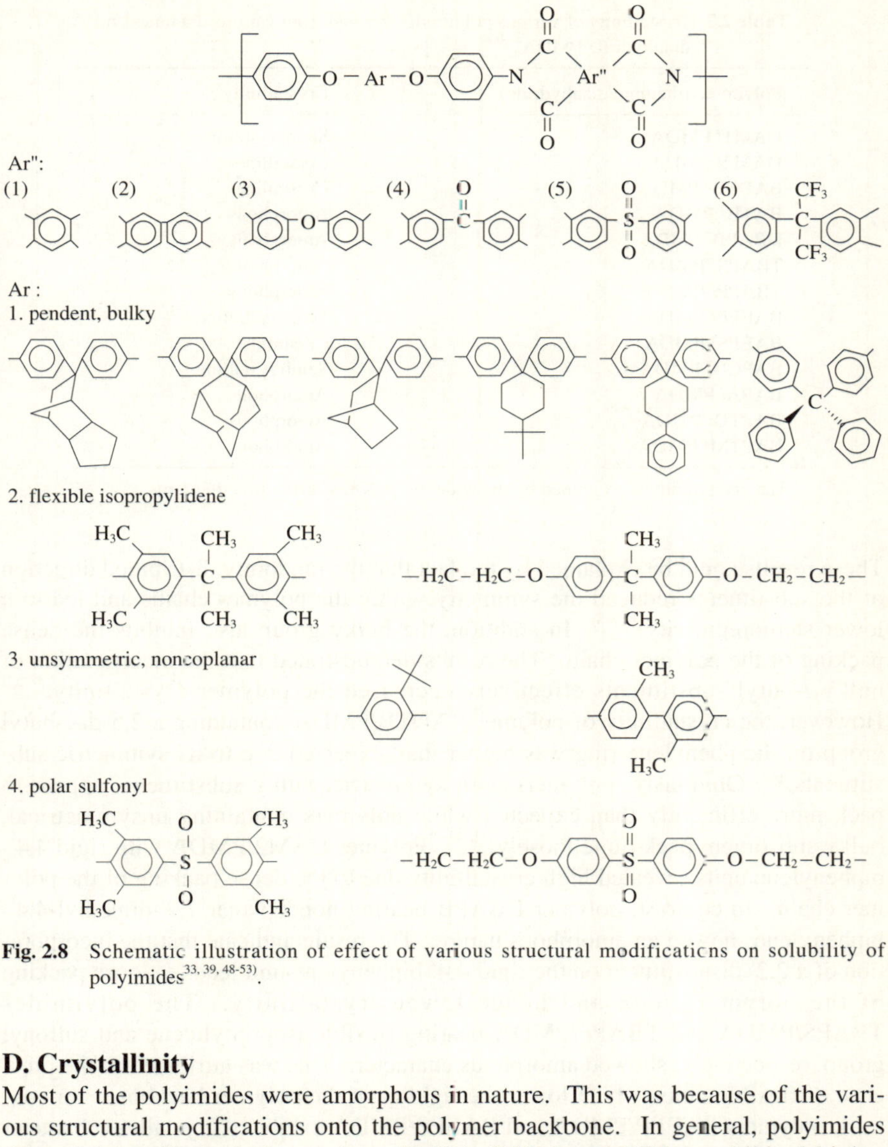


**Fig. 2.8** Schematic illustration of effect of various structural modifications on solubility of polyimides[33, 39, 48-53].

## D. Crystallinity

Most of the polyimides were amorphous in nature. This was because of the various structural modifications onto the polymer backbone. In general, polyimides derived from dianhydrides such as PMDA, BPDA and BTDA exhibited higher crystallinity than the other dianhydrides such as OPDA, DSDA and 6FDA. The higher crystallinity reflected on their poor solubility.[48)]

Crystallinity of the polyimides derived from various diamines with PMDA is summarized in Table 2.7.[48)] It was observed that the polymer DAM1/PMDA having phenylene rings without substituents showed a higher crystallinity than the polymer BATB/PMDA containing *t*-butyl substituent on the phenylene ring.[48-53)]

**Table 2.7** Crystallinity of various polyimides derived from various diamines and dianhydride PMDA[a, 48)]

| Polyimide (diamine/dianhydride) | Crystallinity |
|---|---|
| DAM1/ PMDA | Semicrystalline |
| DAM2/ PMDA | Crystalline |
| BADTB/ PMDA | Crystalline |
| BATB/ PMDA | Amorphous |
| DBAPB/ PMDA | Amorphous |
| TBAPS/ PMDA | Amorphous |
| TBAPP/ PMDA | Amorphous |
| BAEPP/ PMDA | Semicrystalline |
| BAEPS/ PMDA | Crystalline |
| BAPCD/ PMDA | Semicrystalline |
| BAPA/ PMDA | Amorphous |
| BAPTD/ PMDA | Amorphous |
| BAPTM/ PMDA | Amorphous |

[a] The crystallinity is examined by the wide-angle X-ray diffraction diagrams.

These results could be explained by the fact that the randomly distributed direction of the substituents reduced the symmetry within the polymer chains and led to a lower stereoregularity.[42, 48)] In addition, the bulky group also inhibits the dense packing of the polymer chain. The results demonstrated that the incorporation of bulky *t*-butyl substituents effectively decreased the polymer crystallinity.[48-53)] However, the crystallinity of polymer BADTB/PMDA containing a 2,5-di-*t*-butyl group on the phenylene ring was higher than expected due to its symmetric substituents.[48)] Obviously, polymers with symmetric, bulky substituents appear to pack more efficiently than expected while polymers containing unsymmetrical, bulky substituent pack quite loosely.[42, 48)] Polymer DAM2/PMDA with rigid 4,4'-biphenylene units revealed high crystallinity due to the dense packing of the polymer chain. In contrast, polymer DBAPB bearing noncoplanar 2,2'-dimethyl-4,4'-biphenylene showed an amorphous nature. The results indicate that the incorporation of a 2,2'-disubstituent on the rigid 4,4'-biphenylene unit leads to loose packing of the polymer chain and hence lower crystallinity. The polyimides TBAPS/PMDA and TBAPP/PMDA bearing flexible isopropylidene and sulfonyl group, respectively, showed amorphous character. This was attributed to the presence of flexible units, which lower the rigidity of the polymer backbone. In contrast, polymers BAEPP/PMDA and BAEPS/PMDA exhibited crystalline character although they contained flexible oxyethylene units in the backbone. Obviously, the introduction of a flexible and linear oxyethylene group into the polymer backbone led to regular packing of the polymer chain and hence to a high crystallinity.[48)] Owing to the introduction of bulky pendant groups such as adamantane (BAPA/PMDA), tricyclodecane (BAPTD/PMDA) and diphenylmethylene (BAPTM/PMDA), the polymers bearing bulky pendant groups showed amorphous character. It is quite reasonable to assume that the pendant group inhibits the close packing of the polymer chain. In addition, the bulky group also reduces the interaction between polymer chains.[58] However, the polymer containing a pendant cyclododecylidene group exhibited crystalline character. Apparently, the flexibility of the cyclododecane group allows a tight packing than those with stiffer groups such as adamantane, tricyclodecane and diphenyl groups.[58] Comparing the crystallinity of these polyimides, it was concluded that the incorporation of bulky pendant groups along the polymer chain most reduced the crystallinity of polymers.

## E. Thermal Property

Thermogravimetric (TG) analysis revealed good thermal stability for poly(ether imide)s. They were stable up to a temperature of 440 °C. The temperature at 10% weight loss ($Td_{10}$) for these polymers ranged between 441 and 535 °C in nitrogen atmosphere. It was observed that polymers with 2,2'-dimethyl-4,4'-biphenylene units showed a higher $Td_{10}$ value than the others. For example, the $Td_{10}$ value of BATB/DBDPB was observed at 518 °C in nitrogen, which was higher than that of BATB/BDPDM (501 °C, in nitrogen), BATB/BDPTBC (504 °C, in nitrogen), BATB/TBDPP (474 °C, in nitrogen), BATB/BDPCD (441 °C, in nitrogen) and BATB/BDPPC (495 °C, in nitrogen). This is because the polymers with biphenylene units are more rigid than the others. Among the polymers derived from the diamines [Scheme 2.8, (a) and (b)], it was observed that the polymer containing 2,2'-dimethyl-4,4'-bipheylene unit (DBAPB/DBDPB) exhibited a higher $Td_{10}$ value than that bearing an asymmetric *t*-butyl substitution group (BATB/ DBDPB). The noncoplanar conformation can impart greater thermal stability to the polymer than the *t*-butyl substitution.[48] Solubility of polymer was improved by the noncoplanar conformation of 2,2'-dimethyl-substituted biphenylene in the polymer backbone, reducing the polymer chain interaction and rigidity.[33, 48] Similarly, the incorporation of the methyl substituent on the 2,2'-dimethyl-4,4'-biphenylene unit did not significantly sacrifice the thermal stability of the polymer but enhanced the processability.[34] TGA data of these polymers indicated that polymers with noncoplanar biphenylene groups exhibited higher thermal stability than those with kink diphenylmethylene linkages.[54, 55] Isothermal gravimetric analysis (IGA) results demonstrated the high thermo-oxidative stability of these poly(ether imide)s. In general, the IGA results were in agreement with the TGA data. Specifically, following 20 h isothermal aging at 350 °C in static air, the polymer showed weight losses of 7.0–10.5% (e.g., 7.0% for BATB/DBDPB polyimide and 7.9% for DBAPB/DBDPB polyimide). On comparing the weight loss, polyimides containing 2,2'-dimethyl-4,4'-biphenylene units (7.0% for BATB/DBDPB polyimide and 7.9% for DBAPB/DBDPB polyimide) were somewhat more stable thermally than those containing diphenylmethylene units (8.8% for BATB /BDPDM polyimide and 9.0% for DBAPB/BDPDM polyimide).[54, 55] The lower weight loss of 2,2'-dimethyl-4,4'-biphenylene unit-containing poly(ether imide)s in air may reflect the oxidation of the methyl group of the polymer backbone to form a carbonyl group (C=O), leading to weight gain.[34, 69]

Glass transition temperatures ($T_g$'s) of the poly(ether imide)s were found to be in the range between 226 and 262 °C. No melting endotherm peak was observed in the DSC traces. This also verified the amorphous nature of the poly(ether imide)s. All the polyimides exhibited higher $T_g$ values than the commercially available polyimide Ultem 1000® ($T_g$ = 215 °C). These modified polyimides could be considered as new processable high-performance polymeric materials.[69-73) Poly(ether imide) with cyclododecane groups (BATB/BDPCD) showed a higher $T_g$ value than BATB/BDPDM and BATB/BDPTBC containing kink and *t*-butylcyclohexane alicyclic groups, respectively.[54-56) This shows that the introduction of the cyclododecane group into the polymer backbone can enhance the thermal property significantly.

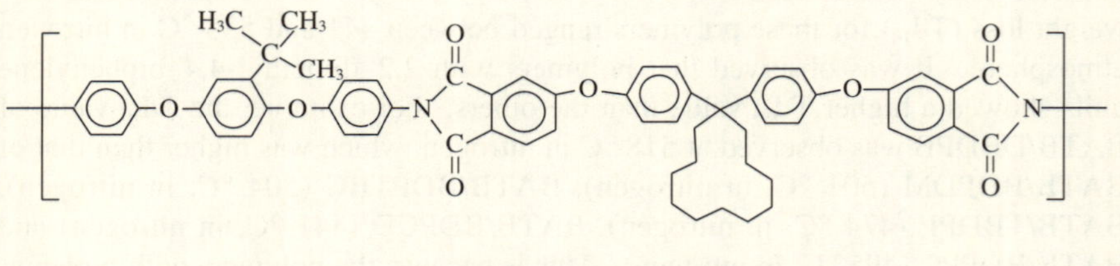

BATB/BDPCD : $T_g$ = 262 °C

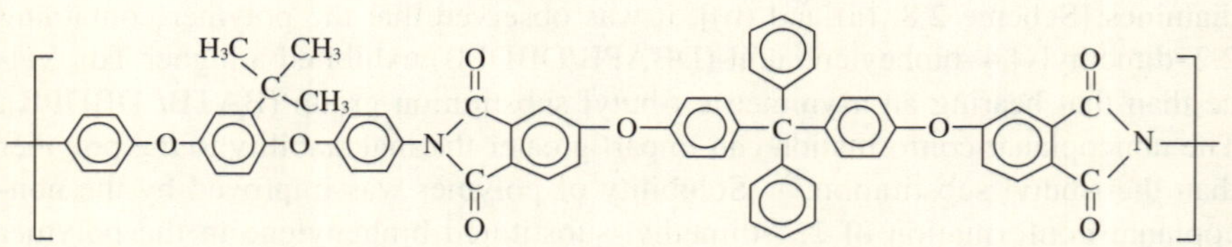

BATB/BDPDM : $T_g$ = 229 °C

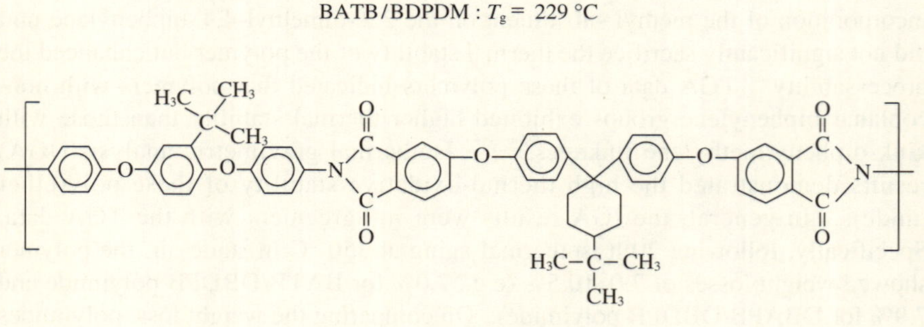

BATB/BDPTBC : $T_g$ = 233 °C

Representative TG curves of polymer BAPDM/BPDA in nitrogen and air atmosphere are shown in Fig. 2.9. In the figure it is apparent that the polymer possesses excellent thermal stability with decomposition temperature in excess of 500 °C. Decomposition temperatures at 10% weight loss of these polyimides reach 542–564 °C and 549–572 °C in nitrogen and air, respectively. It was

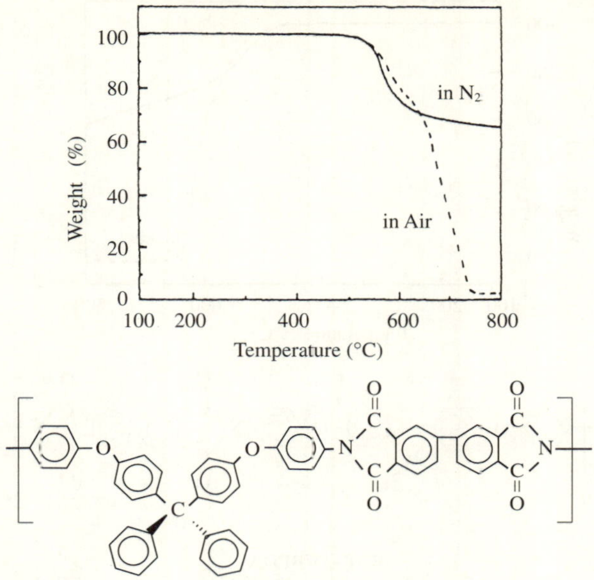


BAPDM/BPDA

**Fig. 2.9** TG curves of BAPDM/BPDA in nitrogen (——) and air ( -------- ) atmosphere at a heating rate of 20 °C · min$^{-1}$ [45, 46].

observed that these polymers had almost the same initial decomposition temperature both in air and nitrogen atmosphere. However, most of these polyimides suffered more weight loss in nitrogen than in air atmosphere below 640 °C, indicating that these diphenylmethylene-containing polyimides possessed good thermo-oxidative stability. The residual weights in nitrogen ranged from 58–67% at 800 °C.[45, 46] Representative TGA curves of polyimide BAPTD/BPDA in nitrogen and air atmospheres are shown in Fig. 2.10.[52] The initial decomposition of polyimide BAPTD/BPDA was almost the same, in both air and nitrogen atmosphere. However, it underwent more weight loss in nitrogen than in air atmosphere below 580 °C, indicating its high thermo-oxidative stability in air at elevated temperatures. Thermal properties of analogous cardo polyimides, according to the contribution of different cardo groups, can be arranged in the following order:

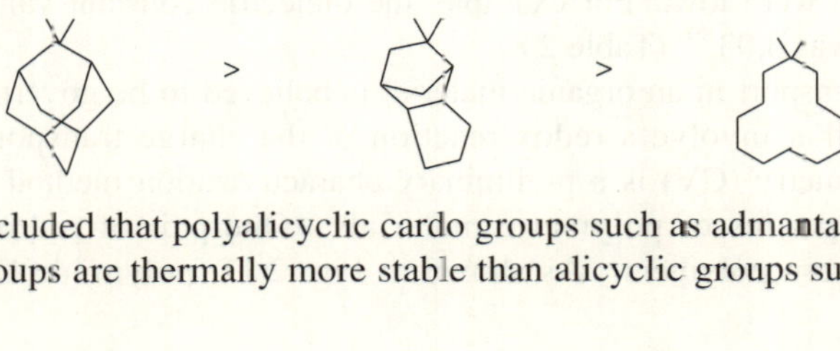

It can be concluded that polyalicyclic cardo groups such as admantane and norbornane-type groups are thermally more stable than alicyclic groups such as cyclododecane.

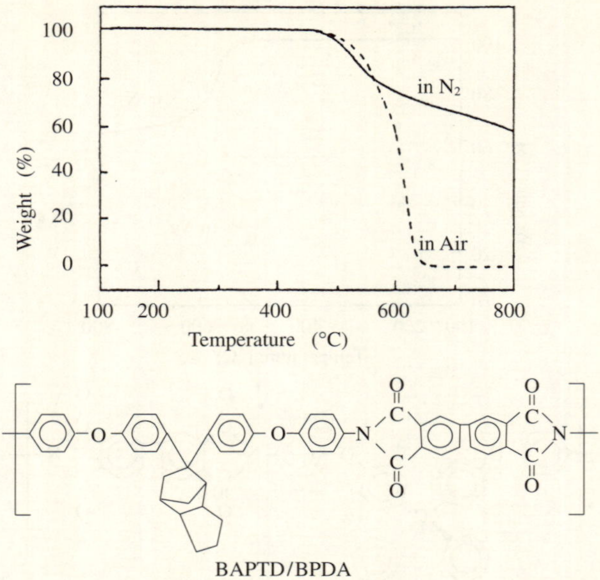


**Fig. 2.10** TGA curves of polyimide BAPTD/BPDA in nitrogen (———) and air (--------) atmosphere at a heating rate of 20 °C · min$^{-1}$ [52].

## F. Electrical and Optical Properties

Polyimides have CTEs of 55.8–75.5 (ppm · °C$^{-1}$), because they have bent ether units.[74] These bent units loosen the packing of films.[74] In addition, a film cast from polyimide solution generally has a lower CTE than that cast from the corresponding poly(amic acid).[74] These polymers have a dielectric constant of 3.57–3.89.[74] The dielectric constant of polyimides [DADT/DSDA (dielectric constant of 3.62) and DADT/6FDA (dielectric constant of 3.89)] is similar to that of the polyimide derived from polyalicyclic bicyclo 〔2,2,2〕 octane-2,3,5,6-tetracarboxylic 2,3:5,6-dianhydride.[74] Optical properties of the polyimides, particularly their UV-visible absorptions in NMP solution, were investigated. Most of the polyimides had absorption maxima at 300–309 and 320–345 nm, which can be ascribed to the $\pi \rightarrow \pi^*$ transition resulting from the conjugation between the aromatic ring and nitrogen atoms.[74] Dielectric constants of trifluoromethyl-containing polyimides were low. For example, the dielectric constant values for DBT-FAPB/6FDA was 3.03 [68] (Table 2.8).

Charge transport in an organic material is believed to be governed by a hopping process that involve a redox reaction of the charge transport molecules. Cyclic voltammetry (CV) is a preliminary characterization method to determine the redox properties of polymeric materials. One pair of redox waves was observed in these polymers. As shown in Fig. 2.11, polyimide DADT/DSDA

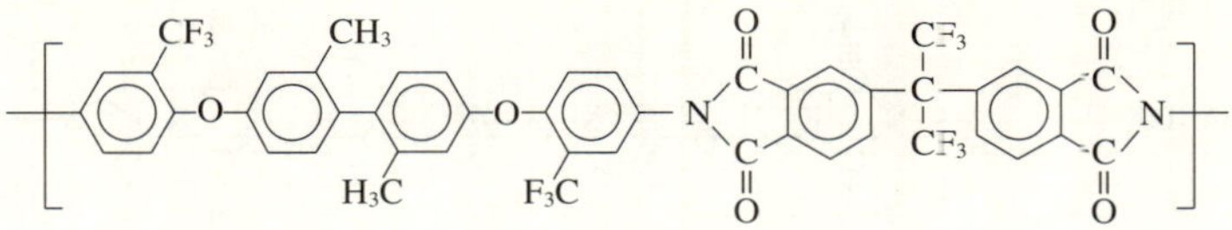

DBTFAPB/6FDA

showed a high oxidation potential with a peak at 1.18 V. The intensity of redox peaks decreased as cycle time increased, probably due to the dissolution of ionized polymer in the solvent. The polymer displayed an electro-chromic property, as they transformed from the orange neutral form to the blue oxidized form.[74)]

The fluoro-PI films were very light-colored to almost colorless, whereas Ultem 1000® and Kapton H films were, respectively, light amber and amber colored, as listed in Table 2.8. Many non-fluorinated polyimide films are known to be yellow to dark amber in color, whereas, fluorinated polyimide films are almost colorless. The incorporation of bulky fluorine-containing groups in the polyimide structure would reduce the refractive index and optical loss. Optical transparency to visible light is affected by the intra- and intermolecular interaction of $\pi$ electrons between the monomer moieties in the polymer chain, including a charge transfer complex (CTC). The $\pi$ electron is transferred from the electron-donating diamine to the electron-acceptor dianhydride moiety. The polyimide chain is basically composed of alternating donor and acceptor moieties, which can interact with each other, inducing interchain CTCs.[75-79)] Therefore, it is possible to reduce CTCs by incorporating electronegative fluorine groups on the polymer backbone or incorporating bulky electron-withdrawing substituent groups, as they restrict the inter-chain conformational mobility and thus lower the formation of CTCs.[75-79)]

**Table 2.8** Dielectric constants and film characteristics of trifluoromethyl-containing polyimides[68, 75)]

| Polyimide code | Dielectric constant (dry, 1 kHz) | Film characteristics |
|---|---|---|
| DBTFAPB/PMDA | —[a] | N/A[b] |
| DBTFAPB/BPDA | —[a] | N/A[b] |
| DBTFAPB/BTDA | 3.32 | Very light yellow to colorless, transparent, flexible |
| DBTFAPB/6FDA | 3.03 | Very light yellow to colorless, transparent, flexible |
| Ultem 1000® | 3.18 | Light amber, flexible |
| Upilex-S | 3.47 | N/A[b] |
| Kapton H | 3.50 | Amber, flexible |

[a] Polymer could not be cast into film from DMAc solution.
[b] N/A: not available

Ultem 1000®

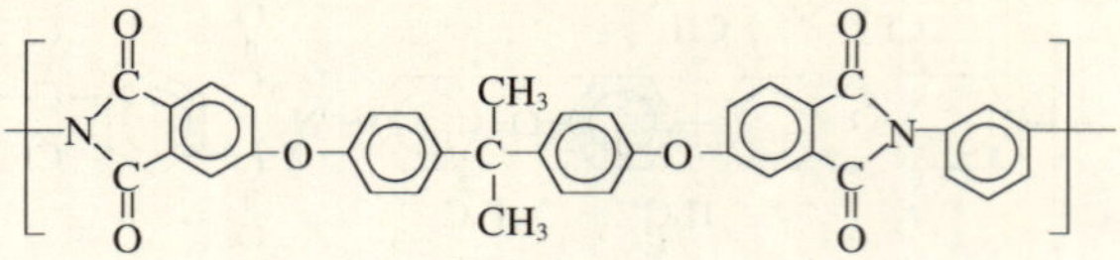

Kapton H

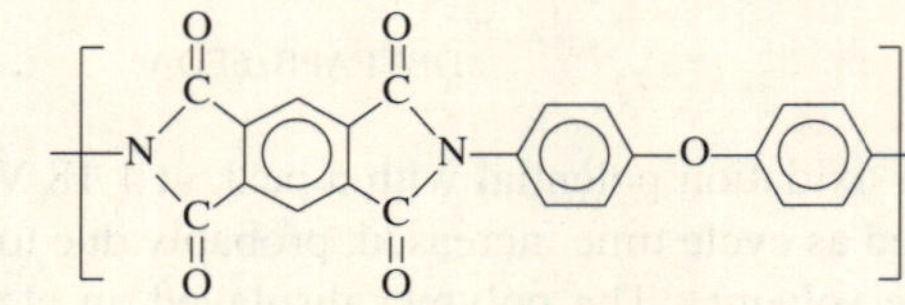

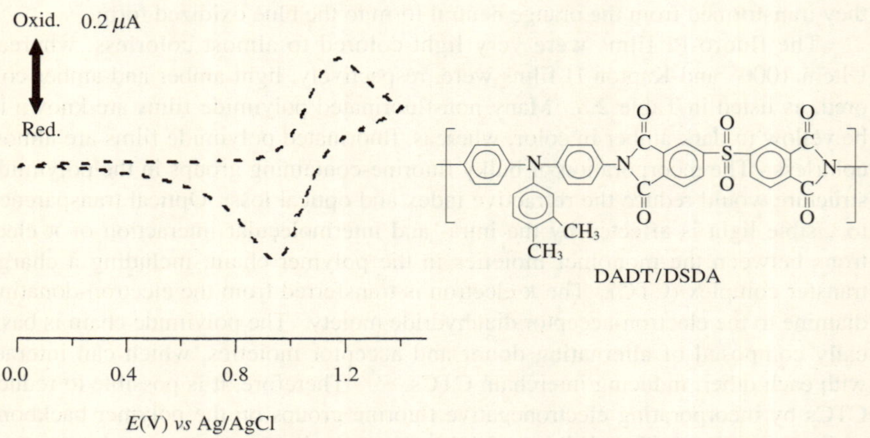


**Fig. 2.11** Cyclic voltammogram of DADT/DSDA in acetonitrile with 0.1 M $Bu_4NClO_4$. Scanning rate 150 mV · $s^{-1}$ [74].


# Acknowledgment

The author thanks C. C. Huang and W. H. Chen for their help in the preparation of this manuscript.

# 2.3 "Inorganic" Combinatorial Chemistry Utilizing Sol-Gel Transcription of Gelatinous Organic Superstructures

**Oliver Gronwald and Seiji Shinkai**

## 2.3.1 Introduction

For more than 150 years organic synthesis was a rich source for new molecules and materials with novel properties. The development of increasingly sophisticated methods to control the formation and manipulation of covalent bonds allowed the construction of highly complex molecular structures. The recent years have been witness to a rapidly growing interest in supramolecular chemistry which lies beyond molecular chemistry based on the covalent bond. This concept embodies the next step in complexity and deals with organized polymolecular systems held together by noncovalent intermolecular forces.

Moreover, for the fabrication of novel inorganic materials these supramolecular assemblies opened new and exciting alternatives to conventional synthetic strategies.[80] The use of various defined molecular architectures provided by higher-order molecular aggregates as templates for sol-gel polymerization resulted in new inorganic materials unattainable though any other method. Proteins, multicellular superstructures, surfactants and DNA have been found to serve as efficient templates for the synthesis of new inorganic materials.[81-83]

Organogel systems made up by the self-assembly of low molecular weight organic gelators constitute a further relevant class of functional materials with potential application in templated materials synthesis. The low molecular weight gelling agents employed in the self-assembly process feature enormous structural diversity.[84] Among recently published gelators cholesterol-based structures as well as amino acids and compounds bearing amide and urea moieties were found as dominating structural motifs. They all have in common the fact that their self-aggregation is solely mediated by physical interactions, including hydrogen-bonding, π-π-stacking and solvophobic effects. As a result, single molecules form well-ordered arrays of molecules, which grow to one-dimensional elongated objects and finally entangle to a three-dimensional network structure, capable of encapsulating the liquid component and suppressing its flow. Different network filament architectures like tapes, rods, fibers, sheets and cylinders have been reported. Compared with other superstructures like micelles and membranes the organogel network is much more stable, thus facilitating the transcription process. In summary, the stability and structural diversity of organogels added to the possibility of introducing chirality, enable them to serve as ideal template-superstructures for the sol-gel transcription process.

## 2.3.2 Sol-Gel Transcription with Different Gel Templates

### A. Cholesterol-based Gels as Templates

Among the organogelators cholesterol-based derivatives were first found to gelate various organic solvents efficiently at low concentrations.[85-89] This phenomenon, based on π-π-stacking, van der Waals forces and solvophobic effects, results in fibrous aggregates that can be even chirally orientated since these aggregates impart the characteristic cholesterol skeleton. The first replication of the 3D gel network structure was demonstrated by our group.[90] By utilizing superstructures based on gelator **19** (Scheme 2.10) as the template for sol-gel polymerization of tetraethoxysilane (TEOS), we obtained a novel porous silica with tubular structure. The presence of the quaternary ammonium group was rationalized to be essential for the transcription process, since during the sol-gel polymerization an electrostatic interaction between "anionic" silica and "cationic" organogel fibers ensures the precise replication of the organogel fiber shapes into silica (Fig. 2.12). Transcription of assembled cholesterol gelators bearing a crown moiety as the metal-binding site[91] confirmed the need for cationic charges in the organogel template: TEOS sol-gel polymerization in 1-butanol gels of compound **21** in the presence of equimolar amounts of $KClO_4$ resulted in a tubular silica structure (Table 2.9, run 1). Scanning electron microscopy (SEM) and transmission electron microscopy (TEM) investigations revealed fibrous silica with outer diameters of around 50 nm and inner diameters of about 3 nm. Using a lower concentration of $KClO_4$ (run 2) gave silica structures with outer diameters ranging from 30 to 200 nm and inner diameters around 50 nm. However, if the $KClO_4$ concentration was too low (run 3), conventional granular silica similar in structure to that prepared from a solution in the absence of the gelator **21** was found. Since 1-butanol was not gelated by **21** without $KClO_4$- additive, the TEOS sol-gel polymerization reference experiment was carried out in aniline as the main solvent (run 4). The granular silica observed substantiated the specific $K^+$-benzo-18-crown-6 interaction as being related to the construction of hollow fiber silica.

For the cholesterol derivatives **22** and **23** the transcription into hollow fiber silica occurred even in the absence of metal salt additives.[92] This phenomenon was ascribed to the protonation of the monoaza- and diaza-18-crown-6 moieties due to their relatively high $pK_a$ values,[93] thus providing under the sol-gel polymerization conditions the required cationic charges for the stimulation of the transcription effect.

Tubular shaped superstructures formed by certain amphiphiles in water aggregate through intermediate helical ribbon structures.[94] Since these amphiphiles feature a polar head in combination with a chiral hydrophobic group, the sol-gel transcription process into tubular silica controlled by crown-appended cholesterol assemblies was early considered to proceed through a similar aggregation process. Evidence for this assumption was obtained by observation of the transcription effect mediated by compound **26** gel-superstructures into helical ribbon-type silica.[95] Fig. 2.13(c) shows a TEM picture of silica, obtained from **26** in acidic acid after calcination: both tubular silica (with 560 nm outer diameter) and helical ribbon structures with 1700 – 1300 nm pitches were found. The right-handed helical

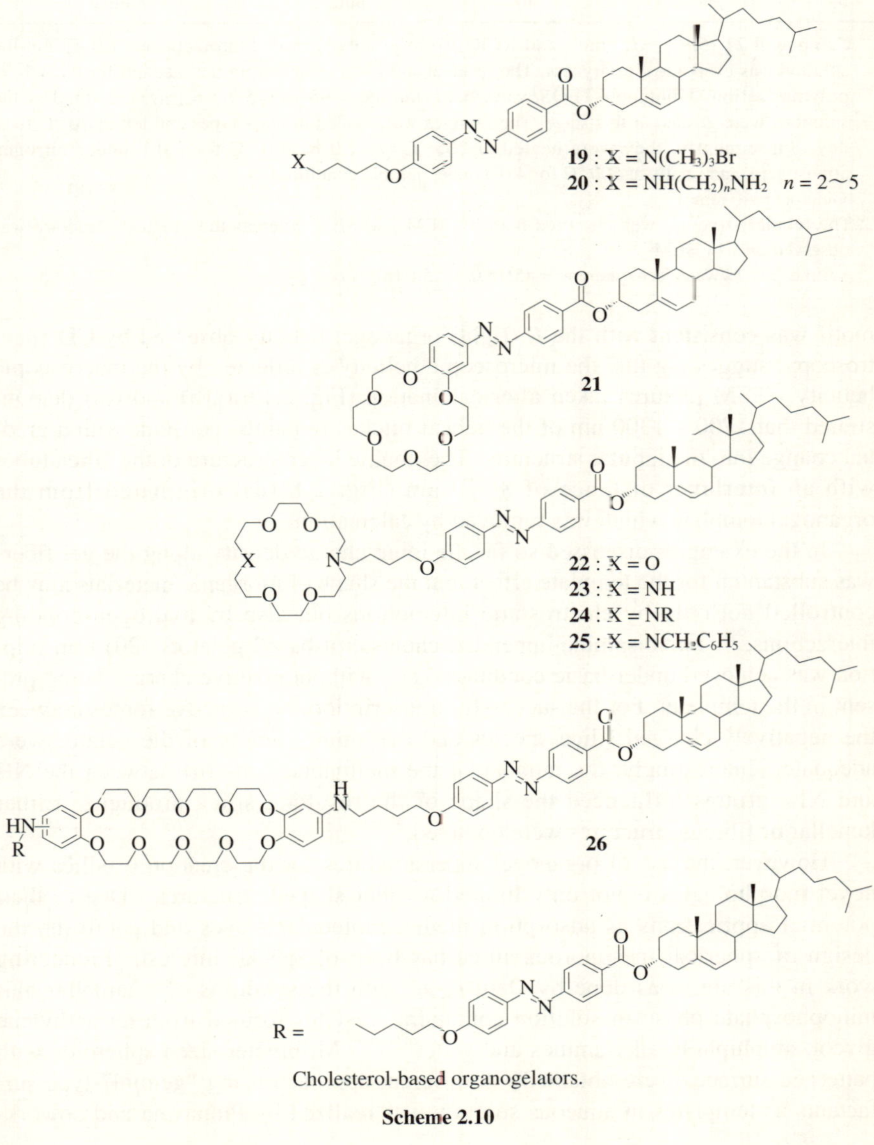

**19** : X = $N(CH_3)_3Br$
**20** : X = $NH(CH_2)_nNH_2$ $n = 2\sim5$

**21**

**22** : X = O
**23** : X = NH
**24** : X = NR
**25** : X = $NCH_2C_6H_5$

**26**

Cholesterol-based organogelators.

**Scheme 2.10**

**Table 2.9** Silica structures resulting from TEOS sol-gel polymerisation[a] in 1-butanol gels of compound **21**

| Run | Gelator[b] | $[KClO_4]$/mmol | Major solvent | Silica structure[c] |
|---|---|---|---|---|
| 1 | **21** | 40 | 1-butanol | tubular |
| 2 | **21** | 20 | 1-butanol | tubular |
| 3 | **21** | 3 | 1-butanol | granular |
| 4 | **21** | 0 | aniline[c] | granular |
| 5 | none | 40 | 1-butanol | granular |

[a] Compound **21** ($5.8 \times 10^{-6}$ mol) and $KClO_4$ were dissolved in dichloromethane (1.0 g) and the solution was evaporated to dryness. The residual solid was dissolved in the medium for the sol-gel polymerization: 1-butanol/ TEOS/water/benzylamine = 95/15.0/5.7/5.6 (mg). In run 1–4 the solutions were gelated at this stage. The samples were sealed in glass tubes and left at 20 °C for a day. Subsequently, they were heated at 200 °C for 1.0 h, 500 °C for 2.0 h under nitrogen atmosphere, and finally at 500 °C for 4.0 h under aerobic conditions.

[b] [Gelator] = 40 mmol

[c] The tubular structure was observed both by TEM and SEM whereas the granular structure was observed only by SEM

[d] Aniline/ TEOS/water/benzylamine = 45/15.0/5.7/5.6 (mg).

motif was consistent with the *R* (right) organogel helicity observed by CD spectroscopy, suggesting that the microscopic helicity is reflected by the macroscopic helicity. TEM pictures taken after calcination (Fig. 2.13(a),(b) and (c)) demonstrated that 1700 – 1300 nm of the helical pitches remained constant with a gradual change into the tubular structure. The double layer structure of the silica tubes with an interlayer distance of 8 – 9 nm (Fig. 2.13(d)) originated from the organogel template which was removed by calcination.

In the examples discussed so far the ionic charge density along the gel fibers was substantial for the template effect, but the shape of inorganic materials may be controlled not only by electrostatic interactions but also by hydrogen-bonding interactions.[96)] For bisamino-appended cholesterol-based gelators (**20**) transcription was achieved under basic conditions, i.e., without positive charges being present in the template. For the successful transcription the attractive forces between the negatively charged silica species and the amino moiety of the gelator were adequate. Interestingly, the number of the methylene units (*n*) between the NH and $NH_2$ groups influenced the shape of the resultant silica products. Either lamellar or fibrous structures were obtained.[97,98)]

However, the use of organogel superstructures for the creation of silica with novel morphologies is not only limited to fiber shaped structures. Due to their potential applications as adsorption media, molecular sieves and catalysts, the design of spherical mesoporous silica has been of special interest. Pioneering work in this area was done by Ozin *et al.* with the synthesis of a lamellar aluminophosphate phase in solution containing vesicles formed from tetraethylene glycol, amphiphilic alkylamines and water.[99,100)] Millimeter-sized spheroids with patterned surfaces were obtained. A different approach using "gemini"-type surfactants as templates in aqueous solution was realized by Pinnavaia and cowork-

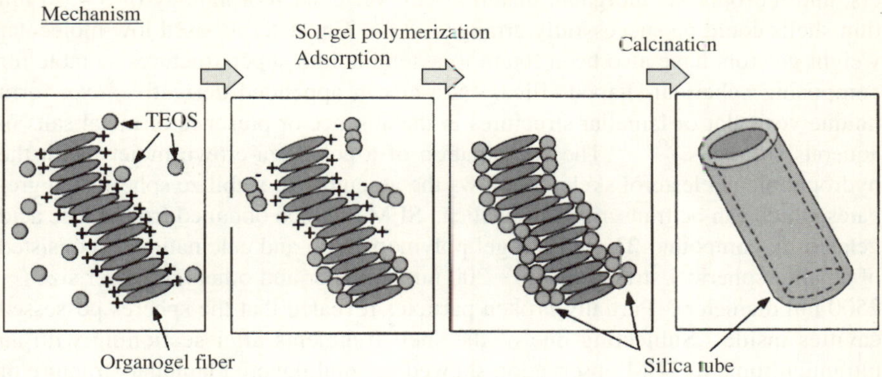


**Fig. 2.12** Schematic representation for the creation of tubular silica structures using electrostatic interactions between the organogel-template and TEOS.

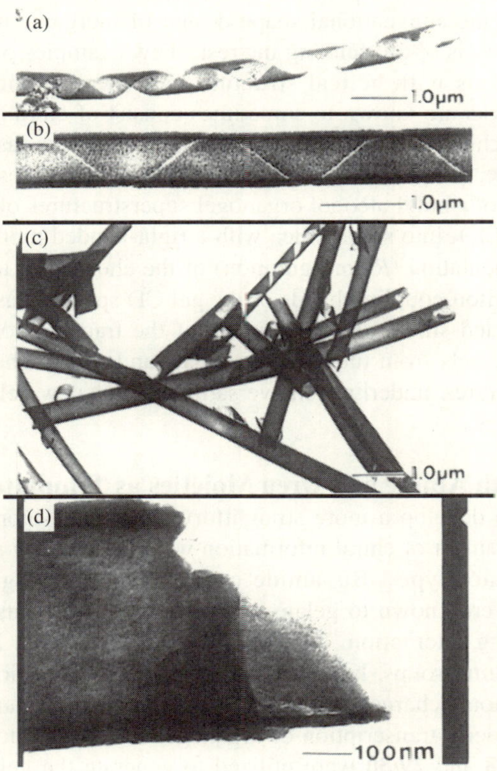


**Fig. 2.13** TEM images of the silica obtained from **26** + acetic acid after calcination. [Reprinted with permission from Jung, J.H. *et al.*, *J.Am.Chem.Soc.*, 2001, **123**, 8785. Copyright 2001 American Chemical Society.]

ers, and mesoporous inorganic material with vesicular morphology of 3 – 70 nm thin shells could be successfully arranged.[101,102] Cholesterol-based low molecular weight gelators have also been found to assemble into superstructures suitable for composing spherical-shaped silica. Aza-crown appended derivatives can form unique vesicular or lamellar structures in the absence or presence of metal salts in aqueous solutions.[103-105] The combination of a polar aza-crown moiety with the hydrophobic cholesterol skeleton allows the assembly to stabilize spherical aggregates which can be transcribed into silica. SEM of silica obtained from acetic acid gelated by compound **23** after sol-gel polymerization and calcination[106] consisted of smaller spherical structures of ~ 200 nm diameter and others of larger size (~ 2500 nm diameter). Partially broken particles revealed that the spheres possessed cavities inside. Subjecting one of the shell fragments after sectioning with an ultramicrotome to TEM observation showed the multilayered lamellar structure of the wall having 5-nm spacing. Interestingly, the smaller vesicles were connected one to another like a pearl necklace, thus disproving the general assumption that only fibrous aggregates induce the gelation effect.

Apart from the conventional shape design of inorganic materials, the introduction of chirality is of outstanding interest. Few examples of the preparation of inorganic materials with helical structure have been reported so far. Since organogel networks are known to sometimes consist of chirally orientated fibers, transferring this chirality into inorganic material presented a fascinating challenge. The first example for this new chiral template method was established based on the transcription of *t*-butyl alcohol organogel superstructures of **25** in the presence of $AgNO_3$ or $CsClO_4$ into spiral silica with a right-handed motif.[107] Surprisingly, the clockwise orientation (*R* configuration) of the cholesterol aggregates as determined by the exciton-coupling bands of the gel CD spectra was transferred directly into right-handed silica. This extension of the transcription methodology for cholesterol-based gels from tubular-, vesicular- and ribbon-shaped silica towards chiral silica structures, underlines the versatility of this new gel template for novel synthetic strategies.

### B. Gelators with Amide and Urea Moieties as Templates

Improvements to develop a more straightforward organic-inorganic transcription process for the transfer of chiral information were made using another low molecular weight gelator type. Bis-amide cyclohexane-based gelators **27** and **29** (Scheme 2.11) were known to gelate a wide range of solvents mediated through hydrogen-bonding interaction.[108] Cyclohexane derivatives **28** and **30** bearing tetralkylammonium groups, however, showed reduced gelation ability. As discussed above cationic charges on the organogel fibers surface are required in order to accomplish orderly transcription of organogel structures into silica. Therefore, mixtures of **27/28** and **29/30** were utilized to generate the gel templates and the aggregate chiralities could be characterized by CD spectroscopy. Analysis of the exiton-coupling bands for mixtures of **27+28** and **29+30** in acetonitrile proved their cooperative organization into aggregates with left- and right-handed helical

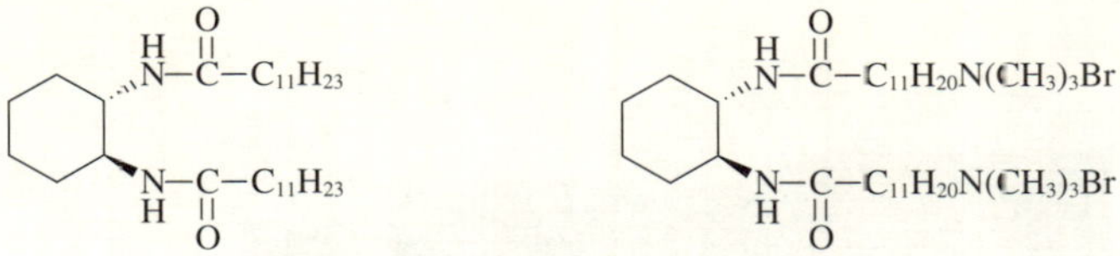

**27** trans (1 *R*, 2 *R*)
**29** trans (1 *S*, 2 *S*)

**28** trans (1 *R*, 2 *R*)
**30** trans (1 *S*, 2 *S*)

Bisurea cyclohexane-based organogelators.

**Scheme 2.11**

structures, respectively. This microscopic chirality was also expressed in the xerogel morphology which was inspected by SEM. Helical fibers (20–60 nm) with left-handed orientation for the (*R*)-enantiomer mixtures **27+28** and right-handed helicity for the (*S*)-enantiomer mixture **29+30** were observed. The mismatch between the CD sign and the fiber helicity was ascribed to the aggregation of the incipient organogel fibers to bundled fibers which resulted in reversed chirality. After transcription with TEOS followed by calcination of (*R*,*R*) **27+28** gels left-handed silica (Fig. 2.14(a)) and for (*S*,*S*) **29+30** mixtures right-handed helical silica (Fig. 2.14(b)) was obtained, having an inner tubular structure with the same helicity and dimension as the organogel fiber itself (Fig. 2.14(c) and (d)).[109,110] Fig. 2.15 depicts a schematic representation of the sol-gel process. This high standard of manipulating the organization of inorganic materials on a nanoscale level is a striking example of the potential of organogels as templates for the synthesis of new inorganic structures.

Extension to different organogelators and inorganic materials demonstrated the universality of the new transcription concept. As an example fibrous structures of porous $TiO_2$ were aggregated by means of sol-gel polymerization of titanium tetraisopropoxide $Ti[OCH(CH_3)_2]_4$ gelated by *N*-carbobenzyloxy-L-isoleucylaminooctadecane mediated through hydrogen bonding interaction of the fibrous gelator aggregates and oligomeric titanias.[111]

### 2.3.3 Combinatorial Approach to Sol-Gel Transcription

So far, novel organogelator structures have been discovered rather by serendipity than by design. Therefore, the attempt to overcome this "empirical" approach and move towards a rational design has been a key objective for many years. Once the design principles of a class of gelator types are understood, novel organogelators and thus a variety of new template superstructures useful for transcription will be accessible. Using combinatorial library methodology can accelerate the discovery of new gelling agents in a resource-effective manner. Designing a range of structurally distinct analogues and analyzing their application as templates in the sol-gel transcription process will lead to a large variety of shapely materials. Therefore, one may refer to this new concept as "inorganic" combinatorial chemistry.

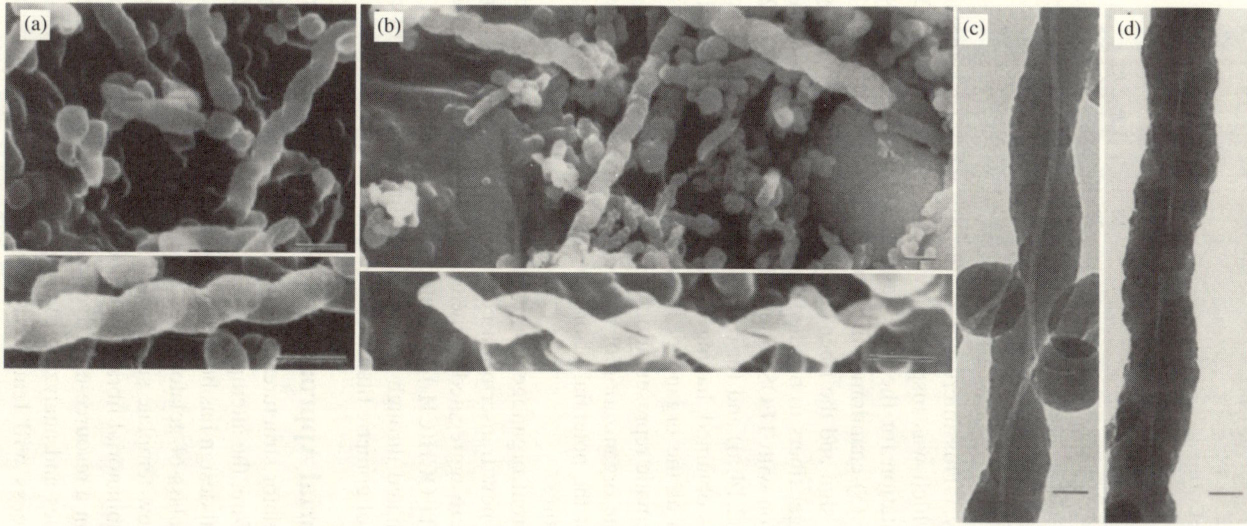


**Fig. 2.14** SEM pictures of silica obtained from (a) **27** + **28** (1 : 1 wt%); (b) **29** + **30** (1 : 1 wt%) after calcination (bar = 200 nm). TEM pictures of silica obtained from (c) **27** + **28** (1 : 1 wt%); (d) **29** + **30** (1 : 1 wt%) after calcination (bar = 50 nm). [Reprinted with permission from Jung, J.H. *et al.*, *Chem. Eur. J.*, 2000, **6**, 4552. Copyright 2000 Wiley-VCH.]

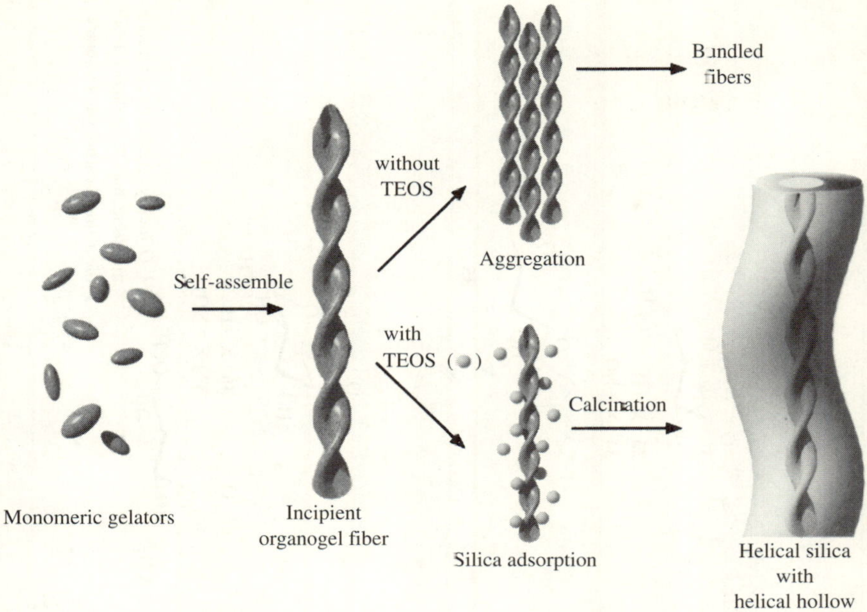


**Fig. 2.15** Schematic representation of the creation of chiral silica structures with inner tubular structures from the organogel state of **27/28** and **29/30** by TEOS sol-gel polymerization.

### A. Carbohydrate-based Derivatives as a Source for Novel Gel Templates

As a relatively new member among the family of low molecular weight gelators 4,6-*O*-benzylidene derivatives of monosaccharides proved to be advantageous to study the structural prerequisites for gelation ability as well as the various morphologies of their gel assemblies in organic solvents.[112] Their facile molecular design and simple synthetic access from building blocks provided by the rich natural carbohydrate library favors this gelator type for combinatorial library methods. The structural features of this new gelator type are i) unmodified 2-OH and 3-OH groups; ii) protection of the 1-OH with a methyl or phenyl group; and iii) protected 4-OH and 6-OH with a benzylidene group (Scheme 2.12). FT-IR and temperature-dependent $^{1}H$ NMR spectroscopy investigations of the gels in organic solvents identified a hydrogen-bond network originated by 2- and 3-OH groups as the driving force for the self-aggregation.[113,114] As a first step to correlate the effects of configurational isomerism with the ability for self-aggregation, a series of eleven 1-*O*-methyl-4,6-*O*-benzylidene derivatives (R = $CH_3$, Scheme 2.12) was investigated. As a result, a significant correlation of the monomer structure with the gelling ability was found. In general, an optimal gelation effect for a wide range of solvents (apolar hydrocarbon and aromatic solvents, carbon disulfide,

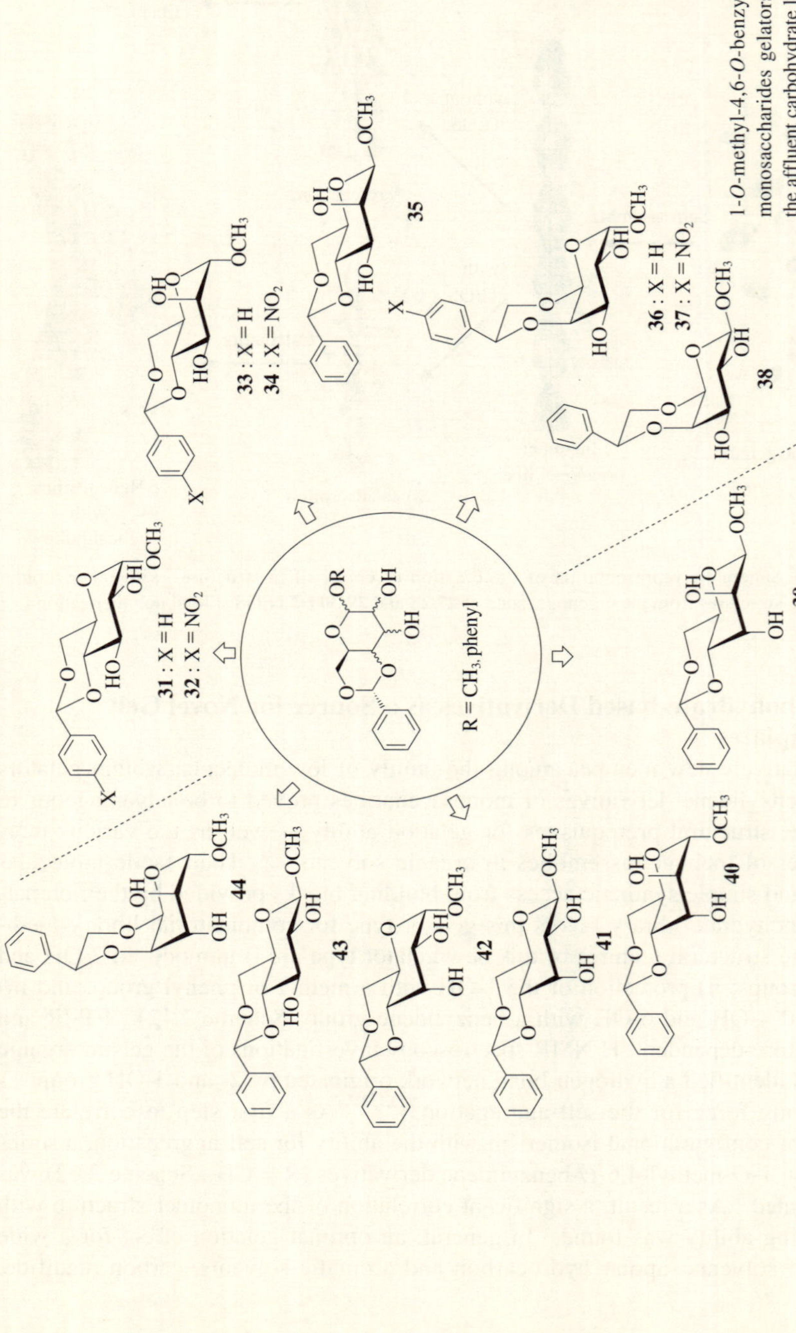


**Scheme 2.12** 1-*O*-methyl-4,6-*O*-benzylidene monosaccharides gelators from the affluent carbohydrate library.

diphenyl ether, triethylsilane, etc.) was achieved with derivatives of D-mannose (**33**, **35**) and D-galactose (**36**, **38**).[115] For the α-anomer of D-glucose (**31**) a reduced gelation potential for a smaller number of solvents was recorded. In contrast, among the derivatives of D-altrose (**39**, **40**), D-allose (**41**, **42**), β-D-glucose (**43**) and α-D-idopyranoside (**44**) no capability for self-assembly was found.

To elucidate the origin of these differences, the molecular arrangement in a single crystal was taken into account. In the history of gel research the question of whether the molecular packing in the gel fibers reflects the solid state structure remained a highly controversial issue. Compared with the solid state the organogel fibers are supposed to be more disordered,[89] but not too wet with solvent molecules.[87] For a long time no single crystal structure of a low molecular weight gelator has been successfully resolved. The fact that 1-*O*-methyl-4,6-*O*-benzylidene monosaccharides were not prone to polymorphism and organogel fibers predestined this type of gelator for further investigations. Since the examined saccharides tended to grow as needles, only some single crystals suitable for X-ray analysis have been isolated. Among them the solid state structure of the gelator 1-*O*-methyl-4,6-*O*-benzylidene-α-D-glucopyranoside (**31**) displayed interesting one-dimensional zigzag chains, in which the molecules were connected by two hydrogen bonds using 2-OH and 3-OH (Fig. 2.16).[116] Phenyl groups positioned at the edge of this one-dimensional chain can form by π-π interaction in other one-dimensional chains. Although this interaction is weaker than hydrogen

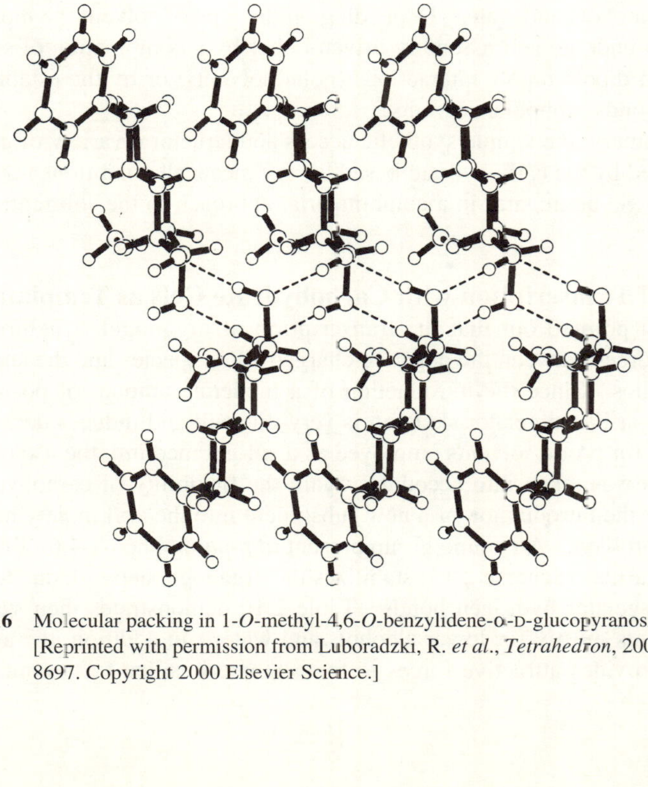

**Fig. 2.16** Molecular packing in 1-*O*-methyl-4,6-*O*-benzylidene-α-D-glucopyranoside (**31**). [Reprinted with permission from Luboradzki, R. *et al.*, *Tetrahedron*, 2000, **56**, 8697. Copyright 2000 Elsevier Science.]

bonding, it may play an important role when chains grow as a bundle. Since amide-based and urea-based gelators tend to form one-dimensional hydrogen-bond arrays,[108,117)] the molecular packing of **31** accounts for a good gelator. Therefore, exploration of the molecular arrangement in the single crystal can help rationalize the different tendencies of saccharide monomers to assemble in one-dimensional aggregates. Assuming, therefore, that these gel-assemblies imply to a certain extent "crystal-qualities", they would provide higher template stability, thus enhancing the quality of the transcription effect.[105)]

Furthermore, the aryl moiety attached to the carbohydrate-based gelators offers the unique possibility of introducing a variety of substituents suitable for promoting the efficiency of the transcription process or additional stability of the gel template. In particular, the presence of a nitro-substituent in the *para*-position of the 4,6-*O*-benzylidene group resulted in a remarkable improvement of the gelation properties. 1-*O*-Methyl-4,6-*O*-*p*-nitrobenzylidene-α-D-galactopyranoside (**37**) acts as "supergelator" for apolar solvents (*n*-hexane, *n*-heptane, *n*-octane, *n*-decane and cyclohexane) with a low minimum gelator concentration range of 0.03–0.05 wt% (0.9–2.48 mM) compared with that of typical gelators (3–15 mM).[118)] Also, for 4,6-*O*-benzylidene derivatives of α-D-glucose (**32**) and α-D-mannose (**34**) the introduction of a nitro-substituent enables these compounds to gelate organic solvents and water in contrast to their non-nitro analogues.[119)] They establish a new class of "bifunctional" low molecular weight gelators capable of forming gels in both nonpolar and aromatic solvents and polar protic solvents including alcohols and water. Depending on the type of solvent the monomers are assumed to undergo self-assembly driven either by a combination of solvophobic effects with dipole-dipole interactions (polar solvents) or by the establishment of hydrogen bonds (nonpolar solvents).

In summary, the simple synthetic access and structural variety of gelator entities provided by the carbohydrate-based library meets all conditions for employing their stable gel aggregates in a combinatorial approach to the sol-gel transcription process.

### B. Sol-Gel Transcription with Carbohydrate Gels as Templates

It has been pointed out that the transcription of organogel structures requires attractive forces between the negative charged silica species and the supramolecular assemblies. Since the introduction of a moderate amount of positive charge into the sugar-based gelator skeleton is very difficult and induces decreased gelation ability for protic solvents employed as a sol-gel medium, the use of non-ionic interactions was taken into account. Again the flexibility of carbohydrate-based gelators for the introduction of a new substituent into the aryl moiety helped overcome this problem. An amino group present in *p*-aminophenyl-4,6-*O*-benzylidene monosaccharides (Scheme 2.13) stabilizes the organogel network due to the intensified inter-gelator hydrogen bonds. Table 2.10 demonstrates their versatility as gelling agents mostly for lower alcohols and water. In addition, the amino function also provides attractive forces to the oligomeric silica for a controlled tran-

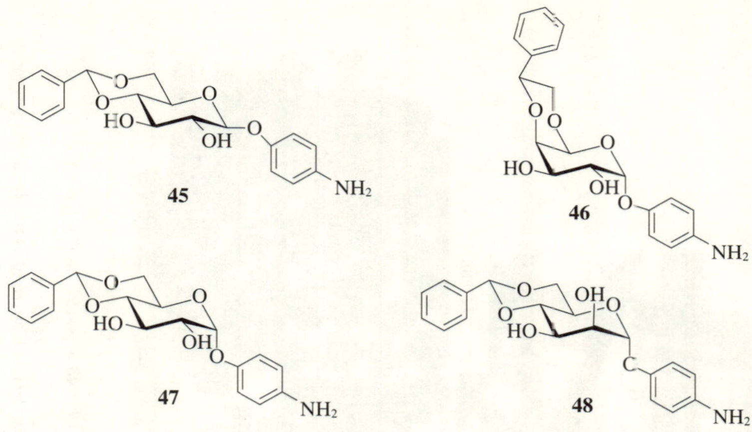

*p*-Aminophenyl-4,6-*O*-benzylindene monosaccharides as gelatops.

**Scheme 2.13**

scription.[120] TEM observations of their xerogels derived from ethanol and water gels revealed various and novel morphologies suitable for composing new silica structures (Fig. 2.17).[121,122] Ethanol gelled by α-gluco (**47**) consisted of a three-dimensional network with 5 – 20 nm frizzled fibrils (Fig. 2.17(a)). In contrast, the organogel of **45** showed a much straighter and larger fiber structure with 50 – 150 -nm diameters(Fig. 2.17(b)). These fibers constitute the integral elements of larger (giant) fibers (Fig. 2.17(c)) which could be seen by a staining method using $OsO_4$ (2.0 wt% solution) applied prior to the organogel fiber growth. When the staining was carried out after the giant fibers were full-grown, no substructure was observable. In water α-galacto (**46**) and α-mannose (**48**) gelators formed spherical structures with 200 – 350 -nm outer diameters (Fig. 2.17(d) and (e)), which

**Table 2.10** Gelation ability of **45**–**48**[a] in organic solvents

| Solvent | 45 | 46 | 47 | 48 |
|---|---|---|---|---|
| Hexane | I | I | I | I |
| Carbon tetrachloride | I | I | I | I |
| Benzene | I | I | I | I |
| Chloroform | I | I | G | I |
| Ethyl acetate | I | R | G | I |
| Methanol | S | G | G | G |
| Ethanol | G | G | G | G |
| Propan-1-ol | G | R | G | R |
| Butan-1-ol | G | R | G | R |
| Acetic acid | S | S | S | G |
| Water | R | G | G | G |

[a] Gelator = 3 wt%, G = stable gel, S = solution, R = precipitation, I = insoluble

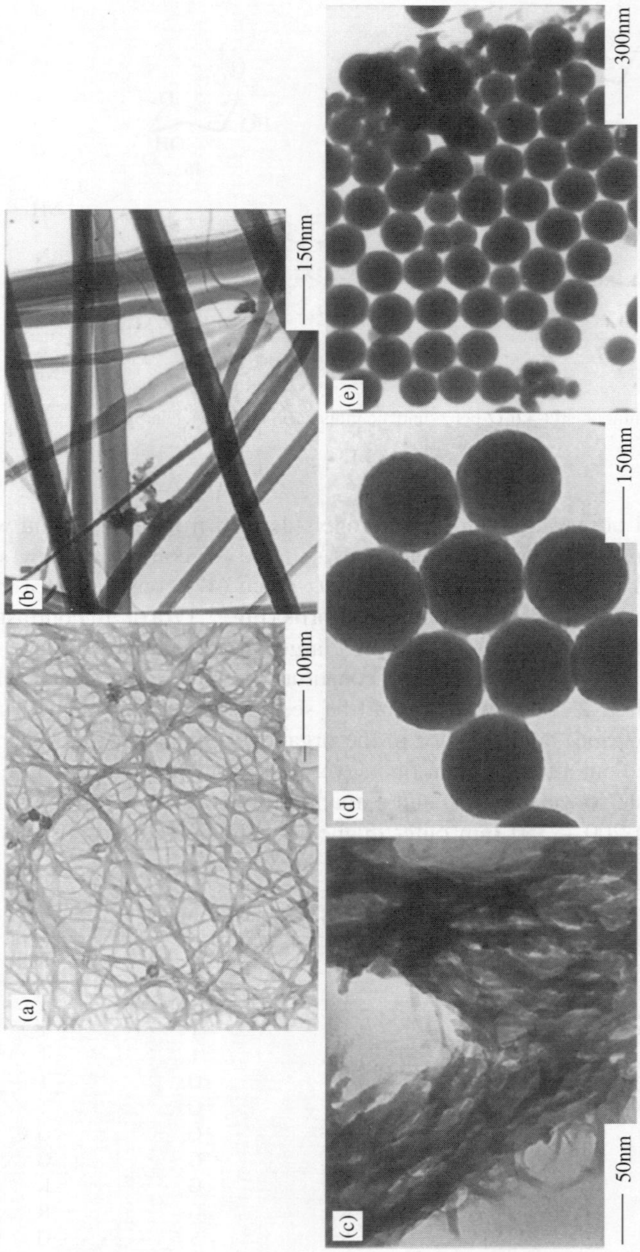


**Fig. 2.17** TEM pictures of xerogels: (a) **47** + ethanol, (b) and (c) **45** + ethanol, (d) **46** + water and (e) **48** + water. In (c), the organogel was stained with $OsO_4$ before the organogel fiber developed.
[Reprinted with permission from Jung, J. H. *et al.*, *J. Chem. Soc.*, *Perkin Trans.* 2, **2001**, 1938. Copyright 2001 Royal Society of Chemistry.]

were considered to be aggregations of colloidal dimension. The strong influence of solvent type on the organogel morphology was demonstrated by observing the fiber structure of **46** in ethanol.

Transcription of the organogel aggregates was carried out in ethanol or water gel phase with TEOS and the presence of benzylamine as catalyst. After calcination the silica obtained from **47**/ethanol gels showed tubular structures of 10 – 30 -nm diameter and 350 – 700 nm length under SEM observation. TEM observations allowed an inside view of the tubular structure and confirmed that the inner diameter of the tube of 5 – 10 nm corresponded to the gel fiber dimensions providing further evidence of a precise transcription effect. Granular silica was found in a reference experiment using *p*-nitrophenyl-4,6-*O*-benzylidene-α-D-glucopyranoside, thus proving the relevance of the amino group for stimulating the transcription. Due to the presence of benzylamine, protonation could be excluded so the interaction of the silica particles with the fiber surface proceeded driven by nonelectrostatic interactions.

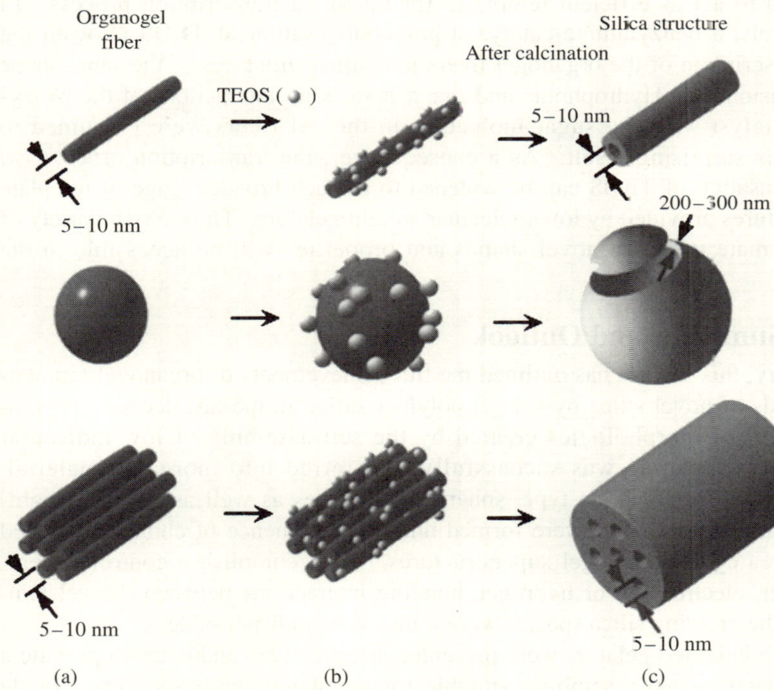


**Fig. 2.18** Schematic representation of the creation of silica structures from the organogel state of **47** (upper), **45** (middle) and **46** (lower) by sol-gel polymerization. (a) gelators, (b) sol-gel polymerization of TEOS and adsorption onto the gelator superstructures and (c) single hollow fiber structure (upper), spherical structure (middle) and lotus-like structure (lower) of the silica materials formed after calcination.

[Reprinted with permission from Jung, J.H. *et al.*, *J.Chem.Soc.*, *Perkin Trans. 2*, **2001**, 1938. Copyright 2001 Royal Society of Chemistry.]

Similar sol-gel polymerization experiments with **45**/ethanol gels and aqueous gels of **46** and **48** succeeded in precise reproduction of the organogel morphologies observed by TEM. With β-gluco (**45**) templates lotus-type silica was created mediated by the absorption of oligomeric silica on the surface of the bundled fiber structures followed by polycondensation along these fibrils. Spherical silicas with 700 – 1300 nm outer and 500 – 1000 nm inner diameters and a wall thickness of 200 – 300 nm were found using gel superstructures of α-manno (**48**) and α-galacto-pyranosides (**46**). A schematic representation of the formation of these silica shapes is given in Fig. 2.18. This structural variety underlines the potential of sugar-based gelators derived from the carbohydrate library to serve as novel templates for inorganic material synthesis.

So far all examples of silica shape design discussed above required the presence of positive charges or H-bonding sites (e.g., amino-groups) on the gelator molecules. However, gel superstructures of 1-*O*-methyl-4,6-*O*-*p*-nitrobenzylidene-α-D-glucopyranoside (**32**),[119] a gelator without binding sites for TEOS, were also found to act as efficient templates for the silica transcription process. In aqueous gels, a benzylamine-catalyzed polycondensation of TEOS allowed the direct transcription of the organogel fibers into silica structures of the same shape and dimensions.[123] Hydrophobic and / or π-π stacking interactions of the benzylamine catalyst with the sugar molecules in the gel fibers were presumed to explain this surprising result. As a consequence , the transcription process *via* polycondensation of TEOS can be widened to a much broader range of template superstructures provided by low molecular weight gelators. Thus, a rich variety of inorganic materials with novel shapes and properties will be accessible in the future.

### 2.3.4 Summary and Outlook

In summary, this section has outlined the full achievements of organogel-templated synthesis of novel silica by sol-gel polymerization in the last decade. A stunning variety of morphologies created by the self-assembly of low molecular weight organogelators was successfully transferred into inorganic material. Hollow tube, helical ribbon type, spherical structures as well as left- and right-handed spiral helical silica were formed under the influence of cholesterol-based and bis-urea cyclohexane gel superstructures. To accomplish a controlled template effect, electrostatic or hydrogen-bonding interactions between the gel template and the growing silica species were found to be indispensable.

Sugar integrated gelators were presented as attractive candidates to provide a greater variety of gel assemblies suitable for templated synthesis. Their facile molecular design using the rich carbohydrate library allows a resource-effective combinatorial approach towards the discovery of novel template superstructures. Single-fiber, lotus-like and spherical silica materials have been synthesized based on these novel sugar-gel architectures. In the future, this new class of gelators may open the door to a creative "inorganic" combinatorial chemistry for the fabrication of well defined inorganic structures at the microscopic or even the nanoscopic level. These novel materials are expected to have great potential for application in a large variety of fields such as chiral catalysis, smart fillers and nanotechnological devices.

# 2.4 Development of Organogelators Based on Supramolecular Chemistry

**Kenji Hanabusa**

## 2.4.1 Introduction

When a saturated solution of sodium chloride in hot water is left at room temperature, sodium chloride of the quantity which corresponds to the difference in solubility is deposited as crystal. This process is called recrystallization, and the crystallization of sodium chloride occurs smoothly. In contrast, there is a case in which the supersaturated solution is so stable that crystallization does not occur. In general, when a saturated solution of low molecular weight compounds is cooled, it crystallizes unless a supersaturated solution is formed. However, we rarely observe the phenomenon in which the entire system is solidified like jelly. This phenomenon is called gelation and gels are defined as soft, solid, or solid-like materials consisting of two or more components, one of which is a liquid present in substantial quantity.

Gel is classified into thermally-reversible gel and thermally-irreversible gel. The former is easily changed into solution by heating, and the solution smoothly returns to the original gel on cooling. On the other hand, the latter gel does not revert to solution on heating. For example, the agar made from Japanese seaweed agar gives the former thermally-reversible type of gel and is called *zeri* in Japanese. The polymerization of acrylamide in the presence of *N,N'*-methylenebis(acrylamide) as the crosslinking agent affords the latter thermally-irreversible poly(acrylamide) gel.

The thermally-reversible gel returns to the solution because the droving forces are noncovalent bonds such as hydrogen bonding, hydrophobic interaction, π-π interaction and electrostatic interaction. These noncovalent bonds are broken down by heating, so the gel reverts to the sol. Since the thermally-irreversible gel forms a network structure by firm chemical covalent bonds, the gel formed never changes to the sol. This is why the thermally-irreversible gel is called physical gel and the thermally-irreversible gel is referred to as chemical gel. Moreover, the gelation ability is found in not only biopolymers but also low molecular weight compounds.

This section deals with the development of organogelators that are low molecular weight compounds capable of gelating liquid by the simple operation of cooling after heating. The features of organogelators can be summarized as follows :

1) they are characterized by good solubility upon heating and inducement of smooth gelation of organic fluids, 2) gelation occurs at a considerably low concentration of gelator; less than 50 g $L^{-1}$, 3) the gels formed show thermally-reversible sol-to-gel phase transition, 4) the driving forces of gelation are cooperating noncovalent interactions such as hydrogen bonding, hydrophobic interaction, π-π interaction and electrostatic interaction.

The disposal of crude oil spills and used cooking oil are serious problems of environmental pollution. Organogelators that solidify waste oils upon addition in small amounts will play a role in environmental protection. Moreover, it is not acceptable to discard liquids such as spent organic solvents in industrial waste. If waste organic solvents can be rapidly solidified by adding organogelators, utilization as fuel becomes a possibility.

## 2.4.2 Low Molecular Weight Organogelators

Gelation of low molecular weight compounds was discovered by chance. About one hundred types of low molecular weight organogelators have been reported. Historical or typical organogelators, except those discovered by the present author are summarized in Scheme 2.14 ; 1,3:2,4-dibenzylidene-D-sorbitol (**49**),[124)] 12-(*S*)-hydroxystearic acid (**50**),[125)] *N*-lauroyl-L-glutamic-α,γ-dibutylamide (**51**),[126)] spin-labeled steroid (**52**),[127)] cholesteryl derivatives (**53**, **54**),[128,129)] 2,3-bis-*n*-hexadecyloxyanthracene (**55**),[130)] cyclic depsipeptide (**56**),[131)] L-cystine derivative (**57**),[132)] surfactant containing L-tartrate anion (**58**),[133)] triphenylamine derivative (**59**),[134)] γ-butyrolactone derivative (**60**),[135)] quaternary ammonium salt (**61**),[136)] biotin derivative (**62**),[137)] functionalized gluconamide (**63**),[138)] *N*-isopropylcholamide (**64**),[139)] enzymatic product (**65**),[140)] and *O*-methyl-4,6-benzylidene-D-galactose (**66**).[141)] Other gelators in the reviews are also notable.[142-144)]

1,3:2,4-Dibenzylidene-D-sorbitol **49** is a commercially available gelator for cosmetics as well as a modifier of polypropylene. 12-Hydroxystearic acid **50**, produced by hydrogenation of castor oil, is a gelator for hardening cooking oils and for cosmetics. *N*-Lauroyl-L-glutamic-α,γ-dibutylamide **51** is a gelator for making cosmetics and for hardening crude oil spills in the sea and is kept in large harbors in Japan.

## 2.4.3 Organogelators of Amino Acid Derivatives

The author evaluates gelation ability by using what is called the test tube upside-down method. A typical procedure for the gelation test is as follows: a weighed gelator is mixed with 2 mL of solvent in a test tube with a screw cap. Then the mixture is heated until the solid is dissolved. The resulting solution is cooled at 25°C for 2 hours and the gelation is checked visually. If when the test tube is turned upside down, there is no fluid running down, we judge it as "gel." When the samples caused gelation, we quantitatively evaluate the gelation ability by minimum gel concentration.

The appearance of formed gels is transparent, translucent, or opaque, depending on the kind of gelator and solvent. Although it depends on the combination of

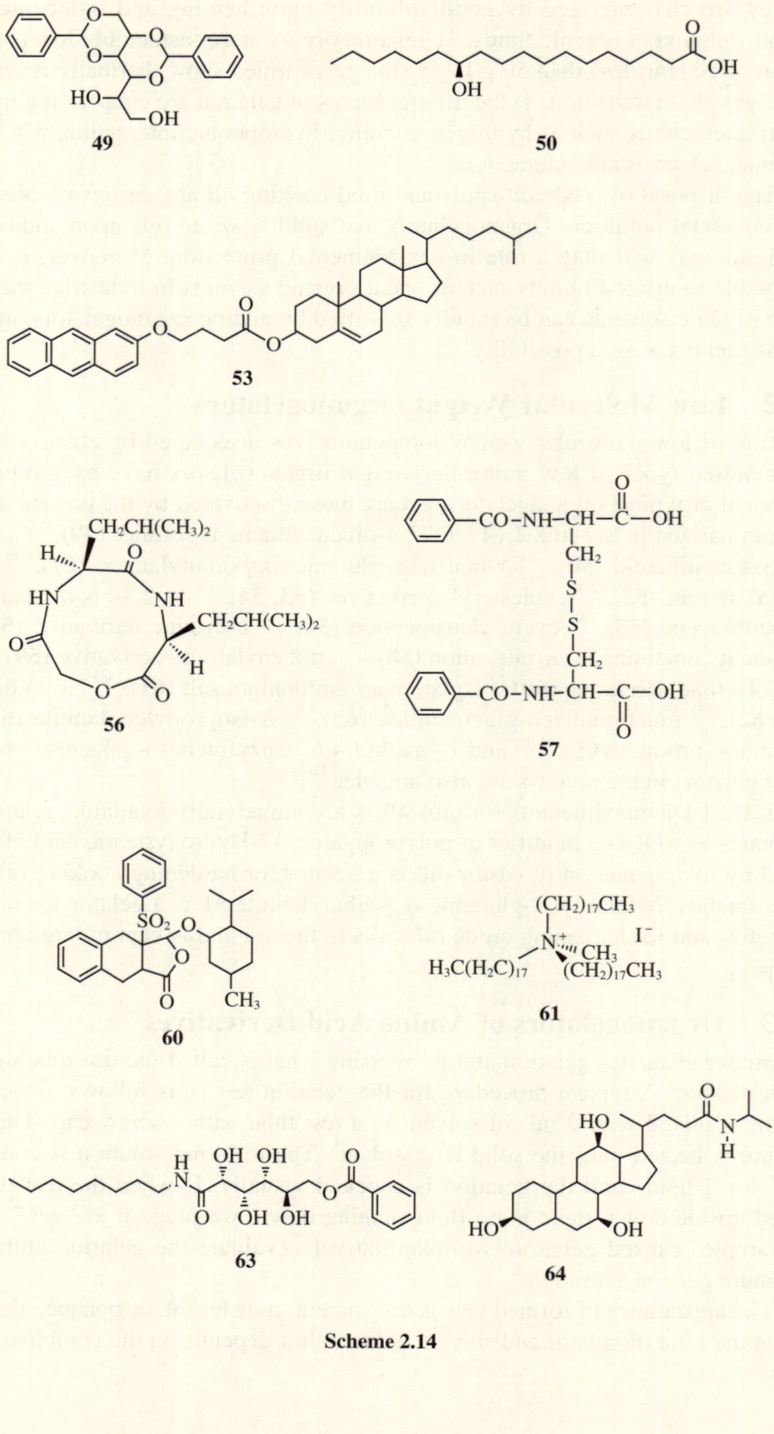


**Scheme 2.14**

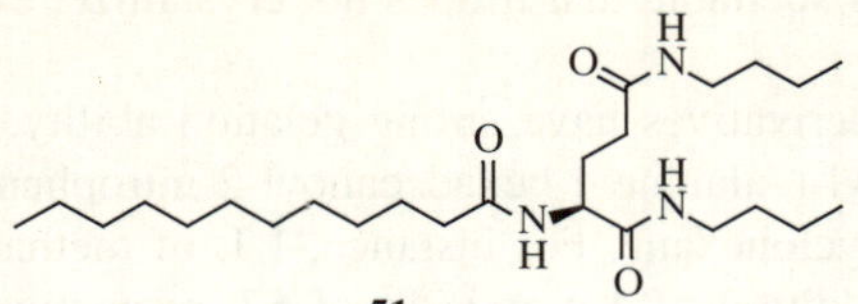

51

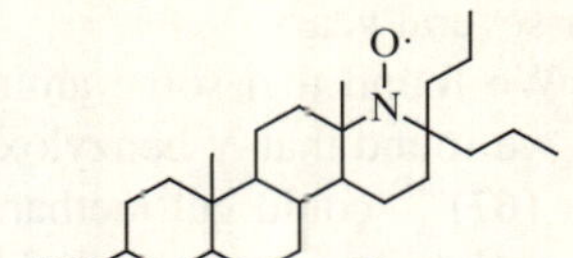

52

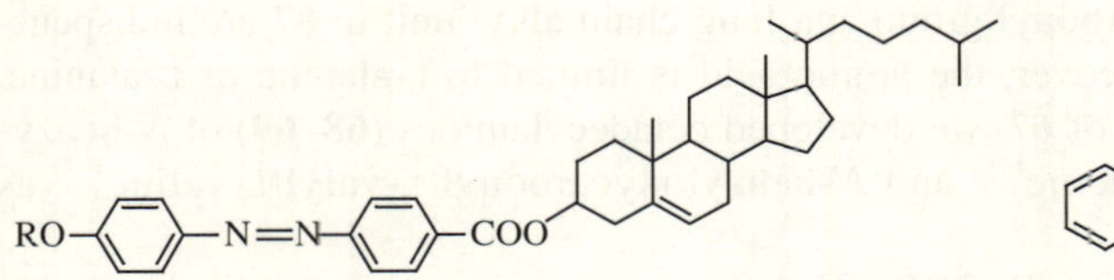

54

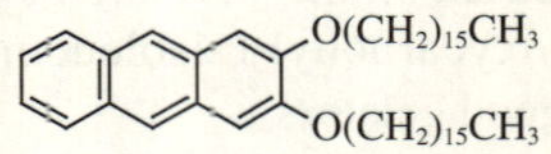

55

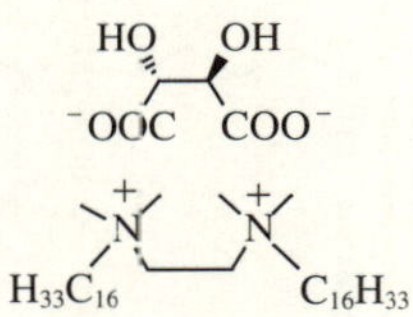

58

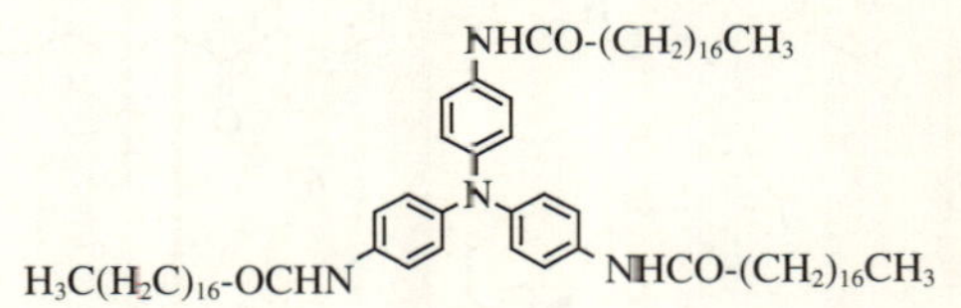

59

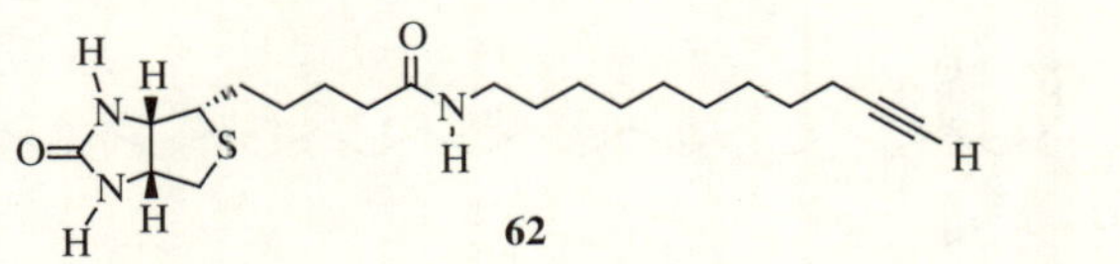

62

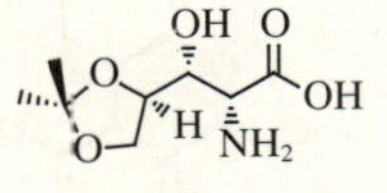

65

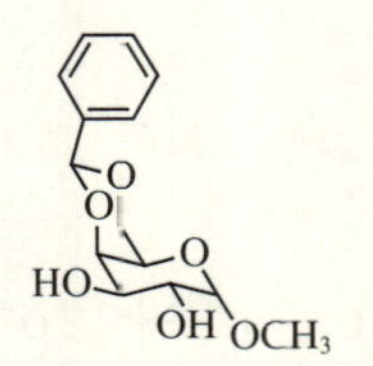

66

**Scheme 2.14** *Continued*

solvents and gelators, the formed gel is so stable that it does not crystallize, even after several years.

We found that some amino acid derivatives have strong gelation ability. At first we found that *N*-benzyloxycarbonyl-L-alanine 4-hexadecanoyl-2-nitrophenyl-ester (**67**)[145] could gel methanol and cyclohexane. For instance, 1 L of methanol and cyclohexane were gelled by the addition of 4 g and 2 g of **67**, respectively. However, **67** could not gel any other solvents except for methanol and cyclohexane. The nitro group carbonyl group and long-chain alkyl unit in **67** are indispensable for gelation; moreover, the amino acid is limited to L-alanine or D-alanine. Based on the structure of **67**, we developed octadecylamides (**68**, **69**) of *N*-benzyloxycarbonyl-L-isoleucine[146] and *N*-benzyloxycarbonyl-L-valyl-L-valine[147] as good gelators.

In particular, **68** (*Z*-L-Ile-$NHC_{18}H_{37}$) was easily prepared from L-isoleucine in moderate yield by the two-step reaction described below.

Results of gelation test of **68** and **69** toward 26 solvents are summarized in Table 2.11, where solvents up to cyclohexane are arranged in dielectric constant

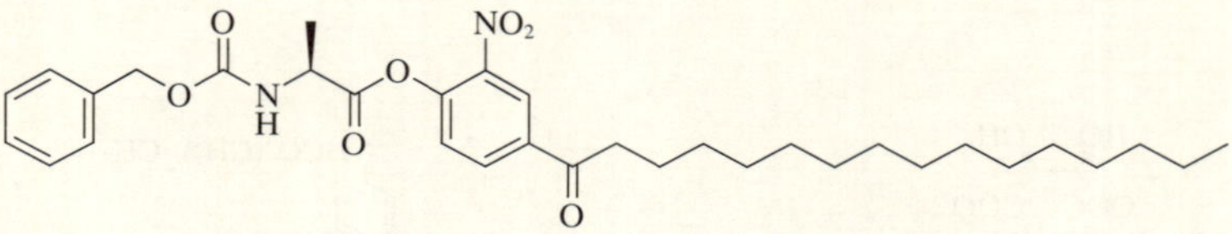

**67**

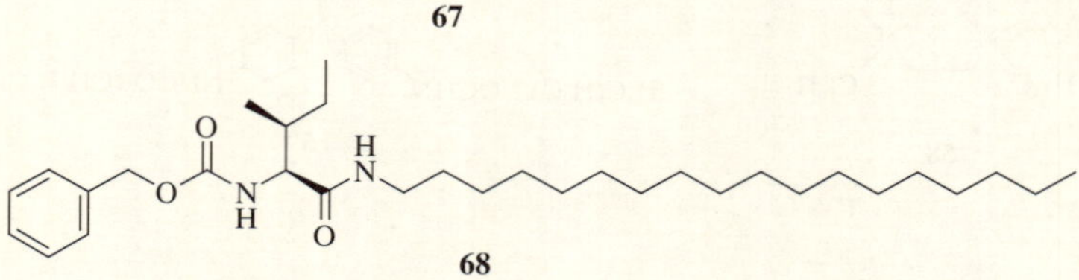

**68**

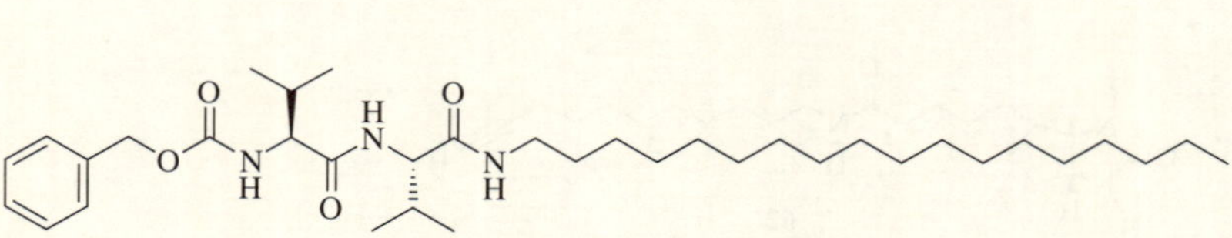

**69**

**Scheme 2.15**

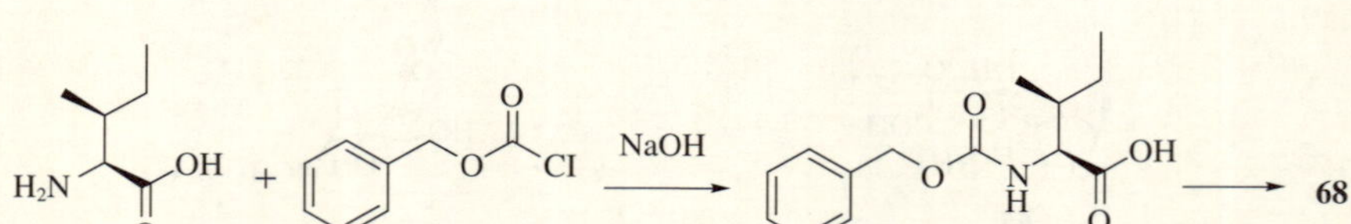

**Scheme 2.16**

**Table 2.11** Gelation test of **68** and **69** and minimum gel concentration necessary for gelation[a]

| Organic fluids | ε | **68** | **69** |
|---|---|---|---|
| Propylene carbonate | 69 | 10 | 7 |
| Dimethyl sulfoxide | 48.9 | 10 | 3 |
| γ-Butyrolactone | 39 | 13 | 7 |
| Acetonitrile | 37.5 | 7 | 8 |
| *N,N*-Dimethylformamide | 36.7 | 20 | 18 |
| Nitrobenzene | 34.8 | 22 | 4 |
| Methanol | 33.1 | 19 | 9 |
| Ethanol | 23.8 | 13 | 15 |
| Polyethylene glycol (400) | – | 7 | 15 |
| 1-Propanol | 22.2 | 20 | 15 |
| Acetone | 20.7 | 7 | 5 |
| Cyclohexanone | 18.3 | 35 | 12 |
| 2-Propanol | 18.3 | 10 | 19 |
| 1-Butanol | 17.1 | 19 | 15 |
| Ethyl acetate | 6.02 | 18 | 8 |
| Chlorobenzene | 5.6 | 30 | 4 |
| Benzene | 2.3 | 10 | 4 |
| Tetrachloromethane | 2.24 | 20 | 15 |
| Toluene | 2.2 | 36 | 14 |
| 1,4-Dioxane | 2.2 | 40 | 18 |
| Cyclohexane | 2.0 | 9 | 2 |
| Silicone oil | – | 5 | 4 |
| Light oil | – | 10 | 3 |
| Kerosene | – | 11 | 4 |
| Salad oil | – | 15 | 2 |
| Soybean oil | – | 23 | 2 |

[a] Gelation test temperature for **68** and **69** are 25°C and 20°C, respectively. ε is dielectric constant. Other values indicate minimum gel concentrations in g $dm^{-3}$ (gelator/organic fluid).

order. Values indicate minimum gel concentrations, whose unit is g $L^{-1}$. These gelators can gel many solvents, including propylene carbonate, DMSO, γ-butyrolactone, acetonitrile, DMF, nitrobenzene, alcohols, ketones, esters, aromatic solvents, silicone oil, mineral oils and edible oils.

It is noteworthy that the gelation ability strongly depends on the kind of amino acid. As long as the structure is fixed to the octadecylamides of benzyloxycarbonyl amino acid like **68**, gelation ability can be observed in only two amino acids, L-valine and L-isoleucine (**68**). There is no gelation ability in compounds including other amino acids, such as glycine, L-alanine, L-leucine, L-*t*-leucine, L-phenylalanine, L-proline, L-methionine, DL-valine, and DL-isoleucine. These results mean that gelation ability is very sensitive to structural factors. Besides **68** and **69**, physical gelation occurs by the following compounds: *Z*-L-Val-$NHC_{18}H_{37}$, *Z*-D-Val-$NHC_{18}H_{37}$, *Z*-D-Val-L-Val-$NHC_{18}H_{37}$, *Z*-L-Val-L-Leu-$NHC_{18}H_{37}$, *Z*-L-Val-β-Ala-$NHC_{18}H_{37}$, Z-L-Leu-β-Ala-$NHC_{18}H_{37}$ and Z-β-Ala-L-Glu$(NHC_{18}H_{37})_2$.

The reason why the amino acid derivatives with simple structures like **68** and

**69** possess gelation ability was studied by FT-IR spectroscopy. Participation of hydrogen bonding with gelation was clarified by FT-IR spectra. The organogelator **69** can gel cyclohexane, but not chloroform. The FT-IR spectrum of the cyclohexane gel of **69** is characterized by bands attributed to intermolecular hydrogen bonding, i.e. 3290 $cm^{-1}$ (ν(N–H) of urethane and amide), 1690 $cm^{-1}$ (ν(C=O) of urethane), and 1640 $cm^{-1}$ (ν(C=O) of amide). On the other hand, the FT-IR spectrum of **69** in chloroform, a homogeneous solution, shows ν(N–H) and ν(C=O) (urethane and amide) at 3437 $cm^{-1}$ 1714 $cm^{-1}$ and 1671 $cm^{-1}$ for non-hydrogen bonding. These results led us to conclude that at least the hydrogen bonding between amide and urethane is a driving force for gelation.

The gelation of the low molecular weight organogelators is sensitive to temperature. In general, the higher the temperature of the solvent, the more organogelator is required to gel. Since the formed gels always show thermally-reversible sol-to-gel transition, we can calculate thermodynamic parameters. Fig. 2.19 shows sol-to-gel phase diagrams of **69** in three solvents in which the minimum gel concentrations are plotted against temperature. The phase above the curve is the gel phase, whereas the phase below the curve is the sol phase.

The thermodynamic parameters for sol-to-gel transition in cyclohexane were found by calculation to be $\Delta H = -8.73$ kcal/mol (20 °C), $\Delta S = -15.9$ cal/K mol. These parameters indicate that the enthalpic contribution compensates for the undesirable entropic change. In other words, the enthalpic contribution is substan-

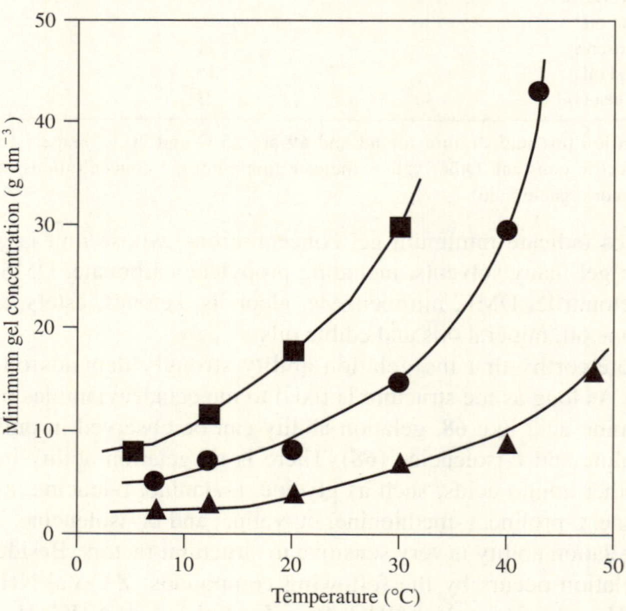


**Fig. 2.19** Sol-to-gel phase diagram; the minimum gel concentration of **69** *vs.* temperature in tetrachloromethane (■), in ethyl acetate (●) and in cyclohexane (▲).

tially important for gelation. In fact, the obtained $\Delta H$ value, –8.73 kcal/mol, suggests that about 2.5 hydrogen bondings were freshly formed when the gelation occurred. Actually, **69** can form three hydrogen bonds between N–H and C=O of a urethane and two amide bonds. From the results of FT-IR spectra, the intermolecular hydrogen bonding of **69** will be formed as shown in Fig. 2.20.

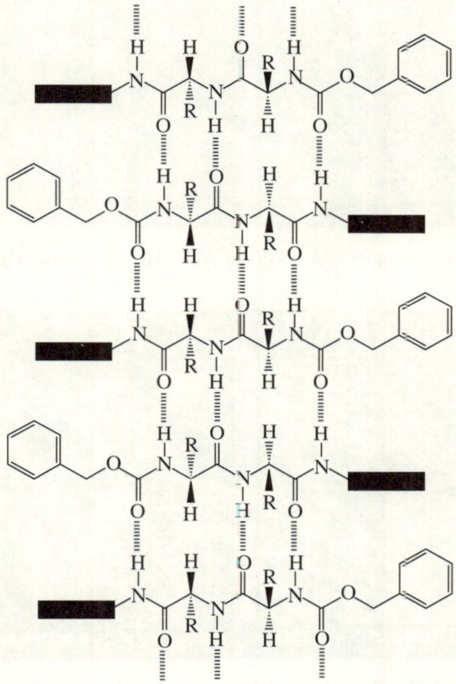

**Fig. 2.20** Molecular aggregate through intermolecular hydrogen bonding of **69**.

Images of supramolecular aggregates can be observed by electron microscope, because the first stage of physical gelation is the self-aggregation of gelator molecules. The TEM of a tetrachloromethane gel formed by *Z*-L-Val-NH$C_{18}H_{37}$, negatively stained by osmic acid, is shown in Fig. 2.21. It is obvious that numerous intertwined fibers were formed from the entanglement of 10–30 nm-long aggregates in the gel.

In the meantime, the octadecyl group of C-ends in **68** and **69** is also important for the gelation. When the octadecyl group shortens like the hexyl one, the gelation ability is remarkably lowered. This suggests that the hydrophobic long alkyl chain as well as hydrogen bonding plays an important part in gelation. The van der Waals interactions between long-chain alkyl units are necessary to gather molecular fibers and form three-dimensional network.

Figure 2.22 shows the SEM of the cyclohexane gel formed by **67**, where the

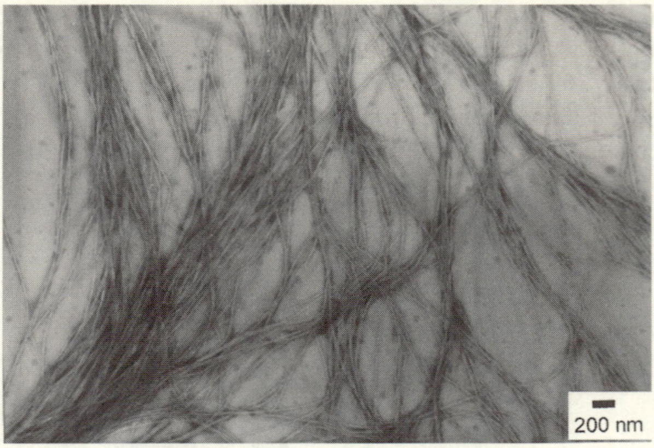


**Fig. 2.21** TEM of a tetrachloromethane gel of *Z*-L-Val-$NHC_{18}H_{37}$ negatively stained by osmic acid.

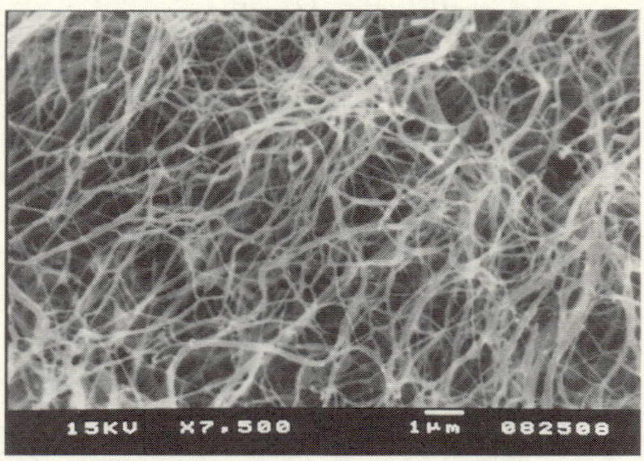


**Fig. 2.22** SEM of a cyclohexane gel of **67** shaded by gold.

three-dimensional image of a huge, steric and deep aggregate (diameter *ca.* 75–300 nm) is observed. Solvent will be trapped in these cavities.

We assume the following gelation mechanism: the fiber-like aggregations are formed from numerous molecules by intermolecular hydrogen bonding, then they are juxtaposed and interlocked by van der Waals interactions, finally immobilizing the organic liquid.

### 2.4.4 Cyclic Dipeptides as Organogelators

We proposed the following working hypotheses for the development of gelators: (i) huge fibrous aggregates are formed through noncovalent-bonding interactions, (ii) the formed huge fibrous aggregates build three-dimensional networks through noncovalent-bonding interactions, (iii) some factors preventing the crystallization from gels exist. We assume that factors (i)–(iii) are necessary for the development of gelators.

With these working hypotheses in mind, we studied the gelation ability of various cyclo(dipeptide)s with the expectation that (a) cyclo(dipeptide)s would form molecular aggregates through intermolecular hydrogen bonding due to four hydrogen bonding sites in each molecule, (b) the random alignment of cyclo(dipeptide)s consisting of different amino acids in the molecular aggregates would prevent the crystallization and consequently stabilize the gel state.

From our detailed study,[148)] it was established that there is no gelation ability for cyclo(glycylglycyl), cyclo(gylcyl-L-alanyl), cyclo(L-alanyl-L-alanyl), cyclo(L-valyl-L-valyl), cyclo(L-leucyl-L-leucyl), and cyclo(L-phenylalanyl-L-phenylalanyl). However, gelation ability was observed with cyclo(L-phenylalanyl-L-leucyl) (**70**), cyclo(L-valyl-L-γ-ethylglutamyl) (**71**), cyclo(L-leucyl-L-γ-ethylglutamyl) (**72**), cyclo(L-β-2-ethylhexylasparaginyl-L-phenylalanyl) (**73**), and cyclo(L-β-3,7-dimethyloctylasparaginyl-L-phenylalanyl) (**74**).

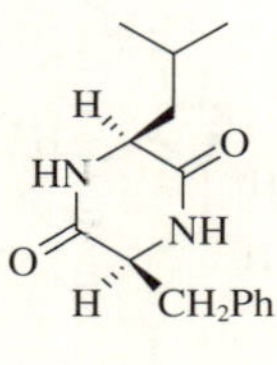

**70**

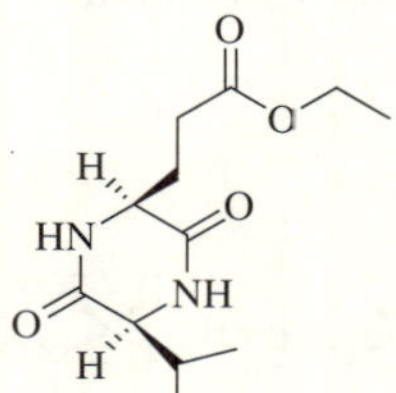

**71**

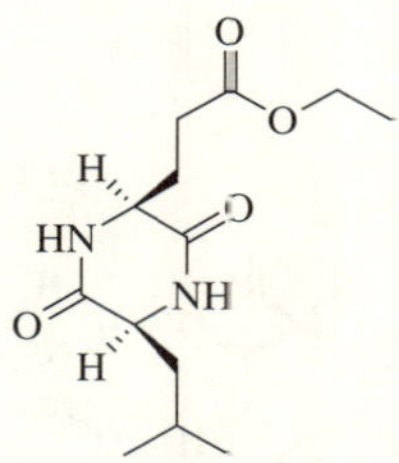

**72**

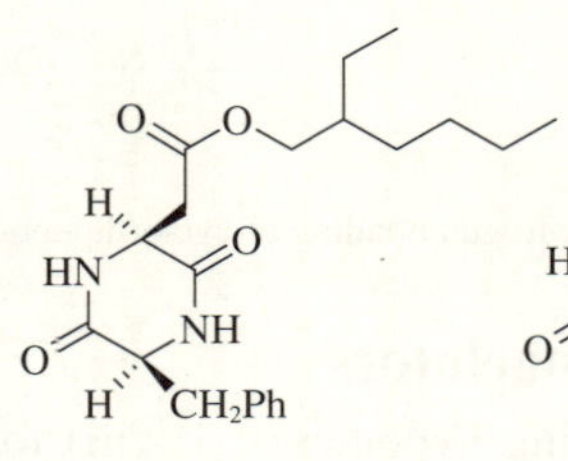

**73**

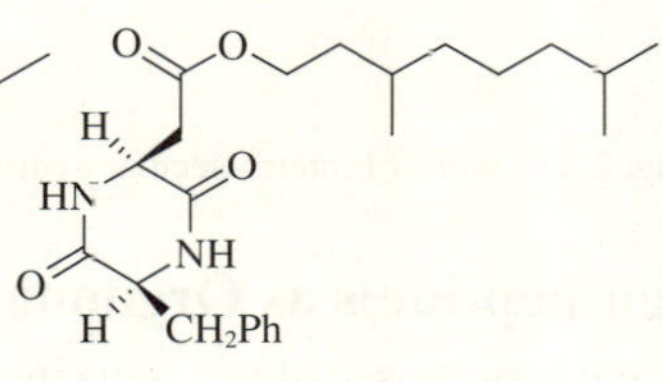

**74**

**Scheme 2.17**

Cyclo(dipeptide)s prepared from the same amino acid always give crystals, but not gels. In general, cyclo(dipeptide)s consisting of different amino acids (e.g., neutral and acidic) are superior as gelators to those containing two neutral amino acids.

The cyclo(dipeptide) **75** was prepared from the commercially available artificial sweetener "Aspartame" by a one-pot reaction according to the following reactions. The amount of **75** necessary to gel one liter of methanol, ethanol, acetone, ethyl acetate, benzene, toluene, tricaprylin, triolenein, soybean oil and silicone oil are 44 g, 28 g, 28 g, 23 g, 10 g, 3 g, 15 g, 15 g, 13 g and 2 g, respectively.

Figure 2.23 illustrates the motif of intermolecular hydrogen bonding between

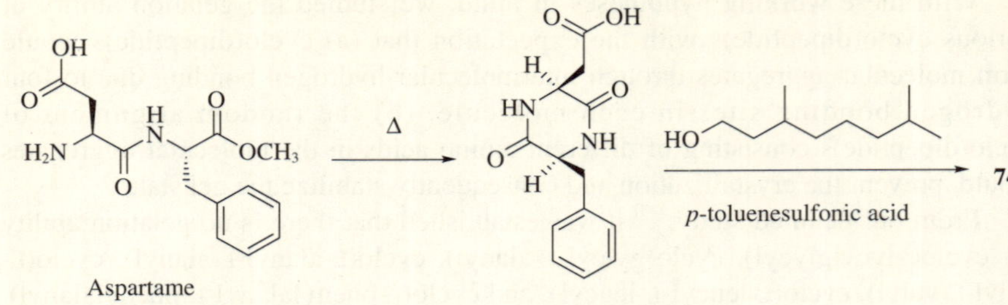

Scheme 2.18

the adjacent molecules of cyclo(dipeptide). The molecules form a ladder-like structure through intermolecular hydrogen bonding between N–H and C=O, and they are interlocked by van der Waals interactions.

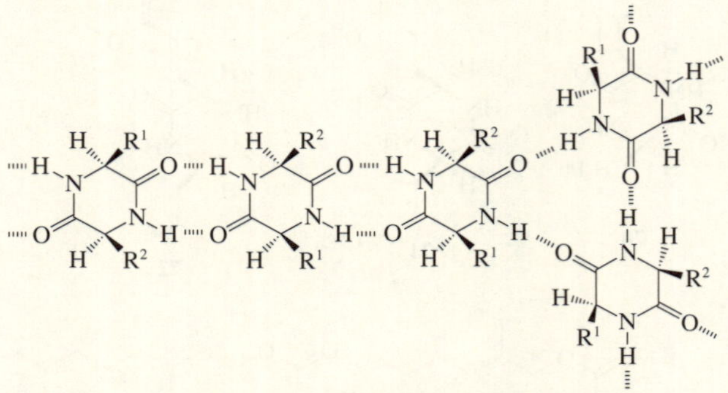

**Fig. 2.23** Motif of intermolecular hydrogen bonding of cyclo(dipeptide).

## 2.4.5 Oligo(peptide)s as Organogelators

Although we have proposed three working hypotheses (i)–(iii) for the development of gelators, the most difficult problem is how to stablize the formed gel, in other words, how to prevent crystallization from the metastable gel to the crystalline state. Actually, most of the compounds we synthesized were precipitated as crystals, but not gels.

Next, we focused on oligo(α-amino acid)s, which are especially promising for new organogelators, because they have plural amide bonds to act as hydrogen bonding sites and they are known not to crystallize owing to both the high molecu-

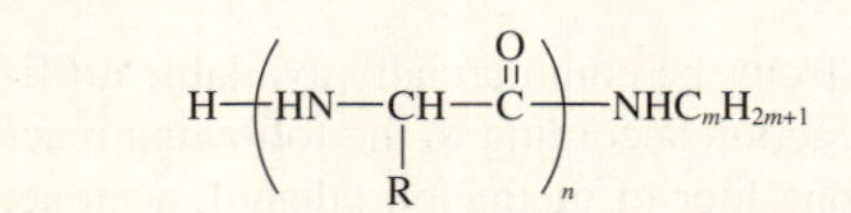

R = $CHMe_2$, $n = ca.$ 5, $m = 18$
R = CHMeEt, $n = ca.$ 5, $m = 18$
R = $CH_2Ph$, $n = ca.$ 5.1, $m = 12$
R = $CH_2CH_2CO_2Me$, $n = ca.$ 4.2, $m = 12$

**75**

Scheme 2.19

lar weight and the molecular weight distribution. Oligo(α-amino acid)s were prepared by decarboxylation of the *N*-carboxyanhydrides of α-amino acids in the presence of alkylamine. We found gelation ability in oligo(α-amino acid)s (**75**)[149].

Results of FT-IR and CD spectroscopy led us to conclude that oligo(α-amino acid)s can cause gelation if they have a β-conformation which is built up through intermolecular hydrogen bonding. When the average molecular weight of oligo(α-amino acid)s is high so that a α-helical conformation is constructed by intramolecular hydrogen bonding, the gel fails to form. In addition, oligomers forming a random coil do not act as gelators. The significance of the β-conformation for gelation was supported by Aggeli *et al.*, who reported gelation by an oligo(peptide) of $NH_2$-Lys-Leu-Glu-Ala-Leu-Tyr-Val-Leu-Gly-Phe-Phe-Gly-Phe-Phe-Thr-Leu-Gly-Ile-Met-Leu-Ser-Tyr-Ile-Arg-COOH.[150]

## 2.4.6 Two-component Type of Organogelators

Based on the three working hypotheses (i)–(iii), we developed a two-component type of gelator. One component is triaminopyrimidine derivatives and the other is barbituric acid derivatives. In fact, we obtained a clue from Lehn,[151] who reported that a 1:1 mixture of a triaminopyrimidine derivative and a barbituric acid derivative formed extended arrays through intermolecular hydrogen bonding and consequently gave the cocrystal of the complementary molecular components. The formation of extended arrays through intermolecular hydrogen bonding satisfies the first necessary factor (i) proposed by us. In order to satisfy the second (ii) and the third factors (iii), we introduced various alkyl groups into triaminopyrimidine and barbituric acid. Finally we developed the component type gelators (**76+77**, **78+79**).[152]

The combination of **78+79** could gel *n*-propanol, chloroform, benzene, ethyl acetate and acetone. The important feature of the two-component type of gelator is the fact that heating is unnecessary for the preparation of the gel, if it dissolves at

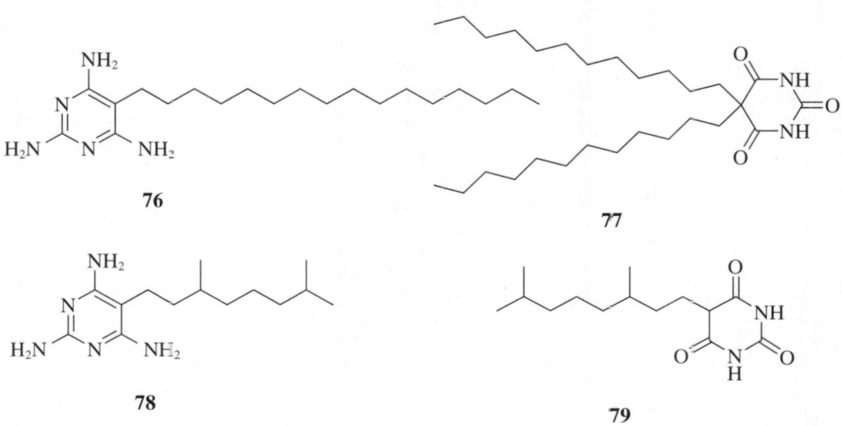


**Scheme 2.20**

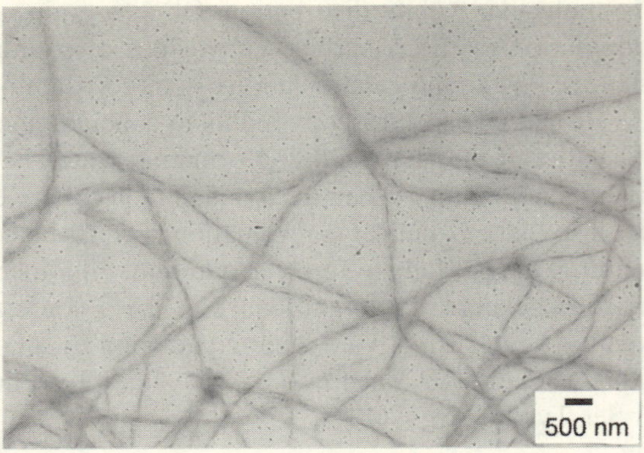


**Fig. 2.24** TEM of an ethyl acetate gel of **78** and 5-(2-ethylhexyl)barbituric acid, negatively stained by osmic acid.

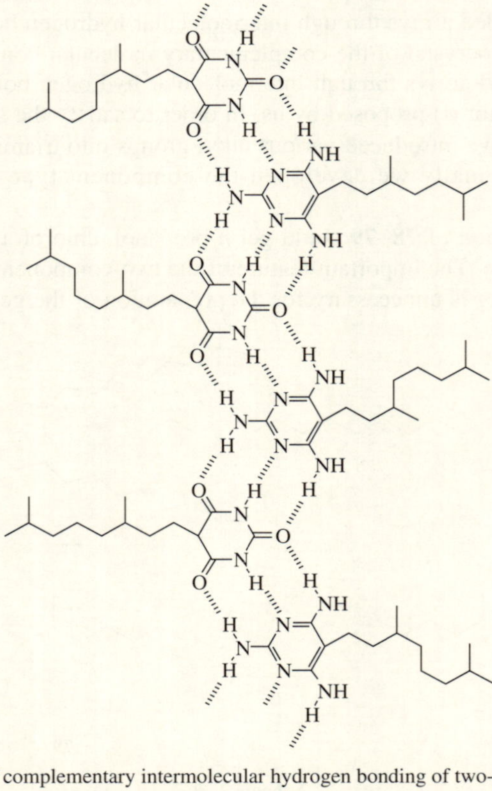

**Fig. 2.25** Motif of complementary intermolecular hydrogen bonding of two-component gelators.

ambient temperature. Namely, when we add two components to solvents individually, the gelation occurs spontaneously at room temperature. Actually, when **78** and **79** were added to chloroform, they naturally dissolved at room temperature, then turned into gel. However, chloroform is one of the more difficult solvents to gel.

TEM image of ethyl acetate gel formed by a 1:1 mixture of **78** and 5-(2-ethylhexyl)barbituric acid is shown in Fig. 2.24. It exhibits the gathering of numerous fibrous twisted aggregates. Constricted ribbons will be formed by twisting of the molecular tape, built up through complementary intermolecular hydrogen bonding.

The motif of complementary intermolecular hydrogen bonding is shown in Fig. 2.25. The molecular arrays based on complementary intermolecular hydrogen bonding is formed, but the branched alkyl chains of the two components are randomly folded and bent, in contrast to the extended alkyl chains of cocrystals. Besides **76+77** and **78+79**, some other two-component gelators have been reported.[153-156)]

## 2.4.7 *trans*-1,2-Cyclohexanediamine Derivatives as Organogelators

The *trans*-(1*R*,2*R*)-bis(undecylcarbonylamino)cyclohexane (**80**)[157)] and *trans*-(1*R*,2*R*)-bis(dodecylureido)cyclohexane (**81**),[158)] prepared from *trans*-(1*R*, 2*R*)-cyclohexanediamine having two amino groups at equatorial positions, can gel a wide variety of fluids. The gelation behavior of **81** is not discussed here, However, the strong gelation ability of **81** indicates that not only the amide group but the ureido group is also good for designing organogelators.

Results of gelation tests are summarized in Table 2.12. The fluids to be gelled range from protic solvents to aprotic ones, from polar solvents to nonpolar ones, and from mineral oils to edible ones. It is obvious that both **80** and **81** are excellent organogelators whose gelation abilities are comparable to those of **68** and **69**. For example, the amount of **80** necessary to gel one liter of acetonitrile, *N*,*N*-dimethylformamide, ethanol, ethyl acetate, tetrachloromethane, toluene and hexane are 5 g, 10 g, 33 g, 8 g, 23 g, 12 g and 6 g, respectively. This means that one molecule of **80** interacts with about 1800 molecules of acetonitrile, 600 molecules of *N*,*N*-dimethylformamide, 200 molecules of ethanol, 600 molecules of ethyl

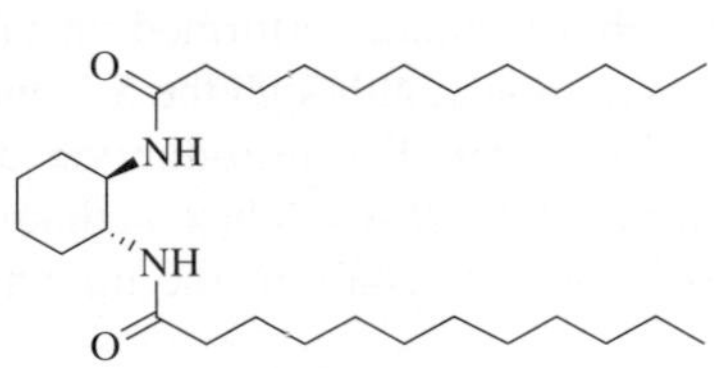


**80**

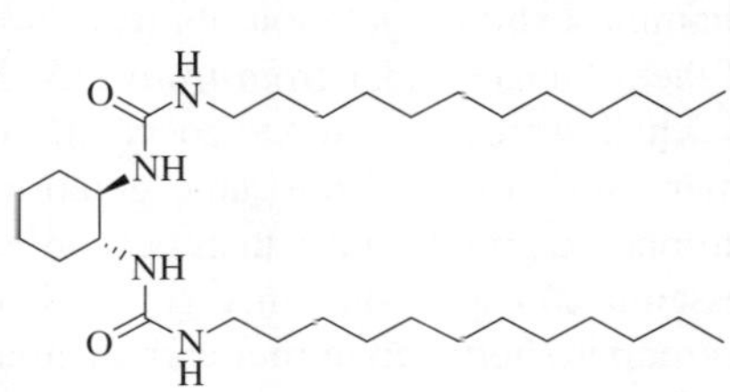


**81**

**Scheme 2.21**

**Table 2.12** Gelation test of **80** and **81** and minimum gel concentration necessary for gelation at 25°C[a]

| Organic fluids | ε | 80 | 81 |
|---|---|---|---|
| Dimethyl sulfoxide | 48.9 | 12 | 5 |
| *N,N*-Dimethylacetamide | 37.8 | 11 | 2 |
| Acetonitrile | 37.5 | 5 | 12 |
| *N,N*-Dimethylformamide | 36.7 | 10 | 2 |
| Methanol | 33.1 | 20 | 3 |
| Ethanol | 23.8 | 33 | 4 |
| Acetone | 20.7 | 10 | insol. |
| 2-Butanone | 18.51 | 15 | 2 |
| Cyclohexanone | 18.3 | 11 | 2 |
| 2-Propanol | 18.3 | 40 | 5 |
| Pyridine | 12.3 | 25 | 2 |
| Tetrahydrofuran | 7.58 | soln. | 2 |
| Ethyl acetate | 6.02 | 8 | insol. |
| Chlorobenzene | 5.6 | 22 | 13 |
| Benzene | 2.3 | 20 | 9 |
| Tetrachloromethane | 2.24 | 23 | 15 |
| Toluene | 2.2 | 12 | 7 |
| 1,4-Dioxane | 2.2 | 12 | 3 |
| Cyclohexane | 2.0 | 11 | 2 |
| Hexane | 1.89 | 6 | insol. |
| Silicone oil | – | 2 | soln. |
| Light oil | – | 8 | 15 |
| Kerosene | – | 7 | 11 |
| Salad oil | – | 6 | 30 |
| Soybean oil | – | 7 | soln. |

[a] ε is dielectric constant. Other values indicate minimum gel concentrations in g $dm^{-3}$ (gelator/organic fluid). insol. = almost insoluble, soln. = solution.

acetate, 200 molecules of tetrachloromethane, 400 molecules of toluene and 600 molecules of hexane.

It is important to note that the corresponding *cis* isomer prepared from *cis*-1,2-cyclohexanediamine, in which two amino groups are located in the axial and equatorial positions, forms no gel with any of the fluids. Neither the corresponding compound from 1,3-cyclohexanediamine nor the compound from 1,4-cyclohexanediamine exhibits gelation ability. Furthermore, the racemate consisting of **80** and the *S,S*-enantiomer from *trans*-(1*S*, 2*S*)-cyclohexanediamine formed unstable gels which were converted to cocrystals after several hours, although the *S,S*-enantiomer itself revealed the same gelation behavior as **80**. The introduction of a hydrophobic group longer than hexanoly into *trans*-(1*R*, 2*R*)-cyclohexanediamine is essential for gelation, since gels fails to form when the chain of the lipophilic tail group is shortened to four carbon atoms.

The thermodynamic parameters for sol-to-gel transition in toluene were found by calculation to be $\Delta H$ = 65 kJ/mol, $\Delta S$ = –160 J/K mol suggesting that the enthalpic contribution is important for gelation. The $\Delta H$ value 65 kJ/mol is

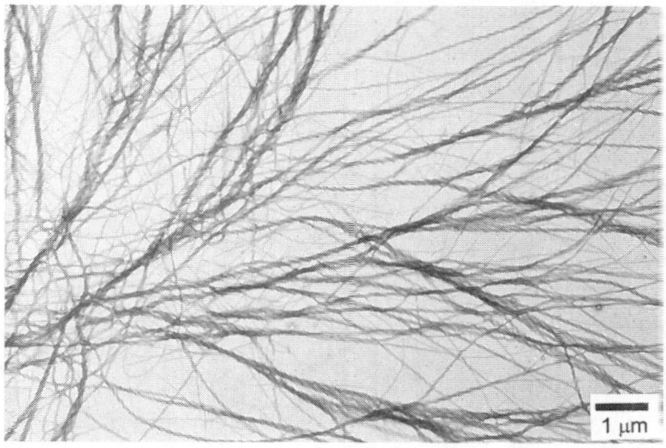


**Fig. 2.26** TEM of an acetonitrile gel of **80** negatively stained by osmic acid.

brought about by two hydrogen bondings when the gelation occurs. Molecular aggregates responsible for gelation were observed by transmission electron microscopy. Fig. 2.26 shows a TEM of loose gel in acetonitrile formed by **80**. Numerous intertwined aggregates were formed from helical fibers with width of 40 nm to 70 nm. It should be noted that the helicity of fibers was always left-handed for the *trans*-(1*R*,2*R*)-isomer, and right-handed for the *trans*-(1*S*,2*S*)-enantiomer.

To study the chiral structure of the aggregate in gel, we measured circular dichroism spectra. The CD spectrum of *trans*-(1*R*,2*R*)-isomer **80** exhibited markedly strong peaks for the amide unit at 200 nm and 213 nm. Their molar ellipticity is more than several tens of thousands. The *S*,*S*-enantiomer gave almost the same spectrum, but the values were opposite in sign. In general, such a much-enhanced CD is derived from exciton coupling among the organized chromophores. Therefore, the observed CD indicates that the aggregate is cooperatively organized for helical stacking. These strong peaks at 20 °C decreased with increasing temperature and disappeared at 50 °C when the gel was transformed to an isotropic solution. The disappearance of the CD signal in the isotropic solution suggests that the observed CD bands originate from a chiral aggregate, but not from a chiral conformation of the molecule itself.

We estimated the motif of intermolecular hydrogen bondings for gel by X-ray crystallographic analysis of the structurally related molecule. Unfortunately, our attempt to make single crystals from the gelators has not yet met with success. However, we obtained the single crystal of *trans*-(1*R*,2*R*)-bis(methylcarbonylamino)cyclohexane. This compound with methylcarbonyl instead of undecylcarbonyl cannot gel any solvent, but the information about the hydrogen bonding motif may help to understand the molecular packing in gels. The molecular packing of *trans*-(1*R*,2*R*)-bis(methylcarbonylamino)cyclohexane determined by X-ray crystallography is shown in Fig. 2.27. It is clear that the molecular packing adopts a parallel coplanar overlap, and the side view reveals that intermolecular hydrogen bondings are formed in the perpendicular direction.

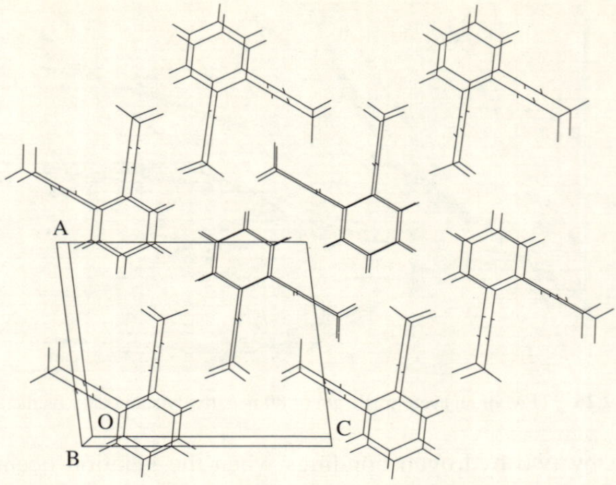


**Fig. 2.27** Molecular packing determined by X-ray crystallography of *trans*-(1*R*,2*R*)-bis(methylcarbonylamino)cyclohexane. $a$ = 10.15 Å, $b$ = 4.838 Å, $c$ = 11.559 Å, β = 96.6° Z = 2.

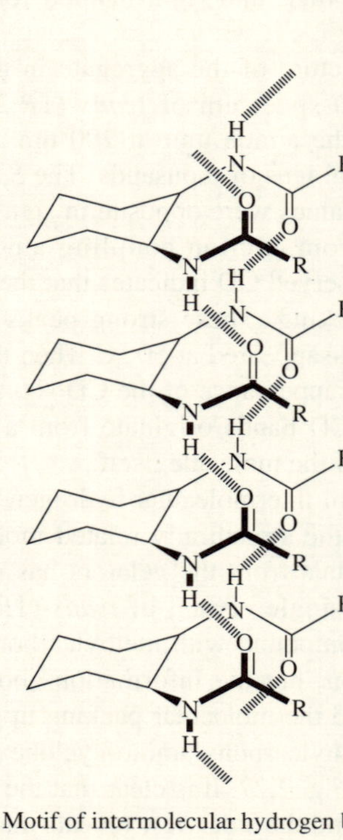

**Fig. 2.28** Motif of intermolecular hydrogen bonding of **80**.

Based on the basis of the molecular packing determined by X-ray crystallography, we assume a motif of intermolecular hydrogen bonding between the adjacent molecules (Fig. 2.28). The extended asymmetric molecular tape is formed through two intermolecular hydrogen bondings between N–H and C=O. The NH and CO can direct themselves antiparallel to each other and perpendicular to the cyclohexyl ring. Therefore, an extended molecular tape was stabilized by two intermolecular hydrogen bonds between each molecule. The resulting asymmetric tape may be responsible for the chiral structure observed in the CD spectra and TEM images.

We assume that, first of all, the tape-like molecular aggregates are formed by intermolecular hydrogen bonding. Subsequently these are interlocked and intertwined through van der Waals interaction leading to the helical fiber aggregates. Then organic fluids are entrapped in the three-dimensional networks and finally gelation occurs.

### 2.4.8 Trialkyl-1,3,5-benzenetricarboxamide and Trialkyl-*cis*-1,3,5-cyclohexanetricarboxamide

We focused on trialkyl-1,3,5-benzenetricarboxamides (**82**, **83**) as organogelators with the expectation that intermolecular hydrogen bonding and $\pi$-$\pi$ stacking play a role in the formation of macromolecule-like aggregates.[159] Actually, trioctadecyl-1,3,5-benzenetricarboxamide was reported to show a discotic liquid crystalline phase and the driving force for liquid crystalline behavior is assumed to be intermolecular hydrogen bonding.[160,161] Although a noteworthy gelation ability was not observed in the family of trialkyl-1,3,5-benzenetricarboxamides, compound **83** was able to gel hexane at a concentration of 55 g $L^{-1}$. The trialkyl-1,3,5-benzenetricarboxamides may be better for thickening rather than as gelators. When adding **82** or **83** to some organic solvents, a marked rise in viscosity was observed. For

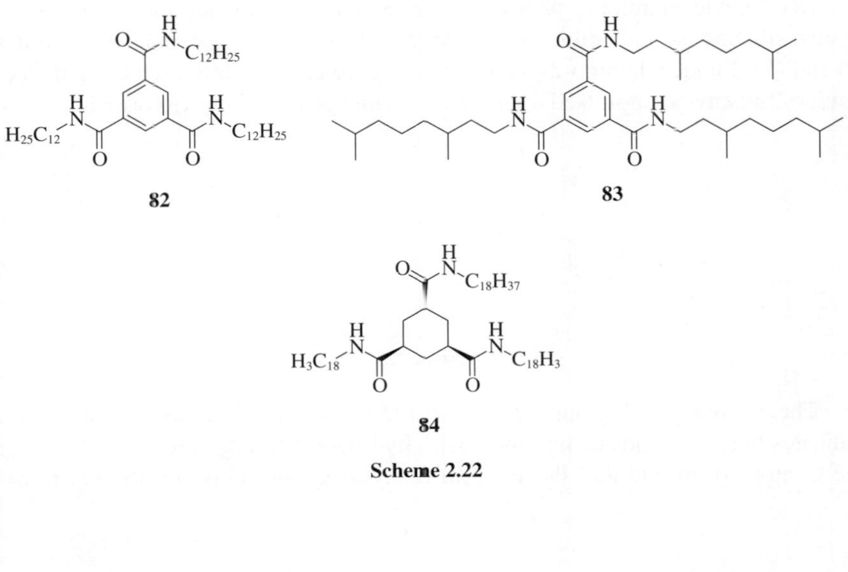

Scheme 2.22

instance, the viscosity of cyclohexane containing **82** (20 g $L^{-1}$) and **83** (30 g $L^{-1}$) increased to 1800 cP and 20000 cP at 20 °C, respectively.

On the other hand, trioctadecyl-*cis*-1,3,5-cyclohexanetricarboxamide (**84**),[162] in which three C=O groups are located at equatorial positions in the cyclohexane ring, is a good organogelator for nonpolar solvents, including hexane (1 g $L^{-1}$), cyclohexane (2 g $L^{-1}$), toluene (5 g $L^{-1}$), tetrachloromethane (5 g $L^{-1}$), benzene (3 g $L^{-1}$), chlorobenzene (3 g $L^{-1}$), tetrahydrofuran (9 g $L^{-1}$), pyridine (1 g $L^{-1}$), kerosene (1 g $L^{-1}$) and salad oil (2 g $L^{-1}$). The values in parentheses indicate minimum gel concentrations.

### 2.4.9 Organogelators Based on 2-Amino-2-phenylethanol

Although a considerable amount of structurally different gelators has been reported so far, there are still several unanswered questions. For example, 1) why do gelators in fluids form metastable gels instead of thermally-stable crystals? and 2) what, if any, are the necessary and sufficient requirements for designing gelators? As yet, we can give no answer to these questions; however, it is interesting to note that most of the reported gelators possess chiral carbon in their chemical structure, as shown in Scheme 2.14, although **55**, **59** and **61** are exceptions. Actually, we have synthesized several types of gelators from chiral compounds as desciribed above and found that racemates tend to crystallize and not form gels. Considering that the most difficult point involved in the development of gelators is how to prevent the transformation from the metastable gel to a crystalline state, we assume that the chirality in gelators probably promotes the formation of chiral aggregates which develop into the network structure for gels. The chiral structure in compounds, although it may be neither a necessary nor a sufficient requirement, will play a vital role in the development of gelators. Following this idea, we attempted to create a new type of organogelators from 2-amino-2-phenylethanol as a chiral starting material.

(*R*)-2-Undecylamino-2-phenylethanols **85** can gel aromatic molecules, mineral oils, silicone oil and edible oils.[163] As we predicted, the racemate consisting of **85** and (*S*)-2-undecylamino-2-phenylethanol gave crystals, although (*S*)-2-undecylamino-2-phenylethanol itself exhibited the same gelation behavior as **85**.

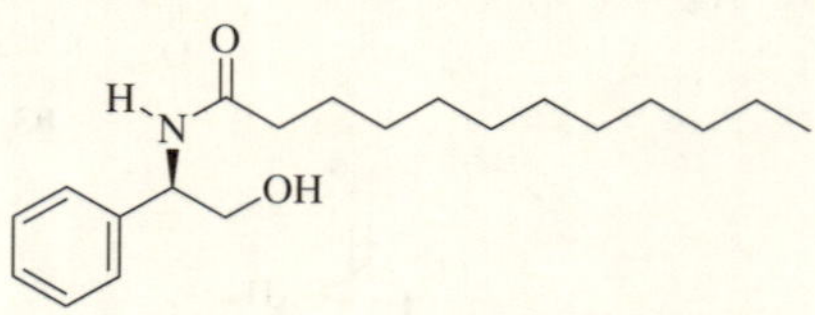


**85**

**Scheme 2.23**

The X-ray crystallographic analysis of (*R*)-(–)-2-acetylamino-2-phenylethanol indicates that two kinds of intermolecular hydrogen bondings are formed between the O atom of amide and the H atom of hydroxy and between the O atom of

**Fig. 2.29** TEM of an ethyl acetate gel of (*S*)-2-heptadecylamino-2-phenylethanol negatively stained by osmic acid.

hydroxy and the H atom of amide. The CD spectra of gel of **85** and its *S*-enantiomer suggest the formation of chiral aggregates. The TEM of an ethyl acetate gel of (*S*)-2-heptadecylamino-2-phenylethanol is shown in Fig. 2.29. The chiral aggregates were found to be large helical ribbons, 500 nm wide and up to 10 μm in length, whose helicity was exclusively right-handed. The helical aggregates account for the formation of metastable gels instead of thermally-stable crystals. When the helical ribbons are sufficiently long and entangled, the fluids are entrapped in the three-dimensional network; consequently, physical gelation occurs.

Finally, we can say with moderate conviction that a chiral compound having multiple hydrogen bonding sites which especially forms helical aggregates in solvents is a candidate for organogelator, although chirality and hydrogen bonding of themselves do not sufficiently fulfill the requirements.

## Acknowledgment

Part of this work is supported by Grant-in-Aid for COE Research (10CE2003) and Grant-in-Aid for Scientific Research on priority Area (A), "Dynamic Control of Strongly Correlated Soft materials" (No.413/13031036) from the Ministry of Education, Science, Sports, Culture, and Technology of Japan.

# 2.5 Synthesis and Self-association of Stimuli-responsive Block Copolymers by Living Cationic Polymerization

**Sadahito Aoshima**

## 2.5.1 Introduction

Well-designed amphiphilic copolymers have recently been synthesized and characterized by many researchers.[164] This field has attracted much attention both academically and for possible practical applications. As selective examples, a number of nano-organized self-assemblies of macromolecules have been prepared using various amphiphilic block copolymers or hydrophobically-modified water-soluble polymers.[165] Furthermore, it has been found that the chemical structure, molecular weight, and hydrophobicity of these amphiphilic polymers controlls the morphology of the self-aggregates. Living polymerization is a very useful and versatile tool for providing this necessary control of structure and molecular weight. However, vinyl monomers having a polar substituent are, in general, known to be very difficult to polymerize into living polymers, because the polar pendants usually cause various side reactions such as chain transfer during polymerization.

Recent progress in living polymerization has given rise to the novel design and synthesis of well-defined amphiphilic block copolymers.[166] In the field of cationic polymerization, living polymerization is currently achievable through the use of nucleophilic counteranions, externally added bases, or added salts. In each case, the key point is the stabilization of propagating carbocations by nucleophilic interaction. If these interactions are sufficient and the growing carbocations become sufficiently stable, then living polymerization will be achieved (Scheme 2.24). Our group has investigated the use of an externally added base as a useful polymerization system for conventional metal halide catalysts at relatively high polymerization temperatures.[167]

In this study, the authors investigate the living cationic polymerization of polar monomers in the presence of added bases to prepare living polymers of various properties and shapes. Based on the results, various types of amphiphilic copolymers are designed, and the stimuli-induced self-association of the products, such as thermosensitive physical gelation, is characterized.

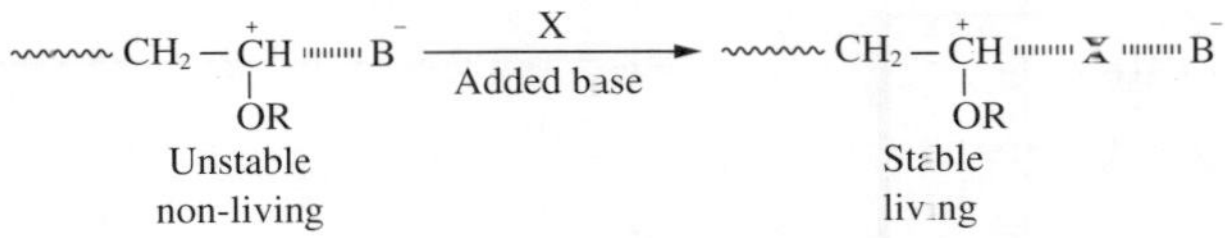


Living cationic polymerization in the presence of added bases (X).

**Scheme 2.24**

## 2.5.2 Results and Discussion

### A. Living Cationic Polymerization in the Presence of Added Bases and Synthesis of Various Functional Polymers

As a typical example of living cationic polymerization in the presence of an added base, isobutyl vinyl ether was polymerized using an $EtAlCl_2$-based initiating system in the presence of ethyl acetate in hexane at 0 °C.[167)] In the absence of ethyl acetate (conventional cationic polymerization), polymerization occurred rapidly to give polymers with broad MWD ($\overline{M}_w/\overline{M}_n = 2.6$). In the presence of a large amount of ethyl acetate (1.0M), the obtained polymers had a very narrow MWD, close to monodisperse ($\overline{M}_w/\overline{M}_n \leq 1.1$). The living nature of the polymerization was confirmed by the linear increase in polymers' $\overline{M}_n$ in direct proportion to the progress of conversion. Thus, the externally added bases stabilized the carbocations, which were unstable under additive-free conditions.

This system is suitable for the polymerization of polar functional monomers and the synthesis of polymers of various shapes. Figure 2.30 shows the living polymerization of polar monomers containing Si, *t*-butyldimethylsilyl vinyl ether. Polymerization was carried out using $Et_{1.5}AlCl_{1.5}$ in toluene at 0 °C in the presence of ethyl acetate.[168)] The C−O−Si bond is known to be readily hydrolyzed by free acid, and the use of added base was effective for inducing living polymerization without such side reactions. The molecular weight increased in direct proportion to the conversion, and the $\overline{M}_w/\overline{M}_n$ was as low as 1.04.

Figure 2.31 shows a typical example of the preparation of diblock copolyalcohol **75**. Block copolymerization was carried out in the presence of an added base by successive second monomer addition. The MWD shifted to higher $\overline{M}_n$, yet remained narrow without tailing. Subsequent acidic hydrolysis resulted in the quantitative conversion of the pendant groups into the polyalcohol.

Scheme 2.25 summarizes the living polymers obtained and the polymer properties (water-soluble polyalcohols and polycarboxylic acid enclosed by a frame were obtained using a corresponding protected monomer). Based on these results, the authors designed block copolymers with various properties such as water-solubility, thermosensitivity and hydrophobicity.

In the course of the investigation illustrated in Scheme 2.25, stimuli-responsive polymers were discovered, in particular, those that exhibit thermally-induced phase separation.[169)] A transparent aqueous solution of a polymer with hydrophilic oxyethylene chains and a hydrophobic ω-alkyl group (POEVE) became cloudy upon heating to a certain critical temperature. This phase separation was found to

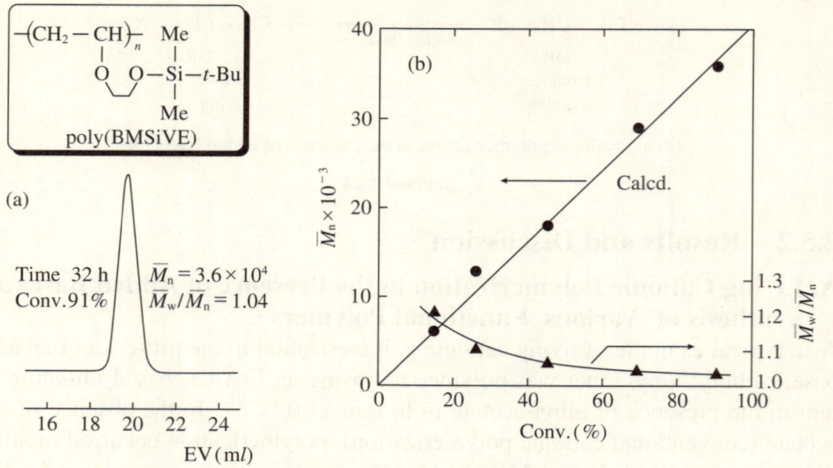


**Fig. 2.30** MWD (a) of poly(BMSiVE) obtained by IBEA/$Et_{1.5}AlCl_{1.5}$ in the presence of ethyl acetate in toluene at 0 °C and relationships between conversion and $\overline{M}_n$ or $\overline{M}_w/\overline{M}_n$ (b): IBEA: $CH_3CH(O\textit{i}Bu)OCOCH_3$, $[BMSiVE]_0$=0.8 M, $[IBEA]_0$=4.0 mM, $[Et_{1.5}AlCl_{1.5}]_0$=20 mM, [ethyl acetate]=1.0 M.

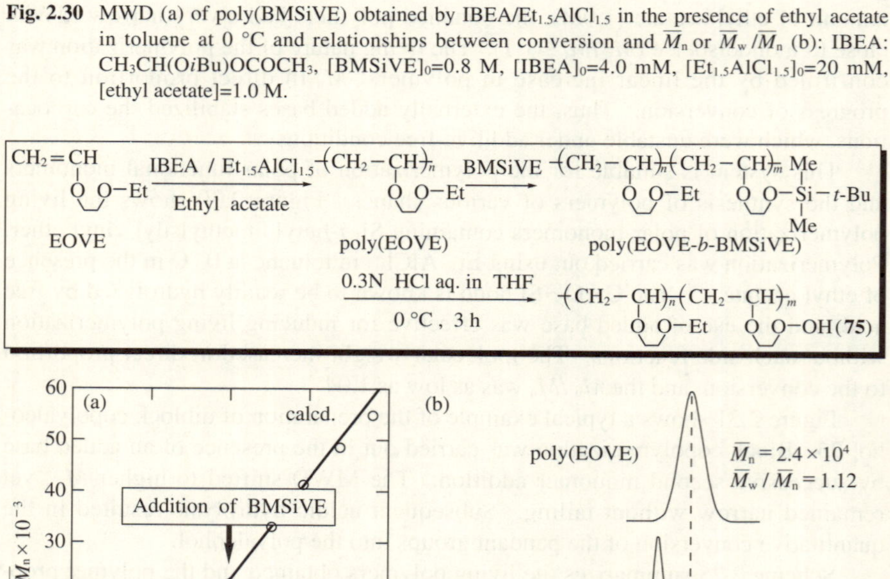


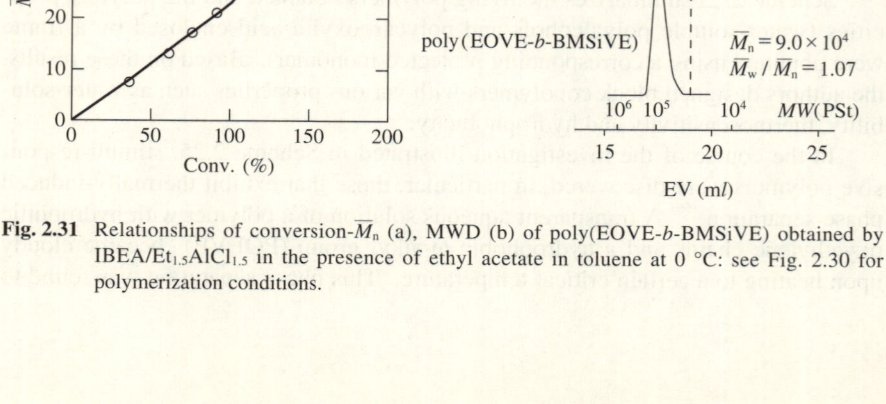


**Fig. 2.31** Relationships of conversion-$\overline{M}_n$ (a), MWD (b) of poly(EOVE-*b*-BMSiVE) obtained by IBEA/$Et_{1.5}AlCl_{1.5}$ in the presence of ethyl acetate in toluene at 0 °C: see Fig. 2.30 for polymerization conditions.

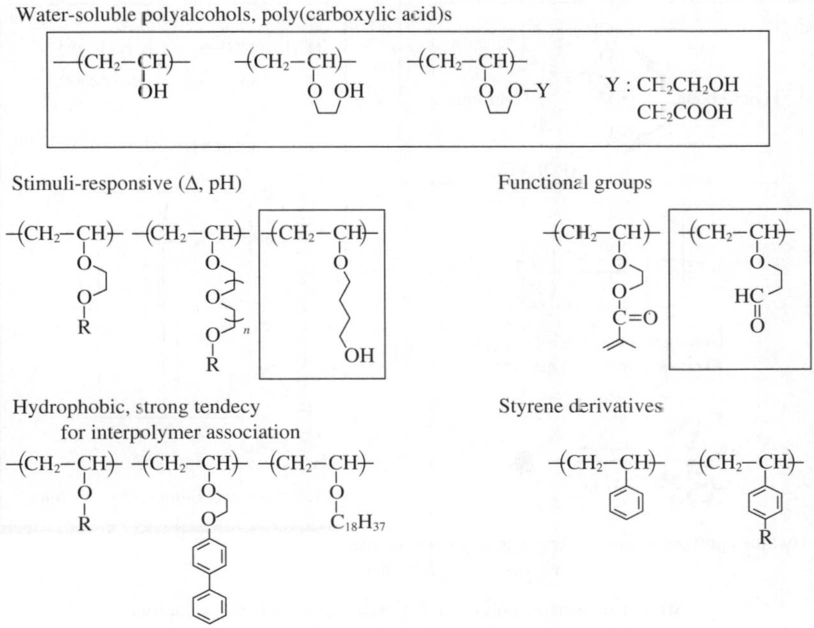


Synthesis of various living polymers with functional groups.

**Scheme 2.25**

be very sensitive and reversible. Poly(EOVE), which has one oxyethylene and one ethyl group, changes from hydrophilic to hydrophobic at 20 °C. This can be readily observed as a change of the solution from transparent to opaque. As can be seen regarding the turbidity results in Scheme 2.26, the phase separation was quite sensitive and reversible on heating and cooling. The mechanism of phase separation is thought to be as follows: at low temperature, the polymers are soluble in water by hydration. When the temperature is increased to the critical point, dehydration occurs and hydrophobic interactions increase, resulting in aggregation (phase separation) .

To the best of our knowledge, POEVEs are the only living polymer system other than poly(ethyleneoxide) that exhibits such phase separation, and as such the authors began a detailed systematic investigation of molecular weight and its distribution, and sequence distribution. The critical temperature $T_{ps}$ was found to change according to the structure of the polymer side-groups, oxyethylene and ω-alkyl groups. The temperatures could also be controlled by random copolymerization of these monomers. A clear relationship between MWD and sensitivity was observed: the $\Delta T_{ps}$ of narrow-MWD polymers was 0.3 °C, whereas that of broad -MWD polymers was 3 °C. The sequence distributions imparted a remarkably effect on the phase separation behavior: in the case of block copolymers, two-step phase separation occurred at critical temperatures for both POEVE segments,

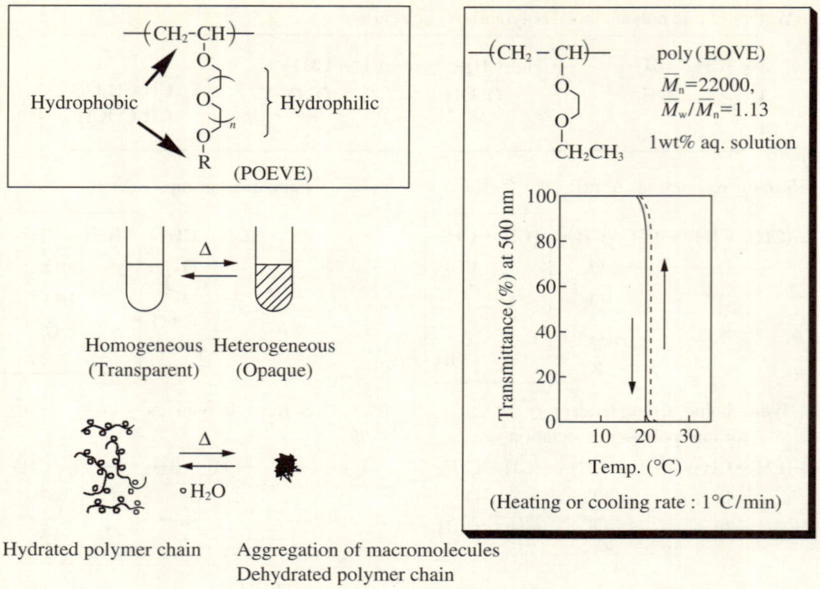


Stimuli-responsive polymers: thermally-induced phase separation.

**Scheme 2.26**

whereas random copolymers having the same MW, composition and narrow MWD exhibited only a single phase separation at the average critical temperature.

As shown in Scheme 2.27, POEVE and PNIPAM have similar structures with both hydrophilic and hydrophobic parts in each monomer unit. Quite recently, our group examined another possibility for thermosensitive phase separation; random copolymers of hydrophilic and hydrophobic monomers.[170] Although random copolymers have been investigated previously, the achieved phase separation was broad, with hysteresis and low turbidity. By living cationic polymerization in the presence of added bases, our group successfully prepared random copolymers of IBVE and HOVE, both of which are typical hydrophobic and hydrophilic monomers, and which are not thermosensitive themselves. At low temperature, the polymers were soluble in water, but when the temperature was increased to a critical point, the transparent solution became opaque. The phase separation was quite sensitive (Scheme 2.27(a)) and the temperature of phase separation was governed by the monomer feed ratio.

## B. Synthesis of Copolymers of Various Shapes

By living cationic polymerization, our group has investigated the synthesis of polymers of various shapes including gradient copolymers,[171] PVA graft copolymers,[172] and end-functional polymers for the modification of liposomes to provide controlled release of drugs.[173]

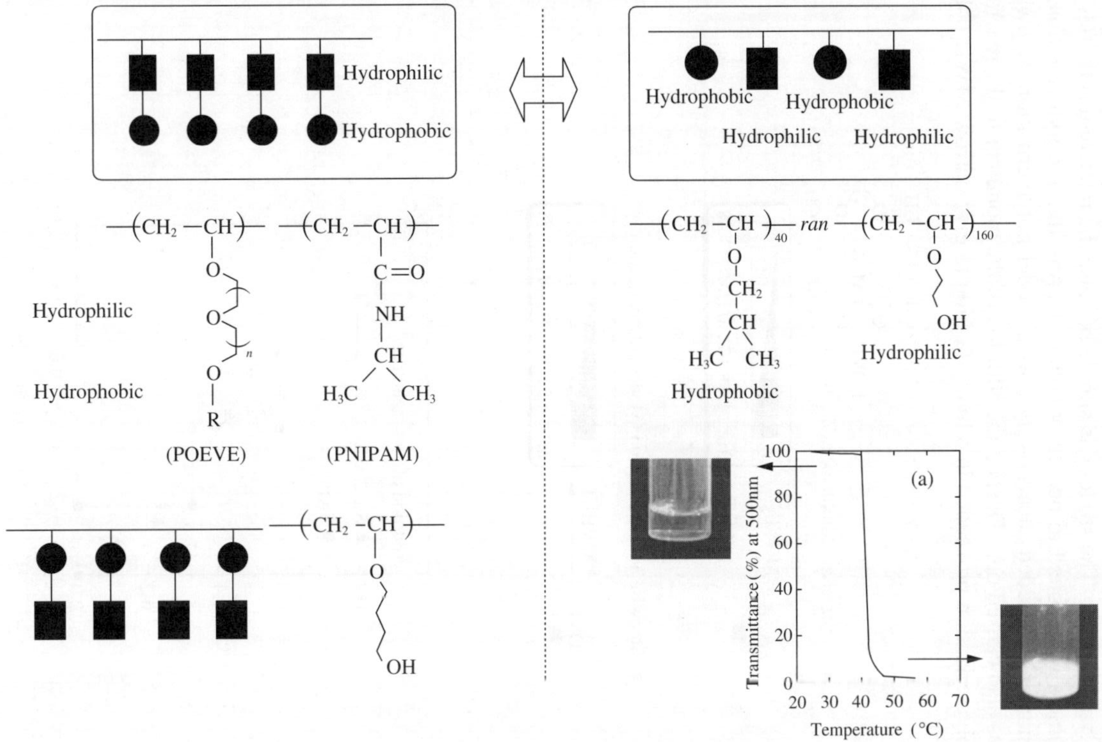


Random copolymers as thermosensitive polymers.

**Scheme 2.27**

Gradient copolymers have a very unusual sequence distribution in which the instantaneous composition changes continuously along each chain.[174] The authours have developed a new concept of thermosensitive polymers based on this polymer.[171] Figure 2.32 shows the synthesis of a gradient copolymer by living cationic polymerization. The strategy is simple because the relative reactivities of these monomers were found to be nearly equal. During living polymerization, a second monomer was added successively. As a model polymerization, stepwise addition was also carried out. Despite the addition of the monomer to the reaction mixture, polymerization proceeded in a living fashion to give narrow-MWD poly-

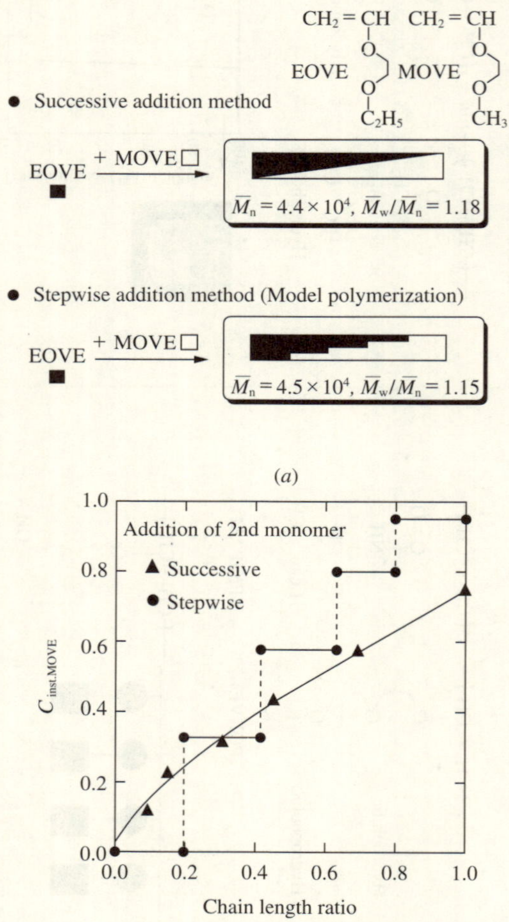


**Fig. 2.32** Schematic representation of monomer addition method (upper) and relationships between instantaneous copolymer composition ($C_{inst.\ MOVE}$) and chain length ratio (*a*) of gradient copolymers obtained by IBEA/$Et_{1.5}AlCl_{1.5}$ in the presence of ethyl acetate in toluene at 0 °C: $[EOVE]_0$=0.8 M, $[MOVE]_{total}$=0.8 M, $[IBEA]_0$=4.0 mM, $[Et_{1.5}AlCl_{1.5}]_0$=20 mM, [ethyl acetate]=1.0 M

mers. Both methods of monomer addition induced changes in composition along the polymer chain, as shown in Fig. 2.32, and confirmed by GPC and NMR analyses. Furthermore, a clear temperature dependence could be seen: in the case of 20 wt% aqueous solution, physical gelation was induced at 30 to 50 °C (sol-gel-sol-precipitation transition on heating). These results indicate that the ratio of hydrophobic segments increased gradually when the polymer solution was warmed. Thus, the changes in the aggregation pattern may be responsible for this complicated phase transition.

Living cationic polymerization has also been employed to prepare PVA graft copolymers consisting of a PVA main chain and poly(vinyl ether)[172] or polystyrene side grafts.[175] As a characteristic feature, various types of well-defined graft chains can be introduced into PVA with nearly quantitative efficiency, and the graft copolymers obtained have the properties of both PVA and the graft chains. The synthetic route (Scheme 2.28) involves the living cationic polymerization of vinyl ethers and its coupling reaction, that is, termination by partially-saponified poly(vinyl acetate) (PVAc-OH). The coupling reagent is soluble in several organic solvents such as toluene, and no side reactions occur in the course of coupling. Furthermore, the polymer main chain can be readily converted to PVA by conventional saponification after the coupling reactions. Using this method, various types of grafts, two different grafts, and diblock grafts etc. were also prepared and the difference in phase separation behavior was investigated.

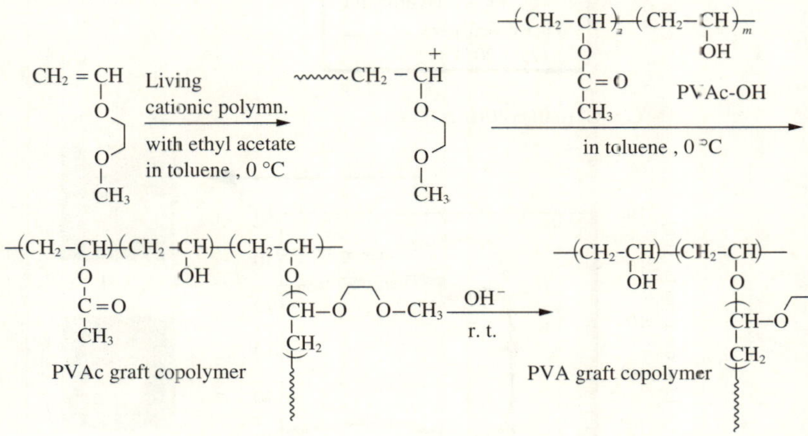


Preparation of PVA graft copolymers by living cationic polymerization.

**Scheme 2.28**

### C. Synthesis and Self-association of Stimuli-responsive Block Copolymers

To examine the characteristic behavior of thermosensitive physical gelation, block copolymer **75** with a hydrophilic polyalcohol segment and a thermosensitive segment was used.[168] As shown in Fig. 2.33, the block copolymer was soluble in water below 20 °C, with low viscosity. When the solution was warmed to above 21 °C, gelation occurred suddenly, yet remained transparent. The transparent gel was relatively stiff at 10% concentration, and could be picked up with tweezers. Upon cooling to below 20 °C, the gel returned sensitively to a solution state with low viscosity.

Measurements of dynamic viscoelasticity below 20 °C (Fig. 2.33) show that the system behaves as a liquid at these temperatures, whereas above 21 °C, G' becomes at least ten times larger than G". This rapid change is indicative of the high sensitivity of this system. Furthermore, separate experiments revealed that G' is much greater than G" over a broad frequency range, indicating the formation of a typical gel. In the case of other block segments with different $T_{ps}$ (other than poly(EOVE)) almost the same physical gelation occurred, but at higher temperatures corresponding to the $T_{ps}$ of each homopolymer.

The estimated mechanism based on our findings is as follows (Scheme 2.29). At below $T_{ps}$ (20 °C in this example), the block copolymer is soluble in water because both constituent segments are hydrophilic. When the solution is warmed

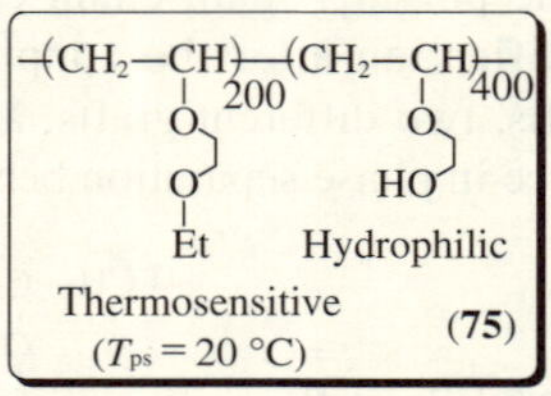


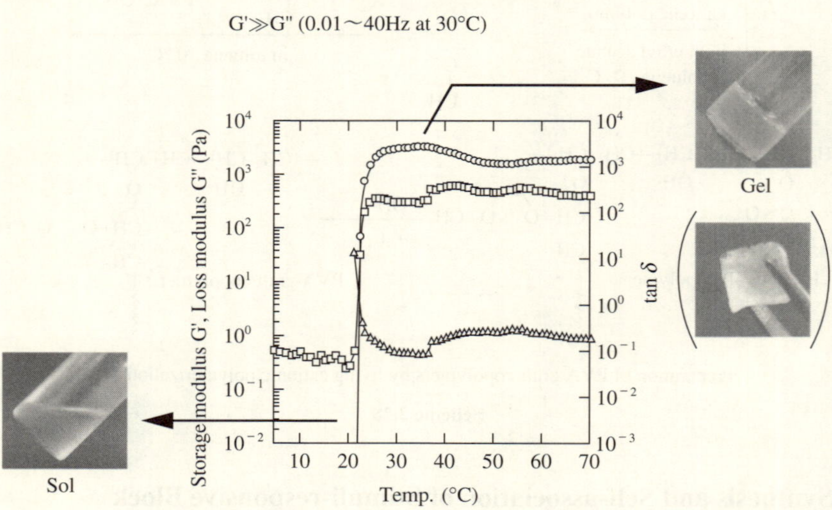


**Fig. 2.33** Temperature dependence of storage modulus G'(○), loss modulus G"(□), and tan δ(△) for 10 wt% $EOVE_{200}$-*b*-$HOVE_{400}$ aqueous solution (frequency=1 Hz) and photographs of aqueous block copolymer solutions when containers were tilted from the upright position.

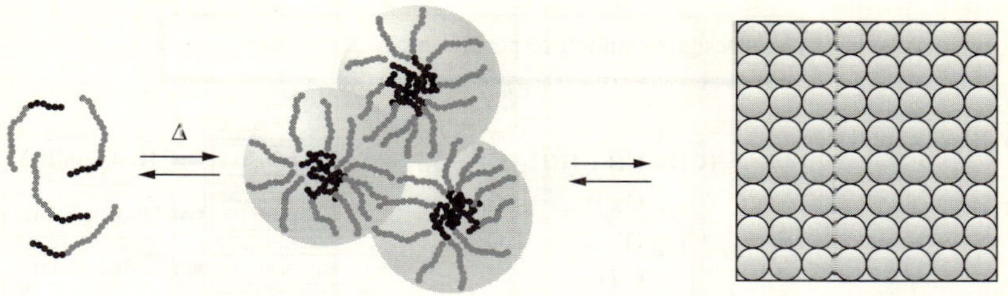


Estimated physical gelation mechanism.

**Scheme 2.29**

to the critical temperature, the hydrophilic poly(EOVE) segment becomes hydrophobic, and forms micelles by hydrophobic interaction. The micelles are composed of a central core of hydrophobic poly(EOVE) and an outer shell of hydrophilic poly(HOVE). The formation of micelles has been confirmed by various measurements,[176] including CMC, SANS, DLS, and TEM, and have been shown to be almost 100 nm in diameter, with a very narrow size distribution. When the polymer concentration is sufficiently high (e.g., 17 wt%), packing of micelles occurs, identified through SANS measurements to form a macrolattice with body-centered cubic symmetry.[177] Physical gelation is therefore considered to occur due to this formation and close packing of micelles. Interestingly, the transition took place in a narrow temperature range of less than 1 °C.

Based on these results, our group has designed many physical gelation systems in which physical gelation is induced by heating or cooling at a certain temperature, or with in a selected temperature range (Scheme 2.30). Instead of heating, other stimuli and interactions can also be used to induce gelation, such as using selective solvents, crystalline-amorphous transitions and hydrogen bonding.

As a typical example, the authors have prepared a diblock copolymer containing two thermosensitive segments of $T_{ps}$=41 and 70 °C.[178] The polymer solution at lower temperature is a clear liquid, and upon heating to around 40 °C, undergoes sudden sensitive physical gelation, while maintaining transparency. Interestingly, further heating induces a decrease in viscosity. In this case, physical gelation is induced in the temperature range of 40 to 50 °C. The mechanism of gelation is similarly thought to be due to micelle formation and packing, and the decrease in viscosity upon further heating is attributed to changes in the size and structure of the micelles. This demonstrates the two-step phase transition associated with each segment. Random copolymers with the same MW and composition behave very differently from the corresponding block copolymers, with no increase in viscosity. This demonstrates the clear influence of the sequence distribution.

Physical gelation by another interaction was investigated, using side groups with a strong tendency for interpolymer association such as long alkyl chains or

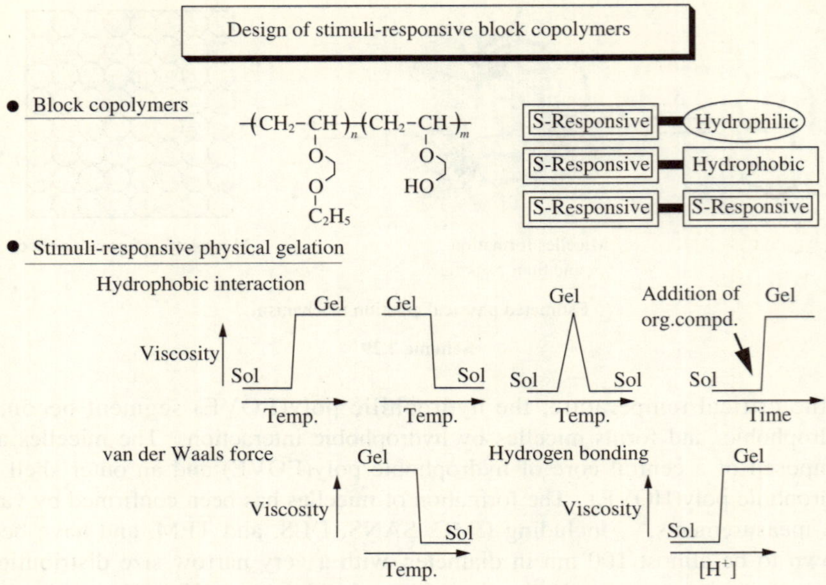


Design of stimuli-responsive block copolymers.

**Scheme 2.30**

biphenyl groups.[179] Van der Waals interaction and close-packed stacking were expected. Several different characteristics were observed. (i) Although a similar gel is obtained, physical gelation is induced on cooling, not heating. The gelation therefore corresponds to the amorphous-crystalline transition of side groups. (ii) The physical gel is obtained for both block copolymers and random copolymers. These differences are attributed to the characteristically strong tendency for interpolymer association of the side groups.

## 2.5.3 Summary

Recent progress in living polymerization has encouraged our group to design various types of amphiphilic copolymers and examine the stimuli-induced self-associations such as thermosensitive physical gelation. The possibility of living cationic polymerization of several functional monomers was first examined in the presence of added bases, carried out using $Et_xAlCl_{3-x}$($x$ =1 or 1.5) in the presence of ethyl acetate or THF. In these systems, the propagating carbocations were stabilized through nucleophilic interaction with added bases to prevent various side reactions. Various shapes of polymers with different sequence distributions were also prepared selectively by living polymerization, including gradient, graft, and end-functionalized polymers. Thermosensitive physical gelation was demonstrated using diblock copolymer **75** having a water-soluble polyalcohol segment [poly(HOVE)] and a thermally-responsive segment [poly(EOVE)]. Upon warming to the LCST of poly EOVE, aqueous solutions of diblock copolymer **75** underwent a sudden transition to a physical gel, and the viscosity change was quite sensitive and reversible. The mechanism of physical gelation in this case is considered to involve the formation of micelles with well-controlled size and structure, followed by close packing of the micelles [superlattice formation (bcc)]. Different types of stimuli-induced sol-gel transition systems have been designed on the basis of these results.

## Acknowledgment

The formation of micelles and the special ordered structure were identified by small-angle neutron scattering measurements in collaboration with Professor M. Shibayama of the University of Tokyo.

## References


< Section 2.1 >

1. Gambardella, P., Dallmeyer, A., Maiti, K., Malagoli, M. C., Eberhardt, W., Kern, K., Carbone, C., *Nature*, 2002, **416**, 301.
2. Sun, S., Murray, C. B., Weller, D., Folks, L., Moser, A., *Science*, 2000, **287**, 1989.
3. Lahti, P. M., *Magnetic Properties of Organic Materials*, Marcel Dekker, New York, 1999.
4. Itoh, K., Kinoshita, M., *Molecular Magnetism-New Magnetic Materials*, Kodansha and Gordon and Breach Science Publishers, Tokyo, Amsterdam, 2000.
5. Freeman, M. R., Choi, B. C., *Science*, 2001, **294**, 1484.
6. Kamachi, M., Tamaki, M., Morishima, Y., Nozakura, S., Mori, W., Kishita, M., *Polym. J.*, 1982, **14**, 363.
7. Nakahara, K., Iwasa, S., Satoh, M., Morioka, Y., Iriyama, J., Suguro, M., Hasegawa, E., *Chem. Phys. Lett.*, 2002, **359**, 351.
8. Korshak, U. V., Madvedeva, T. V., Ovchinnikov, A. A., Spector, V. N., *Nature*, 1987, **326**, 370.
9. Borden, W. T., Davidson, E. R., *J. Am. Chem. Soc.*, 1977, **99**, 4587.
10. Rajca, A., Utamapanya, S., *J. Am. Chem. Soc.*, 1993, **115**, 10688.
11. Rajca, A., Rajca, S., Desai, S. R., *J. Am. Chem. Soc.*, 1995, **117**, 806.
12. Rajca, A., Rajca, S., Wongsriratanakul, J., *J. Am. Chem. Soc.*, 1999, **121**, 6308.
13. Rajca, A., Wongsriratanakul, J., Rajca, S., *Science*, 2001, **294**, 1503.
14. Nishide, H., Kaneko, T., Nii, T., Katoh, K., Tsuchida, E., Lahti, P. M., *J. Am. Chem. Soc.*, 1996, **118**, 9695.
15. Nishide, H., Kaneko, T., Toriu, S., Kuzumaki, Y., Tsuchida, E., *Bull. Chem. Soc. Jpn.*, 1996, **69**, 499.
16. Nishide, H., Miyasaka, M., Tsuchida, E., *Angew. Chem. Int. Ed.*, 1998, **37**, 2400.
17. Nishide, H., Ozawa, T., Miyasaka, M., Tsuchida, E., *J. Am. Chem. Soc.*, 2001, **123**, 5942.
18. Michinobu, T., Takahashi, M., Tsuchida, E., Nishide, H., *Chem. Mater.*, 1999, **11**, 1969.
19. Goodson, F. E., Hauck, S. I., Hartwig, J. F., *J. Am. Chem. Soc.*, 1999, **121**, 7527.
20. Bushby, R. J., Gooding, D., *J. Chem. Soc., Perkin Trans. 2*, **1998**, 1069.
21. Michinobu, T., Inui, J., Nishide, H., *Org. Lett.*, 2003, **5**, 2165.
22. Bottomley, L. A., *Anal. Chem.*, 1998, **70**, 425R.
23. Hartmann, U., *Ann. Rev. Mater. Sci.*, 1999, **29**. 53.
24. Michinobu, T., Inui, J., Nishide, H., *Polym. J.*, 2003, **35**, 71.

< Section 2.2 >

25. Ghosh, M. K., Mittal, K. L., *Polyimides: Fundamentals and Applications*, Marcel Dekker, New York, 1996.
26. Wilson, D., Stenzenberger, H. D., Hergenrother, P. M. eds., *Polyimdes*, Chapman and Hall, New York, 1990.
27. Hergenrother, P. M., Wakelyn, N. T., Havens, S. J., *J. Polym. Sci., Part A: Polym. Chem.*, 1987, **25**, 1093.
28. Asanuma, T., Oikawa, H., Okawa, Y., Yamasita, W., Matsuo, M., Yamaguchi, A., *J. Polym. Sci., Part A: Polym. Chem.* 1994, **32**, 2111.
29. Matsuo, S., Mitsuhashi, K., *J. Polym. Sci., Part A: Polym. Chem.*, 1994, **32**, 1969.
30. Imai, Y., Malder, N. N., Kakimoto, M., *J. Polym. Sci., Part A: Polym. Chem.* 1984, **22**, 2189.
31. Liaw, D. J., Liaw, B. Y., Li, L. J., Sillion, B., Mercier, R., Thiria, R., Sekiguchi, H., *Chem. Mater.*, 1998, **10**, 734.
32. Wirth, J. G., *Discovery and Development of Polyetherimides In High Performance Polymers: Their Origin and Development*, (Seymour, R. B., Kirshenbaum, G. S. eds.), Elsevier, Amsterdam, 1986.
33. Liaw, D. J., Liaw, B. Y., Jeng, M. Q., *Polymer*, 1998, **39**, 1597.
34. Liaw, D. J., Liaw, B. Y., Chen, J. R., Yang, C. M., *Macromolecules*, 1999, **32**, 6860.
35. Harris, F. W., Lanier, L. H., *Structure- Solubility of Relationship*, (Harris, F. W., Seymour, R. B. eds.), p.183, Academic Press, New York, 1977.
36. Sinta, R., Minns, R. A., Gaudiana, R. A., Rogers, H. G., *Macromolecules*, 1987, **20**, 2374.

37. Becker, K. H., Schmidt, H. W., *Macromolecules*, 1992, **25**, 6784.
38. Liaw, D. J., Liaw, B. Y., Chen, Y. S., *Polymer*, 1999, **40**, 4041.
39. Liaw, D. J., Liaw, B. Y., *Macromol. Chem. Phys.*, 1998, **199**, 1473.
40. Mohanty, D. K., Sachdeva, Y., Hedrick, J. L., Wolfe, J. F., McGrath, J. E., *Polym. Prepr.*, 1984, **25**(2), 19.
41. Misra, A. C., Tesoro, G., Hougham, G., Pendharkar, S. M., *Polymer*, 1992, **33**,1078.
42. Park, J. W., Lee, M., Lee, M. H., Liu, J. W., Kim, S. D., Chang, J. Y., Rhee, S. B., *Macromolecules*, 1994, **27**, 3459.
43. Negi, Y. S., Suzuki, Y., Kawamura, I., Kakimoto, M., Imai, Y., *J. Polym. Sci., Part A: Polym. Chem.*, 1996, **34**, 1663.
44. Liaw, D. J., Wang, K. L., *J. Polym. Sci., Part A: Polym. Chem.*, 1996, **34**, 1209.
45. Liaw, D. J., Liaw, B. Y., Yang, C. M., *Acta Polym.*, 1999, **50**, 332.
46. Liaw, D. J., Liaw, B. Y., Yang, C. M., *Macromolecules*, 1999, **32**, 7248.
47. Jin, M. Y., Choi, K. Y., *Macromolecules*, 1997, **30**, 5606.
48. Liaw, D. J., Liaw, B. Y., *Curr. Trends Polym. Sci. Res. Trends*, 1999, **4**, 225.
49. Liaw, D. J., Liaw, B. Y., *Polymer*, 1999, **40**, 3183.
50. Liaw, D. J., Liaw, B. Y., *Macromol. Chem. Phys.*, 1999, **200**, 1326.
51. Liaw, D. J., Liaw, B. Y., *Polym. J.*, 1999, **31**, 1270.
52. Liaw, D. J., Liaw, B. Y., Chung, C. Y., *J. Polym. Sci.,Part A: Polym. Chem.*, 1999, **37**, 2815.
53. Liaw, D. J., Liaw, B. Y., Chung, C. Y., *Macromol. Chem. Phys.*, 2000, **201**, 1887.
54. Liaw, D. J., Liaw, B. Y., Hsu, P. N., Hwang, C. Y., *Chem. Mater.* 2001, **13**, 1811.
55. Liaw, D. J., Hsu, C. Y., Chen, I. W., Liaw, B. Y. *Polymer Prepr.*, 2001, **42**, 596.
56. Liaw, D. J., Hsu, C. Y., Liaw, B. Y., *Polymer*, 2001, **42**, 7993.
57. Liaw, D. J., Hsu, C. Y., Hsu, P. N., Lin, S. L., *J. Polym. Sci., Part A: Polym. Chem.*, 2002, **40**(12), 2066.
58. Liaw, D. J., Chen, I. W., Yang, M. C., *Macromol. Chem. Phys.*, 2002, **203**, 2170.
59. Liaw, D. J., Chen, I. W., Chen, W. H., Lin, S. L., *J. Polym. Sci., Part A: Polym. Chem.*, 2002, **40(15)**, 2556.
60. Son, J. M., Nakao, M., Ogino, K., Sato, H., *Macromol. Chem. Phys.*, 1999, **200**, 65.
61. Son, J. M., Mori, T., Ogino, K., Sato, H., Ito, Y., *Macromolecules*, 1999, **32**, 4849.
62. Tan, L. S., Srinivasan, K. R., Bai, S. J., *J. Polym. Sci., Part A: Polym. Chem.*,1997, **35**, 1909.
63. Nishikata, Y., Fukui, S., Kakimoto, M., Imai, Y., Nishiyama, K., Fujihira, M., *Thin Solid Films*, 1992, **210/211**, 296.
64. Wu, A., Jikei, M., Kakimoto, M., Imai, Y., Ukishima, S., Takahashi, Y., *Chem. Lett.*, 1994, **23**, 19.
65. Ogino, K., Kanegae, A., Yamaguchi, R., Sato, H., Kurjata, J., *Macromol. Rapid Commun.*, 1999, **20**, 103.
66. Liaw , D. J., Liaw, B. Y., *Macromol. Chem. Phys.*, 1998, **199**, 1473.
67. Liaw , D. J., Tseng, W. T., *Macromol. Symp.*, 2003, **199**, 351.
68. Liaw , D. J., unpublished results.
69. Dine-Hart R. A., Wright, W. W., *Angew. Makromol. Chem.*, 1972, **153**, 237.
70. Liaw , D. J., Liaw, B. Y., Tsai, M. Y., *Eur. Polym. J.*, 1997, **33**, 997.
71. Liaw , D. J., Liaw, B. Y., *J. Polym. Sci., Part A: Polym. Chem.*, 1997, **35**, 1527.
72. Liaw , D. J., Liaw, B. Y., *Eur. Polym. J.*, 1997, **33**, 1423.
73. Liaw , D. J., Liaw, B. Y., *Polym. J.*, 1996, **28**, 970.
74. Liaw, D. J., Hsu, P. N., Chen, W. H., Lin, S. L., *Macromolecules*, 2002, **35**, 4669.
75. Vora, R. H., Krishnan, P. S. G., Goh, S. H., Chung, T. S., *Adv. Funct. Mater.*, 2001, **11**, 361.
76. Greenham, N. C., Moratti, S. C., Bradley, D. D. C., Friend, R. H , Holmes, A. B., *Nature*, 1993, **365**, 628.
77. Adronov, A., Fréchet, J. M. J., *Chem. Commun.*, 2000, **18**, 1701.
78. Wang, P. W., Liu, Y. J., Devadoss, C., Bharathi, P., Moore, J. S., *Adv. Mater.*, 1996, **8**, 237.
79. Choong, V., Shi, S., Curloess, J., Shieh, C. L., Lee, H. C., So, F., Shen, J., Yang, J., *Appl. Phys. Lett.*, 1999, **75**, 172.

**< Section 2.3 >**

80. Mann, S., *Biomimetic Materials Chemistry*, (Mann, S., ed.), VCH, New York, 1996.
81. Shenton, W., Douglas, T., Young, M., Stubbs, G., Mann,S., *Adv. Mater.*, 1999, **11**, 253.
82. Douglas, T., Young, M., *Nature,* 1998, **393**, 152; Shenton, W., Pum, D., Sleytr, U., Mann, S., *Nature,* 1997, **389**, 585.
83. Davis, S. A. Burkett, S. L., Mendelson, N. H., Mann, S., *Nature,* 1997, **385** 420.

84. Recent reviews:
a) Terech, P., Weiss, R.G., *Chem. Rev.*, 1997, **97**, 3133.
b) Van Esch, J., Schoonbeck, F., De Loos, M., Veen, E. M., Kellogg, R. M., Feringa, B. L., *Nato ASI Ser. C*, 1999, **527**, 233.
c) Shinkai, S., Murata, K., *J. Mater. Chem.*, 1998, **8**, 485.
85. Otsuni, E., Kamaras, P., Weiss, R. G., *Angew. Chem., Int. Ed. Engl.*, 1996, **35**, 1324 and references therein.
86. Terech, P., Furman, I., Weiss, R. G., *J. Phys. Chem.*, 1995, **99**, 9558.
87. Murata, K., Aoki, M., Suzuki, T., Harada, T., Kawabata, H., Komori, T., Ohseto, F., Ueda, K., Shinkai, S., *J. Am. Chem. Soc.*, 1994, **116**, 6664 and references therein.
88. James, T. D., Murata, K., Harada, T., Ueda, K., Shinkai, S., *Chem. Lett.*, **1994**, 273.
89. Jeong, S. W., Murata, K., Shinkai, S., *Supramol. Sci.*, 1996, **3**, 83.
90. Ono, Y., Nakashima, K., Sano, M., Kanekiyo, Y., Inoue, K., Hojo, J., Shinkai, S., *J. Chem. Soc., Chem. Commun.*, **1998**, 1477.
91. Ono, Y., Kanekiyo, Y., Inoue, K., Hojo, J., Shinkai, S., *Chem. Lett.*, **1999**, 23.
92. Jung, J. H., Ono, Y., Shinkai, S., *Langmuir*, 2000, **16**, 1643.
93. Sakamoto, H., Kimura, K., Shono, T., *J. Org. Chem.*, 1986, **51**, 5974.
94. Nakashima, N., Asakuma, S., Kunitake, T., *J. Am. Chem. Soc.*, 1985, **107**, 509.
95. Jung, J. H., Kobayashi, H., Masuda, M., Shimizu, T., Shinkai, S., *J. Am. Chem. Soc.*, 2001, **123**, 8785.
96. Kim, S. S., Zhang, W., Pinnavaia, T. J., *Science,* 1998, **282**, 1302.
97. Jung, J. H., Ono, Y., Shinkai, S., *Chem. Lett.*, **2000**, 636.
98. Jung, J. H., Shinkai, S., *J. Chem. Soc. , Perkin Trans. 2*, **2000**, 2393.
99. Ozin, G. A., *Acc. Chem. Res.*, 1997, **30**, 313.
100. Oliver, S., Kuperman, A., Coombs, N., Ozin, G. A., *Nature,* 1995, **378**, 47.
101. Tanev, P. T., Ling, Y., Pinnavaia, T. J., *J. Am. Chem. Soc.* 1997, **119**, 8616.
102. Kim, S. S., Zhang, W., Pinnavaia, T. J., *Science,* 1998, **282**, 1302.
103. Echegoyen, L. E., Hernandez, J., Kaifer, A. E., Gokel, G. W., *J. Chem. Soc., Chem. Commun.*, **1988**, 836.
104. Nakano, A., Hernandez, J. C., Dewall, S. L., Berger, D. R., Gokel, G. W., *Supramol. Sci.*, 1997, **8**, 21.
105. Echegoyen, L. E., Portugal, L., Miller, S. R., Echegoyen, L., Gokel, G. W., *Tetrahedron Lett.*, 1998, **29**, 4065.
106. Jung, J. H., Ono, Y., Sakurai, K., Masahito, M., Shinkai, S., *J. Am. Chem. Soc.*, 2000, **122**, 8648.
107. Jung, J. H., Ono, Y., Shinkai, S., *Angew. Chem. Int. Ed.*, 2000, **39**, 1862.
108. Hanabusa, K., Shimura, K., Hirose, K., Kimura, M., Shirai, H., *Chem. Lett.*, **1996**, 885.
109. Jung, J. H., Ono, Y., Hanabusa, K., Shinkai, S., *J. Am. Chem. Soc.*, 2000, **122**, 5008.
110. Jung, J. H., Ono, Y., Shinkai, S., *Chem. Eur. J.*, 2000, **6**, 4552.
111. Kobayashi, S., Hanabusa, K., Suzuki, M., Kimura, M., Shirai, H., *Chem. Lett.*, **1999**, 1077.
112. Gronwald, O., Shinkai, S., *Chem. Eur. J.*, 2001, **7**, 4328.
113. Yoza, K., Ono, Y., Yoshihara, K., Akao, T., Shinmori, H., Takeuchi, M., Shinkai, S., Reinhoudt, D. N., *J. Chem. Soc., Chem. Commun.*, **1998**, 907.
114. Yoza, K., Amanokura, N., Ono, Y., Akao, T., Shinmori, H., Takeuchi, M., Shinkai, S., Reinhoudt, D. N., *Chem. Eur. J.*, 1999, **5**, 2722.
115. Gronwald, O., Sakurai, K., Luboradzki, R., Kimura, T., Shinkai, S., *Carbohydr. Res.*, 2001, **331**, 307.
116. Luboradzki, R., Gronwald, O., Ikeda, M., Shinkai, S., Reinhoudt, D. N., *Tetrahedron,* 2000, **56**, 8697.
117. Schoonbeeck, F. S., Van Esch, J. H., Hulst, R., Kellogg, R. M., Feringa, B. L., *Chem. Eur. J.*, 2000, **6**, 2633.
118. Luboradzki, R., Gronwald, O., Ikeda, M., Shinkai, S., *Chem. Lett.*, **2000**, 1148.
119. Gronwald, O., Shinkai, S., *J. Chem. Soc., Perkin Trans. 2*, **2001**, 1933.
120. Amanokura, N., Kanekiyo, Y., Shinkai, S., Reinhoudt, D. N., *J. Chem. Soc., Perkin Trans. 2*, **1999**, 1995.
121. Jung, J. H., Amaike, M., Shinkai, S., *J. Chem. Soc., Chem. Commun.*, **2000**, 2343.
122. Jung, J. H., Amaike, M., Nakashima, K., Shinkai, S., *J. Chem. Soc., Perkin Trans. 2*, **2001**, 1938.
123. Friggeri, A., Gronwald, O., Van Bommel, K. J. C., Shinkai, S., Reinhoudt, D. N., *J. Chem. Soc., Chem. Commun.*, **2001**, 2434.

**< Section 2.4 >**


124. Yamamoto, S., *J. Chem. Soc. Jpn. Ind. Chem. Soc.*, 1943, **46**, 779. (*Chem. Abstr.*, 1952, **46**, 7047i)
125. Tachibana, T., Mori, T., Hori, K., *Bull. Chem. Soc. Jpn.*, 1980, **53**, 1714.
126. Honma, M., *Gendaikagaku*, **1987**, 54.
127. Terech, P., Wade, R. H., *J. Colloid Interface Sci.*, 1988, **125**, 542.
128. a) Lin, Y., Kachar, B., Weiss, R. G., *J. Am. Chem. Soc.*, 1989, **111**, 5542.
b) Mukkamala. R., Weiss, R. G., *Langmuir*, 1996, **12**, 1474.
129. Murata, K., Aoki, M., Suzuki, T., Harada, T., Kawabata, H., Komori, T., Ohseto, F., Ueda, K., Shinkai, S., *J. Am. Chem. Soc.*, 1994, **116**, 6664.
130. Brotin, T., Utermöhlen, R., Fages, F., Bouas-Laurent, H., Desvergne, J., *J. Chem. Soc., Chem. Commun.*, **1991**, 416.
131. Vries, E. J., Kellogg, R. M., *J. Chem. Soc., Chem. Commun.*, **1993**, 238.
132. Menger, F. M.. Yamasaki, Y., Catlin, K. K., Nishimi, T., *Angew. Chem. Int Ed. Engl.*, 1995, **34**, 585.
133. Oda, R., Huc, I., Candau, S. J., *Angew. Chem. Int. Ed.*, 1998, **37**, 2689.
134. Yasuda, Y., Takebe, Y., Fukumoto, M., Inada, H., Shirota, Y., *Adv. Mater.*, 1996, **8**, 740.
135. Snijder, C. S., Jong, J. C., Meetsma, A., Bolhuis, F., Feringa, B. L., *Chem. Eur. J.*, 1995, **1**, 594.
136. Lu, L., Weiss, R. G., *J. Chem. Soc., Chem. Commun.*, **1996**, 2029.
137. Crisp, G. T., Gore, J., *Synthetic Commun.*, 1997, **27**, 2203.
138. Hafkamp, R. J. H., Kokke, B. P. A., Danke, I. M., Geurts, H. P. M., Rowan, A. E., Feiters, M. C., Nolte, R. J. M., *J. Chem. Soc., Chem. Commun.*, **1997**, 545.
139. Hishikawa, Y., Sada, K., Watanabe, R., Miyata, M., Hanabusa, K. *Chem. Lett.*, **1998**, 795.
140. Vassilev, V. P., Simanek, E. E., Wood, M. R., Wong, C-H., *J. Chem. Soc., Chem. Commun.*, **1998**, 1865.
141. Yoza, K., Amanokura, N., Ono, Y., Akao, T., Shinmori, H., Takeuchi, M., Shinkai, S., Reinhoudt, D. N., *Chem. Eur. J.*, 1999, **5**, 2722.
142. Terech, P., Weiss, R. G., *Chem. Rev.*, 1997, **97**, 3133.
143. Abdallah, D. J., Weiss, R. G., *Adv. Mater.*, 2000, **12**, 1237.
144. Esch, J. H., Feringa, B. L., *Angew. Chem. Int. Ed.*, 2000, **39**, 2263.
145. Hanabusa, K., Okui, K., Karaki, K., Kimura, M., Shirai, H., *J. Colloid Interface Sci.*, 1997, **195**, 86.
146. Hanabusa, K., Hiratsuka, K., Kimura, M., Shirai, H., *Chem. Mater.*, 1999, **11**, 649.
147. Hanabusa, K., Tange, J., Taguchi, Y., Koyama, T., Shirai, H., *J. Chem. Soc., Chem. Commun.*, **1993**, 390.
148. Hanabusa, K., Matsumoto, M., Kimura, M., Kakehi, A., Shirai. H., *J. Colloid Interface Sci.*, 2000, **224**, 231.
149. Hanabusa, K., Naka, Y., Koyama, T., Shirai, H., *J. Chem. Soc., Chem. Commun.*, **1994**, 2683.
150. Aggeli, A., Bell, M., Boden, N., Keen, J. N., Knowles, P. F., McLeish, T. C. B., Pitkeathly, M., Radford, S. E., *Nature*, 1997, **386**, 259.
151. Lehn, J-M., Mascal, M., DeCian, A., Fischer, J., *J. Chem. Soc., Chem. Commun.*, **1990**, 479.
152. Hanabusa, K., Miki, T., Taguchi, Y., Koyama, T., Shirai, H., *J. Chem. Soc., Chem. Commun.*, **1993**, 1382.
153. Danda, H., Nishikawa, H., Otaka, K., *J. Org. Chem.*, 1991, **56**, 6740.
154. Tata, M., John, V. T., Waguespack, Y. Y., McPherson, G. L., *J. Am. Chem. Soc.*, 1994, **116**, 9464.
155. Nakano, K., Hishikawa, Y., Sada, K., Miyata, M., Hanabusa, K., *Chem. Lett.*, **2000**, 1170.
156. Bauer, T., Thomann, R., Mulhaupt, R., *Macromolecules*, 1998, **31**, 7651.
157. Hanabusa, K., Yamada, M., Kimura, M., Shirai, H., *Angew. Chem. Int. Ed. Engl.*, 1996, **35**, 1949.
158. Hanabusa, K., Shimura, K., Hirose, K., Kimura, M., Shirai, H., *Chem. Lett.*, **1996**, 885.
159. Hanabusa, K., Koto, C., Kimura, M., Shirai, H., Kakehi, A., *Chem. Lett.*, **1997**, 429.
160. Matsunaga, Y., Miyajima, N., Nakayasu, Y., Sakai, S., Yonenaga, M., *Bull. Chem. Soc. Jpn.*, 1988, **61**, 207.
161. Malthéte, J., Levelut, A.-M., Liébert, L., *Adv. Mater.*, 1992, **4**, 37.
162. Hanabusa, K., Kawakami, A., Kimura, M., Shirai, H., *Chem. Lett.*, **1997**, 191
163. Hanabusa, K., Maesaka. Y., Kimura, M., Shirai, H., *Tetrahedron Lett.*, 1999, **40**, 2385.

**< Section 2.5 >**


164. As a review, see: MacCormick C. L. ed., *Stimuli-responsive Water-soluble and Amphiphilic Polymers, ACS Symp. Ser.*, **780**, 1, 2000.
165. a) Zhang, L., Eisenberg, A., *Science*, 1995, **268**, 1728.
    b) Schillen, K., Bryskhe, K., Mel'nikova, Y. S., *Macromolecules*, 1999, **32**, 6885.
    c) Cornelissen, J. J. L. M., Fischer, M., Sommerdijk, N. A. J. M., Nolte, R. J. M., *Science*, 1998, **280**, 1427.
    d) Won, Y. -Y., Davis, H. T., Bates, F. S., *Science*, 1999, **283**, 960.
166. Matyjaszewski K. ed., *Controlled /Living Radical Polymerization, ACS Symp. Ser.*, **768**, 1, 2000.
167. Aoshima, S., Higashimura, T., *Macromolecules*, 1989, **22**, 1009.
168. Aoshima, S., Hashimoto, K., *J. Polym. Sci., Part A: Polym. Chem.*, 2001, **39**, 746.
169. a) Aoshima, S., Oda, H., Kobayashi, E., *J. Polym. Sci., Part A: Polym. Chem.*, 1992, **30**, 2407.
    b) Aoshima, S., Kobayashi, E., *Macromol. Symp.*, 1995, **95**, 91.
170. Sugihara, S., Aoshima, S., *Kagaku To Kogyo*, 2001, **75**, 510.
171. a) Aoshima, S., Kikuchi, T., Matsuzono, S., Miyazawa, H., *Polym. Prep. Jpn.*, 2000, **49**, 389.
    b) Aoshima, S., Tsujimoto, I., Kikuchi, T., Yoshida, T., *Polym. Prep. Jpn.*, 2001, **50**, 1333.
172. Aoshima, S., Ikeda, M., Nakayama, K., Kobayashi, E., Ohgi, H., Sato, T., *Polym. J.*, 2001, **33**, 610 .
173. Aoshima, S., Yoshida, T., Murakami, Y., Kono, K., *Polym. Prep. Jpn.*, 2001, **50**, 1335.
174. Matyjaszewski, K., Zieger, M. J., Arehart, S. V., Greszta, D., Pakula, T., *J. Phys. Org. Chem.*, 2000, **13**, 775.
175. Aoshima, S., Segawa, Y., Okada, Y., *J. Polym. Sci., Part A: Polym. Chem.*, 2001, **39**, 751.
176. Sugihara, S., Matsuzono, S., Sakai, H., Abe, M., Aoshima, S., *J. Polym. Sci., Part A: Polym. Chem.*, 2001, **39**, 3190.
177. Okabe, S., Sugihara, S., Aoshima, S., Shibayama, M., *Macromolecules*, 2002, **35**, 8139.
178. a) Aoshima, S., Sugihara, S., *J. Polym. Sci., Part A: Polym. Chem.*, 2000, **38**, 3962.
    b) Sugihara, S., Aoshima, S., *Kobunshi Ronbunshu*, 2001, **58**, 304.
179. Aoshima, S., Yoshida, T., *J. Jpn. Soc. Colour Mater.*, 2001, **74**, 569.

# 3
# Macromolecular Inorganic Assemblies

# 3.1 Spherical, Fibrous Rock- and Sheet-like Noncovalent Polymers

Guangtao Li, Andreas Klyszcz, Matthias Lauer and Jürgen-Hinrich Fuhrhop

## 3.1.1 Introduction

Noncovalent polymers are defined here as molecular assemblies with a high degree of polymerization (PN > 100) and well-defined molecular arrangements in solution as well as in the dry state. We also include membrane-coated colloidal metal or silicate particles with rigid nanometer gaps. The spherical or planar carrier is needed to fixate the membrane structures, which are not distinguishable by substances in the bulk water or solvent phase from clefts in the surfaces or polymers with an organic core.

## 3.1.2 Spheres

Hemoglobin and many enzymes are covalent polymers with a globular shape. This shape is enforced by the tendency of hydrophobic amino acids to form a hydrophobic droplet in aqueous solutions solubilized by hydrophilic side-chains around them.[1,2] The same is true for synthetic block polymers made of hydrophobic and hydrophilic segments.[3,4] Spherical biopolymers thus usually appear as "micelles," with a core made of organic material.[3] Covalency allows the construction of fully organized micelles, e.g. dendrimeric spheres, where one half has a hydrophilic, the other a hydrophobic surface.[5] Block polymers may not only form micelles, but they may also arrange to form vesicles which entrap a water volume. Such spheres have a thick polymer wall.[4] Both the polymer micelles and vesicles can be removed from solution without collapsing.

Noncovalent spherical assemblies of amphiphilic lipids, on the other hand, are either short-lived already in aqueous solution (micelles) or collapse immediately upon drying (vesicles).[6,7] This behavior is due to the character of the forces which form them. Curvature is retained by repulsive hydration forces. If the hydration sphere is removed from the head groups, the amphiphiles will pack together and form crystalline sheets and 3-D crystals. Neither crystallites nor micelles and vesicles can be considered as noncovalent polymers, because they change their molecular arrangement drastically when going from the dissolved to the dry state. They do not have material properties.

This situation will change drastically if the curvature of the micelles is enforced not by repulsive hydration forces, but by the stereochemistry of the head groups. A hydrophobic droplet of about 100 alkyl chains of moderate length, e.g., $C_{18}$, has a diameter of about 5 nm corresponding to a surface of about 80 $nm^2$. One hundred carboxylate groups of stearate, for example, would only cover about 25 $nm^2$, which means 75% of the surface is covered by hydration spheres. Each carboxylate group would be isolated from the others. If, however, the charges are neutralized by the addition of acids, COOH groups will form and bind to neighbors by hydrogen bonds. Curvature will disappear, vesicles will be formed first and then rearrange to form planar sheets and crystals. If, however, very large planar or spherical head groups are supplied, they will be able to cover the whole surface of the alkane droplet.

Stable spheres with interconnected head groups were obtained for the first time with ruthenium(II)-tris(bipyridyl) complexes. Hexafluorophosphate or perchlorate counterions worked as cement between these building blocks. Ruthenium(II) complexes are kinetically inert compounds and can be substituted in a step-by-step fashion with different ligands. It was thus possible to synthesize a ruthenium(II) complex with one bipyridyl-malonic diester with two long alkyl chains and two naked bipyridyl ligands. With two bromide counterions no defined supramolecular assemblies could be detected. Upon titration with $PF_6^-$-ions, however, the large ruthenium(II) head groups with a width of about 10 Å were glued together and the alkyl chains were forced to form an interdigitated bilayer. Isolable micelles were characterized by transmission electron microscopy (TEM) and atomic force microscopy (AFM)[8)]. TEM showed multilayered micelles in which the ruthenium complexes not only formed a rigid monolayer with the $PF_6^-$-counterions, but also dimerized to a herringbone type structure.

These micelles were isolated as a solid material and could be re-suspended in water. No structural changes occurred within several months. On most solid substrates (gold, silicon, mica, hydrophobic polymers) individual micelles retained

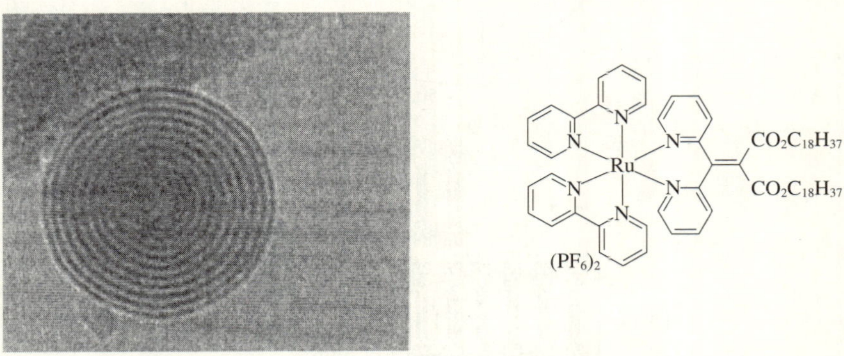


**Fig. 3.1** Transmission electron micrograh (TEM) and model of a multilayered micelle made of the depicted amphiphilic ruthenium complex with $PF_6^-$-counterions. The whole bilayer assembly does not measureably dissociate in aqueous solution and can be isolated in the dry state. It is therefore "a noncovalent polymer."

their spherical shape; only on graphite did they collapse to form monolayers of flat-lying molecules. Each of the individual micelles behaved like a molecule. With respect to tightness and rigid organization the noncovalent assemblies can be compared with the 4th generation dendrimers, only the size of the assemblies is not as uniform. The noncovalent polymer also produces an interesting property, which the monomer does not provide: it fluoresces. The stability of the spheres depends, however, strongly on the chain length of the alcohols in the diester. $C_{18}$-$C_{20}$ micelles are stable on most surfaces except (on) graphite; $C_{16}$-micelles are much more vulnerable and defined $C_{12}$-micelles do not form at all upon addition of $PF_6^-$-ions. Stability of the micellar material will therefore only be guaranteed if several conditions are fulfilled, e.g., at least two $C_{18}$-chains, $PF_6^-$ as counterion, temperatures below 80 °C and no organic solvent. This situation is typical for noncovalent polymers. One may prepare it in large quantities and in quantitative yield under optimal conditions with optimal compounds and one may dissect it quantitatively by changing the conditions.

A fluid micelle which dissolves apolar porphyrins but can nevertheless be isolated as a material and characterized by transmission electron (TEM) and atomic force microscopy (AFM) has kanamycin as a head group. This is a trisaccharide-like cyclohexane derivative containing several OH- and four amino groups. One of them was amidated with stearic acid, and upon sonication spherical micelles with a diameter of about 6 nm were formed. TEM shows a large assembly of spheres which look similar to covalent fullerene spheres, but are ten times larger.[9]

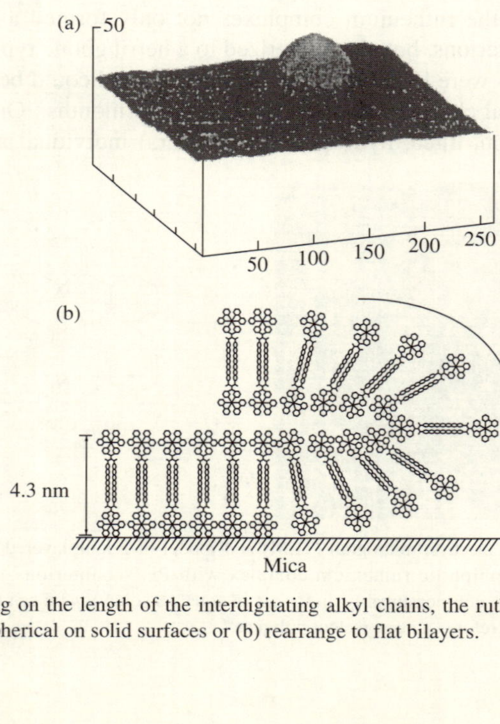


**Fig. 3.2** Depending on the length of the interdigitating alkyl chains, the ruthenium micelle will (a) remain spherical on solid surfaces or (b) rearrange to flat bilayers.

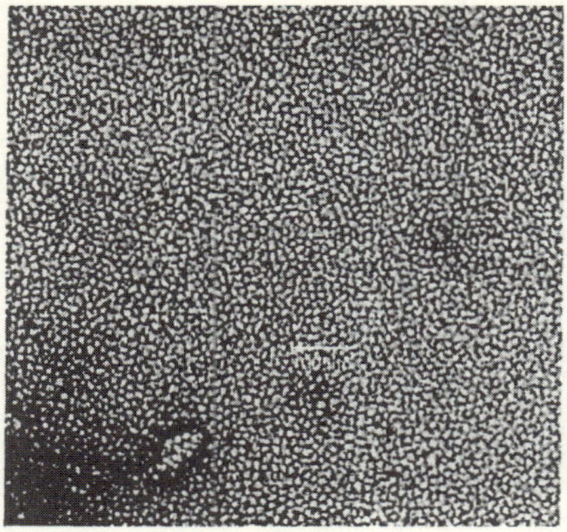

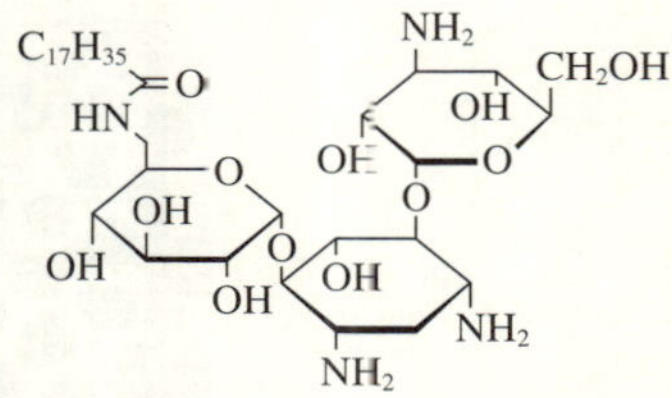

Kanamycinstearoylamide

**Fig. 3.3** TEM and model of an assembly of fluid kanamycin micelles. The picture shows no signs of fusion or collapse. It looks similar to pictures of much smaller covalent dendrimer micelles.

In the case of kanamycin it is hydrogen bonding between $NH_2$-, $NH_3^+$- and OH-groups which ties the head groups together. A micelle, which would collapse immediately upon dehydration, stabilized and was detected at room temperature under the TEM.[9] Upon dilution of the micellar assembly, the micelles form a monolayer of kanamycin stearate on mica (Fig. 3.4) and stable half-micelles deposit on this monolayer. In water the cmc is again smaller than $10^{-5}$ M and the micelle behaves essentially like a polymer molecule. Addition of organic solvents, however, leads to quantitative dissociation and/or rearrangement to crystals. Again it is the cooperativity of the hydrophobic effect with the binding interactions between large head groups which leads to isolable particles.

Enzymes are globular proteins which carry a cleft with a width of 1–2 nm (a 1–2 nm wide cleft) on its surface.[2,7] Most active centers of enzymes are localized in such clefts. The substrate has perfect access and the products are quickly released. In proteins such clefts are usually realized as an intermittent space between three or four helices. Can such clefts also be realized with noncovalent assemblies? The answer is half a yes. The membrane-type walls of the cleft can be made of amphiphilic diamides, the bottom of the cleft can be formed by a non-connected flat molecule, but the whole assembly must be fixed covalently on a colloidal carrier. This has been realized, for example, with citrate gold or silicate spheres covered with a flat-lying porphyrin and a rigid membrane with an oligoethylene (OEG) head group. The thin OEG-coating makes the particle water- and toluene-soluble, and the rigid gap can be used to attract reactants to specific sites within the membrane walls. These gaps have, for example, been functionalized with rings of methylammonium groups.[10]

Again each membrane coat behaves like an isolated polymer molecule. Because of the solid carrier it is now also resistant to organic solvents.

Vesicles with rigid membranes made of carotenoids have also been isolated

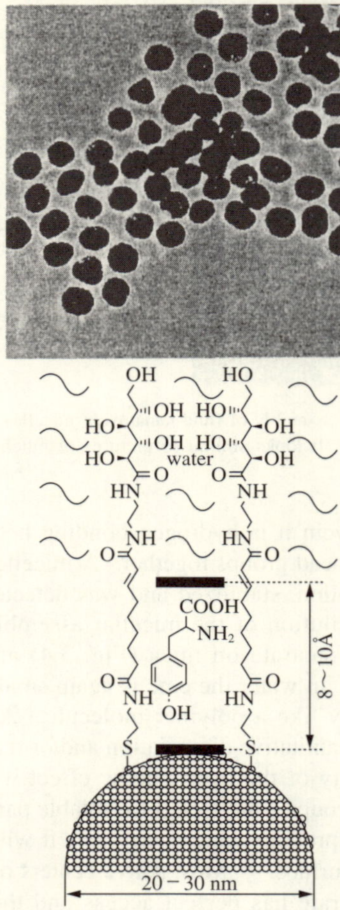


**Fig. 3.4** TEM and model of a spherical gold particle covered with functionalized membranes. Similar systems have also been realized on 100-nm silicate particles.

in the dried solid state. They behave like sea-urchin shells: rigidity and porosity make them completely stable towards electrolytes. No osmotic pressure is experienced by these large fullerene analogues. They are also stable on solid substrates and have been characterized by TEM and AFM.[11)]

## 3.1.3 Fibers

Micellar and vesicular fibers are easy to realize with amphiphiles containing a secondary amide group or with tridentate ligands for ruthenium. At temperatures above 80 ˚C these amphiphiles give standard hydrated micelles in water, upon cooling below 70 ˚C hydrogen bond chains are formed and fibers appear. Again a combination of the hydrophobic effect with hydrogen bonding is important, but

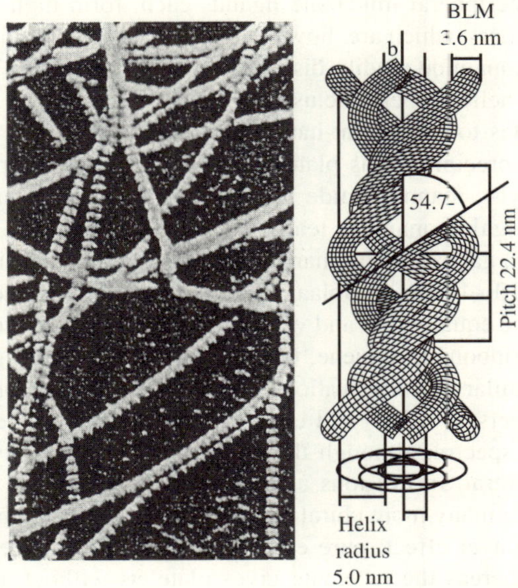


**Fig. 3.5** TEM and model of *N*-octylgluconamide quadruple helices. The fibers look similar to covalent collagen fibers and have been stored for months in the dry state without decomposition.

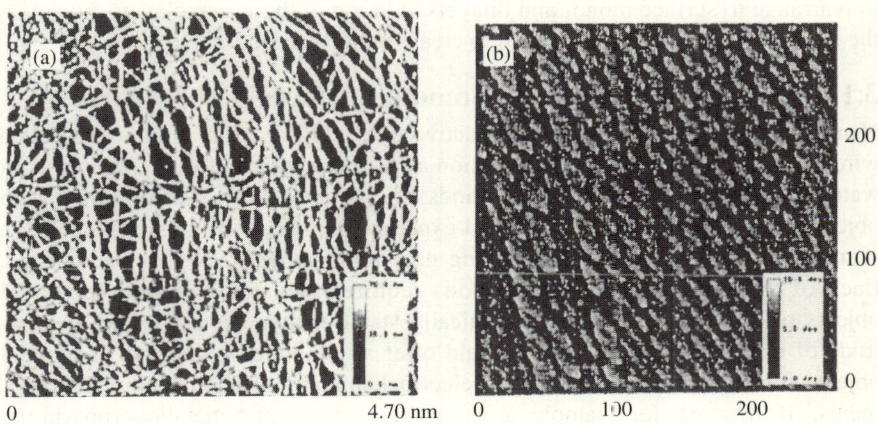


**Fig. 3.6** AFM images of fibrous aggregates of *N*-octylgluconamide lying directly on silicon surfaces: (a) network of fibers; (b) high-resolution phase image of the quadruple helices.

fiber formation can also be enforced in organic solvents.[12] Covalent porphyrin dimers with two peripheral imidazole ligands each, form high molecular weight coordination polymers, which are, however, stable only in organic solvents.[13]

*N*-D-octylgluconamide readily dissolves in water above 80 °C and gives left-handed quadruple helical fibers exclusively upon cooling (Fig. 3.5). *N*-L-octylgluconamide assembles to form right-handed quadruple fibers. A 1 : 1 mixture of both enantiomers precipitates as platelets, whereas the fibers produce gels with water.[14a,b] *N*-octyl-D-mannonamide gives under the same conditions rolled-up sheets, *N*-octyl-D-galactonamide leads to twisted ribbons. 1 : 1 mixtures of diastereomers, e.g., galacton- and mannonamide aggregate to thin tubules with an inner diameter of 10–40 nm.[15] Bolaamphiphiles with an L-lysine and an ammonium head group give equally thin and very long tubules.[16] 11-Hydroxy-stearic acid produces twisted ribbons in toluene, the sense of twisting changes with different counterions.[17] Similar CD observations have been made with protoporphyrin IX-D-gluconamide fibers.[18] Zinc- and copper-porphyrinates, for example, produce mirror-image CD spectra, although the chiral head group is always the same.[19] Counterions of central Sn(IV) ions also strongly influence the helical twist and height.[20] There are many more chiral fibers known now; many but not all of them show the chiral bilayer effect: pure enantiomers produce helices or tubules with high curvature, whereas the racemate gives platelets without any curvature. In organic solvents one does not find manifold fine-tuned structures, which occur in water. Up-rolled sheets, short rods and ill-defined twisted ribbons prevail (Fig. 3.4). The lysine tubules and gluconamide quadruple helices can be stored indefinitely as dry materials.

Upon contact with solid substrates in aqueous dispersions the helices collapse to rearrange to surface mono- and bilayers. On top of these monolayers, however, the fibers are stable and can be used to create molecular landscapes.[21,22]

### 3.1.4 Nanometer Rocks on Smooth Surfaces

Nanometer-size assemblies of redox-active dyes may be used as light controlled wires or switches for electron conduction as well as catalysts for the photolysis of water by sunlight. The analytic methods of choice to study and optimize such objects are electro- and photochemical experiments with a scanning force (SFM), scanning tunneling (STM) or scanning near-field optical (SNOM) microscope. Each of these high resolution methods requires elevated and well-separated objects on smooth materials. Mechanically stable nanometer probes on an electrode of macroscopic dimensions would offer an unlimited number of individual objects for comparative mechanical, electrochemical and photochemical experiments. If there are, for example, about ten rigid stacks of a metalloporphyrin per 100 nm × 100 nm surface area of a silicon or gold electrode one would have $10^{10}$ individual stacks per square centimeter.

Noncovalent assemblies of porphyrins have been realized with four zirconium (IV) phosphonate bonds as a cement between porphyrin planes.[10] Tetraphosphonic acid porphyrins have been reported in the literature.[23,24] An octaphosphonic analogue seemed most promising with respect to the formation of rugged molecular assemblies ("rocks"), since phosphonate and zirconium(IV) ions are known to fit perfectly together in crystalline bilayer structures of biphosphonate-bolaamphiphiles.[25,26)]

Silicon wafers were covered by an α, ω-silylphosphonate in 10 M methanolic HCl solutions (~ $10^{-1}$ M) at 60 °C for 3 days. The phosphonate-covered substrate was then treated with an aqueous solution of $ZrOCl_2 \cdot 8H_2O$ for 15 minutes, followed by an aqueous solution of the octaphosphonate porphyrin. Washing and repeated alternating treatment of the wafer with $ZrOCl_2$ and aqueous solutions of porphyrin yielded ill-defined porphyrin multilayers only. No island-like structures were detected in AFM pictures. Transmission electron micrographs (TEM) of dried aqueous suspensions of the porphyrin on carbon grids before and after addition of $ZrOCl_2 \cdot 8H_2O$ showed only crystallites, no extended fibers. Self-assembly from methanol solutions of porphyrin and aqueous $ZrOCl_2$ solution yielded a few fibers not suitable for the envisaged AFM or STM experiments (not shown). Addition of the voluminous tetrabutylammonium hydroxide (1% w/W) to the methanolic solution of octaphosphonate porphyrin dissolved the fibers, and self-assembly applying this solution yielded isolated porphyrin islands with typical heights of 2.5, 5.0, 7.5 and 10 nm for the first time. After several further trials we found that a $10^{-6}$ M acetonitrile solution of porphyrin containing 1% w/W of tetra-butylammonium-hydroxide was the most suitable. Four self-assembly cycles using this solution alternately with $ZrOCl_2 \cdot 8H_2O$ in water yielded a rocky landscape corresponding to a high density of porphyrin assemblies (Fig. 3.1(b) and (c)). The average height was 10 nm, the maximum height about 30 nm. Since one flatlying porphyrin measures 2 nm in height, the building blocks for the porphyrin rocks were obviously oligomers. Attempts to obtain larger distances between these rocks by dilution of the acetonitrile solution or by shortening the self-assembly time failed. An efficient capping procedure became mandatory. At first phenyl- or tert-butyl phosphonic acid in water were applied as capping agents after the attachment of the first porphyrin islands to the zirconium(IV)-coated wafers. In each self-assembly cycle the density of the porphyrin islands became, however, even higher than without capping (Fig. 3.1(b)). We then applied alizarin S, a good ligand for zirconium(IV). This dye is known to form upright standing crystallites on all kinds of planar substrates.[27)] This time the capping was successful and the desired number of about ten islands per 104 $nm^2$ could easily be established. In the first self-assembly step of the porphyrin a few single cone-shaped islands with a height of 2.5 or 5.0 nm were detectable. Their diameter as determined by scanning electron microscopy (SEM) was 8–20 nm corresponding to 4–10 porphyrins. In further cycles the number of towers increased only marginally, their width seemed to grow slightly, their height averaged about 25 nm after for cycles (Fig. 3.1(c)). The most likely shape of the rocks is that of a cone standing on a monomolecular or somewhat broader base and being surrounded there by an alizarin S monolayer. The shape of individual islands remained, however, unchanged for several months on the silicon wafer's surface. Plunging the silicon platelet for several minutes

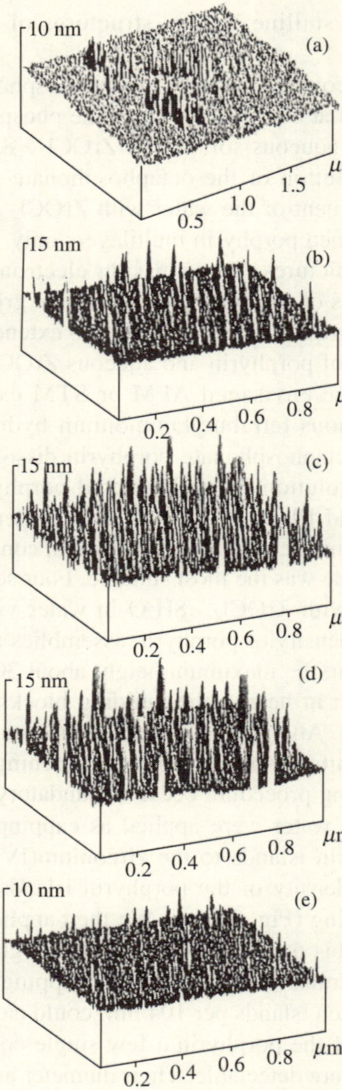


**Fig. 3.7** Porphyrinphosphonate **1** – zirconium(IV) multilayered rocks on silicon substrates. The solvents applied in the self-assembly of **1** were (a) methanol, (b) methanol –1% tert. $BuONH_4$, (c) acetonitrile, (d) acetonitrile plus tert. butylphosphonate capping, (e) acetonitrile plus alizarin S capping (see Fig. 3.8).

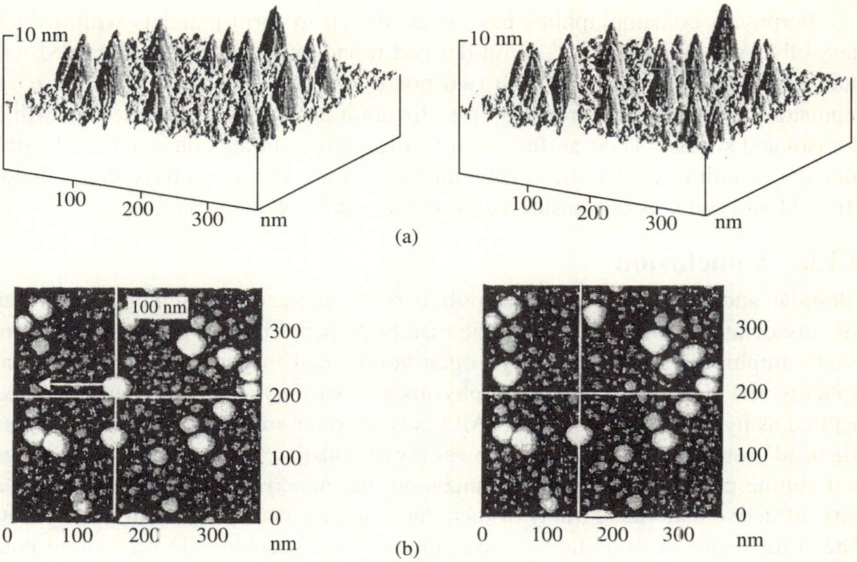


**Fig. 3.8** Two mechanical displacements of a porphyrin rock by a tip in the tapping mode on a zirconium(IV)-alizarin S-covered silicon substrate (a) three-dimensional image, (b) two-dimensional image.

into water or ethanol did not change the appearance of the porphyrin islands.

We then investigated the mechanical stability of individual porphyrin cones by pushing them with an AFM-tip in the tapping mode.[10)] It was found that the towers slid on the alizarin S ground or, in some cases, broke. Fig. 3.2(a), (b), (c) and (d) shows an example of a 100-nm displacement of an individual tower. There neither occurred a change in shape of height of the island, nor is any trace of dragging detectable on the alizarin S ground. None of the neighbor islands had changed its position.

Wasielewski reported on similar porphyrin nanoparticles on mica, which were simply deposited by spin coating. These porphyrins carried four perylene substituents and were shown to be photoconductive.[28)]

## 3.1.5. Planar Sheets

Sheets of noncovalent polymers, which would be analogous to the β-pleated sheets of silk, are commonplace on water and solid surfaces. The molecular weight and extension of these surface mono- and bilayers is unlimited. Of current interest are well-defined rigid gaps within such monolayers. They provide small reactive domains which can be manipulated on a molecular scale. It was found possible, for example, to fixate porphyrin heterodimers with a plane-to-plane distance of 10 or 20 Å within such gaps and to position tyrosine monomers as electron relays between them.[29,30)] These sheets can then be fixated on spherical colloids (see Fig. 3.4) and we are back to spherical polymers.[31a, 31b)]

Porphyrin bolaamphiphiles have been shown to form platelets with a thickness of 8 Å in bulk water.[32] Well-defined monolayers have been obtained with bolaamphiphilic porphyrins with two positive charges on each edge. Charge repulsion obviously first prevents crystallization and second unlimited growth of the isolated sheets. Their surface area is a few $\mu m^2$. In aqueous solution lecithin bilayers of unlimited extension (myelin figures) are known to have cmc's below $10^{-10}$ M and can thus be considered as noncovalent polymers.[6]

### 3.1.6 Conclusion

Globular and fibrous noncovalent polymers are molecular assemblies with very low dissociation constants and can be readily prepared by self-assembly of appropriate amphiphiles in water. Hydrogen bonds lead in general to linear chains. Spheres and stacks of alkanes, porphyrins and various aromatic dyes have been applied as hydrophobic skeletons. All kinds of water-soluble molecules are versatile head groups. With respect to the variety of constituents, high molecular weight and simple preparation by self-organization, the noncovalent polymers appear as very attractive materials. Many of them have already been isolated in the dry state. The components such as fatty acids, glucose, protoporphyrin(IX) are often common products or side products of food processing and therefore low-priced.

Both biopolymers and noncovalent polymers are produced at the borderline between hydrophobic membrane structures and bulk water. The hydrophobic effect enforces curvature in both, hydrogen bonds lead to linear chains, chirality produces helices, sheets are often insoluble and are applied only occasionally. All kinds of planar thin layers occur, however, on solid surfaces, e.g., bones and minerals in nature or gold, graphite and mica in model systems.

If one leaves water and gold to organic solvents as a medium, the situation will change. Biopolymers and synthetic block polymers form ill-defined clusters, nanometer helices are replaced by micrometer twisted ribbons, and sheets immediately crystallize. In the case of micellar assemblies bilayer sheets prevail, showing a tendency to roll up and form bulky scrolls.

Water is clearly the best medium to produce polymer assemblies with stable imprints on their surface. Small clefts do not survive in organic solvents, so swelling crystallization cannot be avoided.

In our context of a comparison between covalent and noncovalent polymers important differences appear in general and should be considered with respect to any application.

(i) Noncovalent polymers are best formed in water, if solvophobic effects play a role. Only some very rigid coordination polymers are uniform in organic solvents. Covalent polymers with hydrogen-bonded secondary structures do, in general, not change their shape in organic solvents, but often do not dissolve there.
(ii) Noncovalent polymers are rapidly formed *in situ* by a change in temperature or pH, and readily dissociate by the reverse treatment. Yields are, after opti-

mizing conditions, always quantitative. Covalent polymers are much more robust against external conditions and therefore more difficult to remove.

(iii) Synthetic noncovalent polymers provide complex and very uniform supramolecular structures (quadruple helices, narrow tubules), which are not accessible from synthetic polymers or copolymers. The shape of noncovalent polymer assemblies can be varied by very simple modifications of the amphiphilic monomers and reaction media. The variety of supramolecular structures approaches that of biopolymers. Dyes as a hydrophobic backbone for high molecular weight fibers are unique to noncovalent fibers. The cmc has not been measured for any of the fibers but is certainly much below $10^{-5}$ M in stable systems.

(iv) Applications of noncovalent fibers are scarce. They are only used in a simple manner to stabilize aqueous or alcoholic gels and as sensitizers in photographic emulsions. To the best of our knowledge, no one has ever taken advantage of the reversibility of fiber formation and dissolution.

What is needed is an intelligent commercial application which utilizes the reversible character of the polymerization by self-assembly. Can one use and sell a paper which dissolves upon heating and reforms readily? Are plugs made of a meltable gel and applicable in micro- or nanometer size of any interest, given their ability to disintegrate and to re-form at given temperature jumps? Are cheap energy transport systems of interest apart from photographic layers? Today's answer is no. The beautiful micrographs and sharp solid state NMR signals, which indicate well-defined and ordered arrangements of the monomers, are of academic interest only. Noncovalent polymers could be produced easily on a large scale but they will be of no use, if no motivation is found to do so.

### 3.1.7 Summary

Noncovalent polymers of lipid and porphyrin amphiphiles were obtained by self-organization in water and isolated as dry materials in quantitative yields. Strong noncovalent interactions between very large head groups were used in spherical particles, amide hydrogen bond chains in rigid fibers and monolayers containing nanometer clefts. Form-stable elevations on smooth substrates were stabilized by phosphonate-phosphonate bonds. Spheres, rods, tubules and sheets with rigid clefts were formed by hydrogen bonds between head groups and characterized by both electron TEM and atomic force microscopy. TEM yielded nanometer resolution, AFM produced the height profile.

### Acknowledgement

Financial support by the Deutsche Forschungsgemeinschaft(SFB 348 “Mesoscopic Systems”), the European TMR research network “Carbohydrate Recognition,” the Fonds der Deutschen Chemischen Industrie and by the FNK of the Free University is gratefully acknowledged.

# 3.2 Macromolecular Conjugated Complexes

**Toshikazu Hirao**

## 3.2.1 Introduction

π-Conjugated polymers and oligomers are subjects of wide interest due to their high potential in a variety of applications to electric materials as conducting compounds.[33] The electric properties are related to their π-conjugated function and structure. π-Conjugated polymers exist in various oxidation states as represented by pernigraniline, emeraldine and leucoemeraldine bases of polyaniline (Scheme 3.1). Protonic acid doping plays an important role in controlling their redox processes.

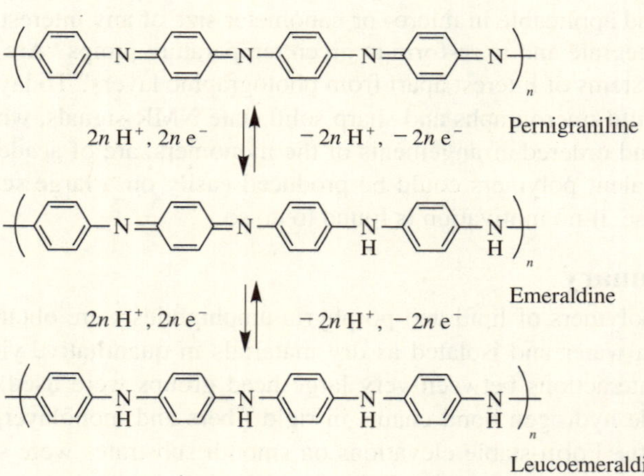


Redox processes of polyaniline

**Scheme 3.1**

These oxidation states interconvert each other, allowing the construction of a catalytic system for oxidation reaction. Actually, polyanilines or polypyrroles have been demonstrated to serve as synthetic metal catalysts under oxygen atmosphere in dehydrogenative oxidation of benzylamines, 2-phenylglycine, and 2,6-di-*t*-butylphenol (Scheme 3.2).[34] A similar catalysis is also observed with the protonated quinonediimine or phenylenediamine derivative, a unit of polyaniline.[35] These catalytic activities are revealed to be controlled by protonic acid doping since the presence of a proton source is required for the formation of a reversible redox cycle under oxygen atmosphere.[36)]

An efficient redox process of transition metal complexes for electron transfer

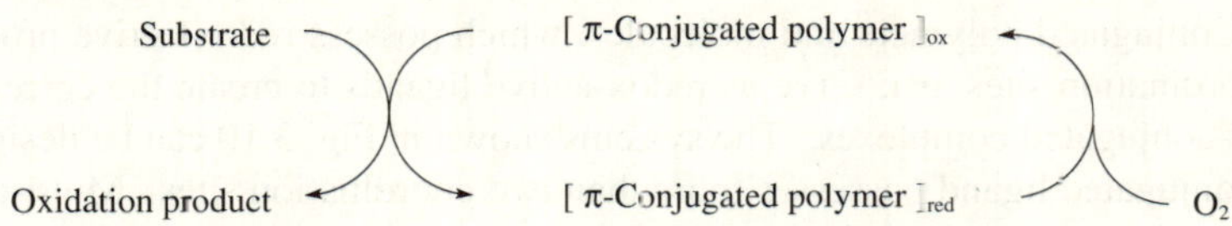


π-Conjugated polymer = polyanilines, polypyrroles

A catalytic scheme for the synthetic metal catalyzed oxidation

**Scheme 3.2**

is essential to the development of functionalized materials and catalysts. Coordination interaction with ligands can control the redox process. If ligands are redox active, a combination of both redox properties is envisioned to provide a multi-redox system as shown in Scheme 3.3. The redox interaction between transition metals and redox moieties of ligands appears to operate through coordination, but if the redox moieties are not coordinated directly, the interaction is also possible through bond or space.[37)] A more efficient redox function through coordination is likely to be realized in both systems.

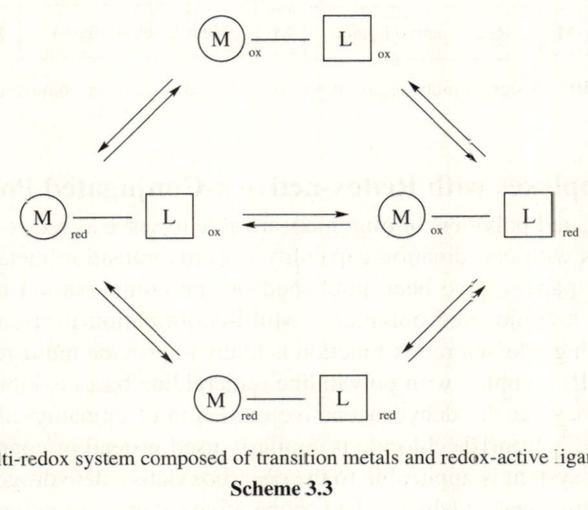

A multi-redox system composed of transition metals and redox-active ligands

**Scheme 3.3**

Furthermore, when ligands bear more than two coordination sites, interaction between metals can occur through the ligand spacer (Fig. 3.9). This design is related to molecular wire systems.[38)]

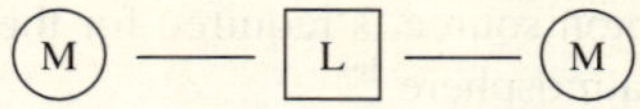


**Fig. 3.9** A transition metal complex bearing a bidentate redox-active ligand.

π-Conjugated polymers and molecules, which possess redox-active properties and coordination sites, can serve as redox-active ligands to create the corresponding d, π-conjugated complexes. The systems shown in Fig. 3.10 can be designed if the π-conjugated ligand possesses more than two coordination sites. Multi-coordination to a redox-active π-conjugated polymer ligand leads to the multi-nuclear complex. In the case of both metals and ligands possessing two coordination sites, these components are arrayed alternatively to give the corresponding block polymer-like complexes. A cyclic skeleton structure is also allowed to be formed, depending on their geometries. Based on the number and geometry of the coordination sites, a variety of dimensionally designed structures may be constructed. Electronic communication is considered to occur between metals through a π-conjugated backbone chain. This review focuses mainly on these kinds of hybrid conjugated systems composed of transition metals and redox-active π-conjugated ligands.

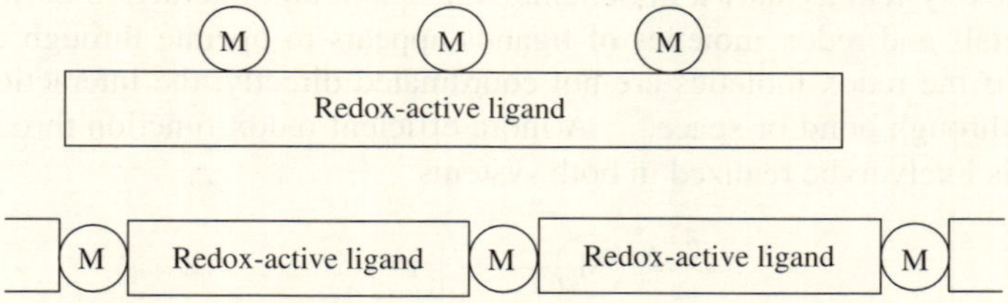


**Fig. 3.10** Design of multi-redox complexes with redox-active π-conjugated ligands.

## 3.2.2 Complexes with Redox-active π-Conjugated Polymers

The π-conjugated polymers or oligomers are able to serve as redox-active π-conjugated ligands with coordination capability towards transition metals as described above. Few papers have been published on the complexation behavior of the redox-active π-conjugated polymers. Multi-coordination to π-conjugated polymers possessing relevant redox function is likely to provide multi-redox systems.

The Cu(II) complex with polyaniline (emeraldine base) exhibits a higher catalytic efficiency for the dehydrogenative oxidation of cinnamyl alcohol into cinnamaldehyde.[34)] Iron(III) chloride is similarly used instead of copper(II) chloride. The catalytic system is applicable to the decarboxylative dehydrogenation of mandelic acid to give benzaldehyde. The cooperative catalysis of polyaniline and copper(II) chloride operates to form a reversible redox cycle under oxygen atmosphere as shown in Scheme 3.4.[39)] The copper salt contributes to not only oxidation process but also metallic doping. The reduced phenylenediamine anionic species appear to be stabilized by the metallic dopants.

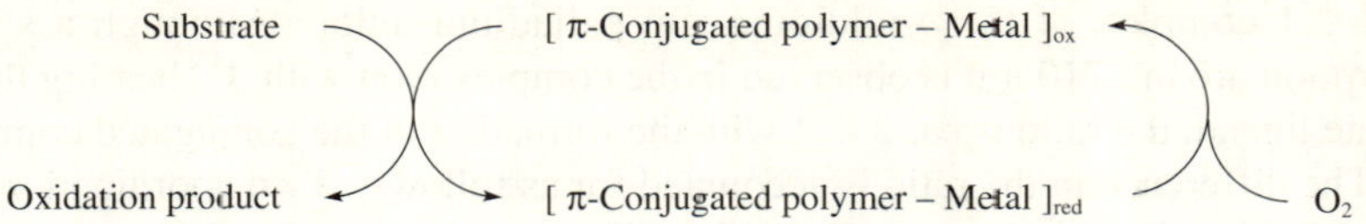


Metal = Cu(II), Fe(III)

A catalytic scheme for the π-conjugated complex induced oxidation

**Scheme 3.4**

The redox interaction between transition metals and redox-active ligands is likely to permit a smooth reversible redox cycle in the transition metal-catalyzed oxidation reactions. Actually, the Wacker oxidation reaction of a terminal olefin proceeds catalytically only in the presence of a catalytic amount of polyaniline or polypyrrole derivative as a cocatalyst in acetonitrile-water under oxygen atmosphere to give 2-alkanone (Scheme 3.5).[40,41] Copper-free catalytic systems are achieved by this method. The catalysis is not observed in the absence of a polyaniline derivative, indicating that the π-conjugated polymer participates in the catalytic cycle of a palladium species. The redox processes of polyaniline are monitored by UV-vis. spectra. π-Conjugated polymers are considered to behave as a redox-active ligand based on their reversible redox function in the transition metal-catalyzed oxidation reaction under oxygen atmosphere.[42] The solubility of the complex catalysts depends on the substituent on the polymer, which can provide both homogeneous and heterogeneous catalysts.

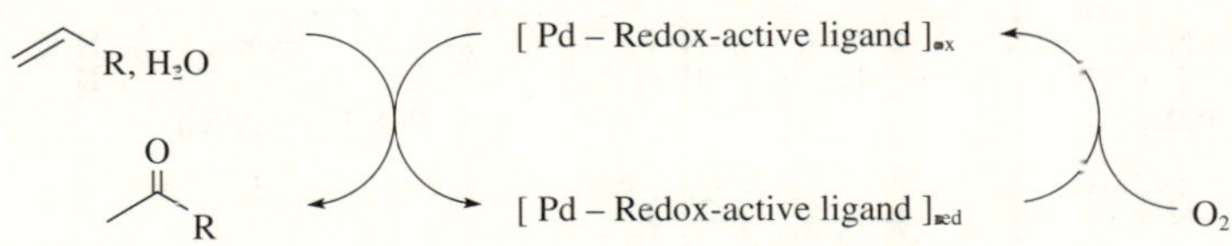


Redox-active ligand = Polyanilines, Polypyrroles

A catalytic scheme for the palladium-catalyzed oxidation in the presence of the redox-active ligand

**Scheme 3.5**

The controlled complexation of a polyaniline derivative is performed in organic solvent to afford the structurally defined conjugated complex **2**.[43] The spectral measurement supports the formation of such a polymer complex. Coordination sites are the quinonediimine nitrogens of the emeraldine base, as confirmed by a model study described below. On treatment of the emeraldine base of poly(*o*-toluidine), possessing the phenylenediamine and quinonediimine moieties at *ca.* 1 : 1 ratio, in THF with $Pd(OAc)_2$ or $PdCl_2(MeCN)_2$, a new absorption appears in the longer wavelength region as compared with the CT band of the polymer around 590 nm. The titration experiment indicates the formation of the *ca.* 4 : 1 complex of the *o*-toluidine and palladium units. Although a similar absorption around 710 nm is observed in the complexation with **3**[44] bearing the tridentate ligand, the ratio is *ca.* 2 : 1 with the formation of the conjugated complex.

The difference in the ratio is accounted for as follows. Two coordination sites are available with $Pd(OAc)_2$ or $PdCl_2(MeCN)_2$, but the complex **3** bearing the tridentate ligand has only one free coordination site. A cross-linked polyaniline network complex **4** is considered to be formed in the former case. On the contrary, the latter complexation affords the single-strand conjugated complex **5**. Thus, a hybrid system composed of transition metals and π-conjugated polymers is controlled structurally (Fig. 3.11).

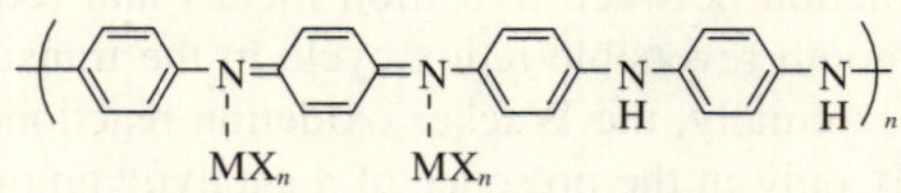

**2**

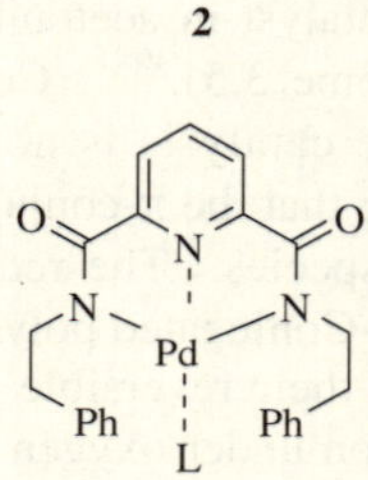

**3**

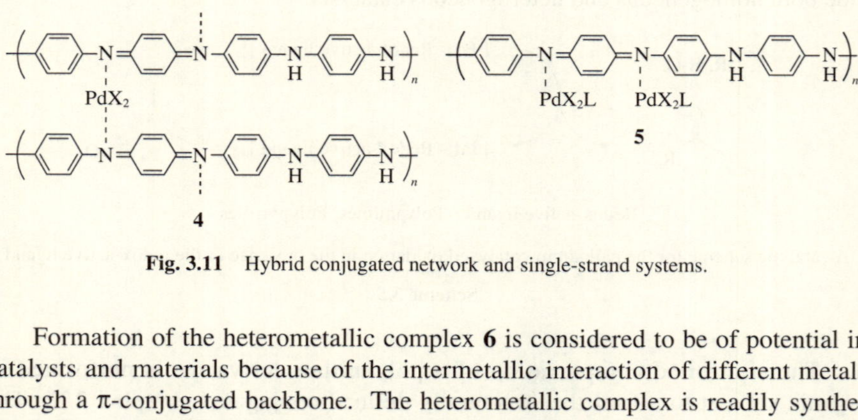

**4** **5**

**Fig. 3.11** Hybrid conjugated network and single-strand systems.

Formation of the heterometallic complex **6** is considered to be of potential in catalysts and materials because of the intermetallic interaction of different metals through a π-conjugated backbone. The heterometallic complex is readily synthesized, for example, by successive treatment of poly(*o*-toluidine) with $Cu(OAc)_2$ and $Pd(OAc)_2$.[45]

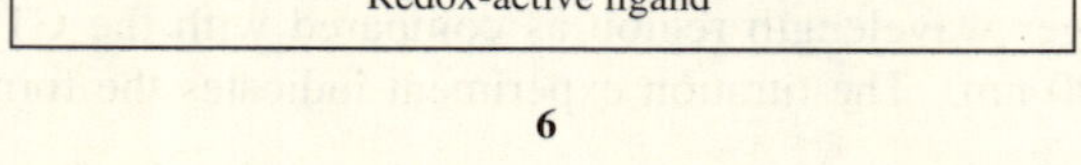


**6**

Based on the redox properties of π-conjugated polymers, complexation may be accompanied by redox reactions. The emeraldine base of poly(*o*-toluidine) undergoes redox complexation with $VCl_3$, affording the corresponding conjugated complexes.[46] In this complexation process, the CT band between the benzenoid and quinoid moieties disappears, being consistent with the reduction of the quinonediimine moieties along with the oxidation of V(III) to V(IV). The vanadium species serves as a one-electron reductant and metallic dopant. A similar complexation is also observed with *N,N'*-bis(4'-dimethylaminophenyl)-1,4-quinonediimine (**7**).[47]

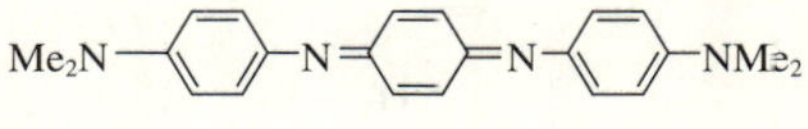

**7**

A photoactive metal center is introduced in these systems. The ruthenium bipyridyl complexes are coordinated to the emeraldine base to form the corresponding polymer complexes as described above.[48] The incorporation of the ruthenium centers to the pyridyl backbone has been also reported to give the ruthenium complexes **8**.[49] Conjugated ruthenium bipyridine complexes thus obtained are evaluated to be photorefractive materials. Other transition metal complexes can be employed to form the corresponding polymer complexes.[50] The pyridine unit is replaced by bithienyl, 1,4-diazabutadiene, ethylene, benzimidazole or thiazole.[51]

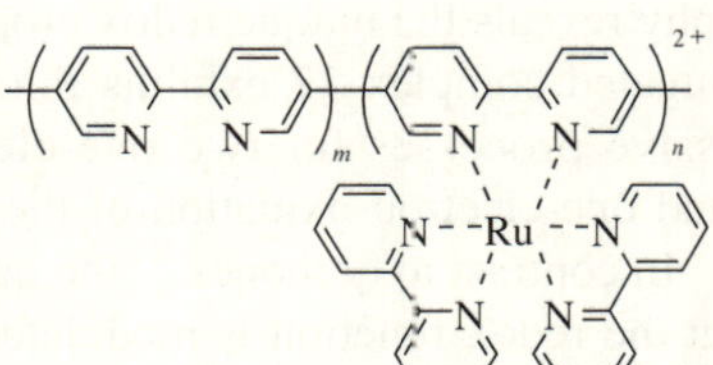

**8**

A polythiophene backbone attached by salen complex moieties is synthesized by anodic polymerization of the terthiophene derivative **9** (Scheme 3.6).[52]

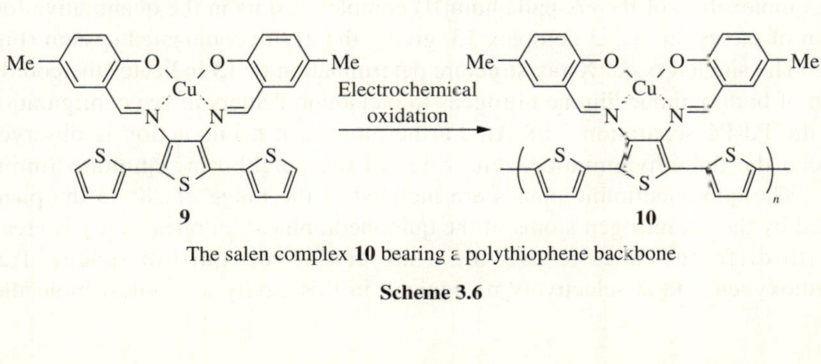

The salen complex **10** bearing a polythiophene backbone

**Scheme 3.6**

### 3.2.3 Complexes with Redox-active π-Conjugated Molecules

A strategy for obtaining block polymer-like conjugated complexes is described here. Complexation of a palladium species like $Pd(OAc)_2$ or $PdCl_2$ with the quinonediimine ligand **7** possessing two coordination sites results in the formation of the polymeric palladium(II) complex **11** with an alternate array of the palladium and ligand units. The structure of the complexes depends on the geometry of both units.[53)]

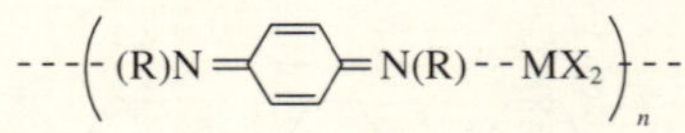


**11**

To elucidate the structures of **4**, **5** and **11**, a model complex is prepared as follows. Complexation of the π-conjugated molecule **7** with the palladium(II) complex **3** affords the 1 : 2 conjugated palladium(II) complex **12** as *syn*- and *anti*-isomers.[54)] The two palladium complex units are bridged by the quinonediimine moiety. The X-ray crystal structural analysis of the isolated $C_2$-symmetrical complex **12** ***anti*** indicates the structure shown in Fig. 3.12. Furthermore, the bridging π-conjugated spacer moieties are aligned along a straight line almost parallel to the *a* axis, to form the columns of the π-conjugated molecules in the molecular packing. A similar coordination to the quinonediimine moieties is also suggested in the case of the above-mentioned conjugated complexes.

Cyclic voltammography reveals the unique redox properties of the quinonediimine ligands. The conjugated complex **12** exhibits three separate redox waves assignable to the successive processes for two one-electron reduction of the quinonediimine moiety and one-electron oxidation of the two terminal dimethylamino groups (Fig. 3.13). In contrast to quinones,[55)] the quinonediimine moiety of **7** is redox-inactive so that the redox function is modulated by complexation with the palladium complex, probably due to the stabilization of the generated anionic species. The added electrons are considered to be delocalized over the Pd(II)-quinonediimine d-$\pi^*$ system. This finding is related to the role of protonic acid doping in the redox process of **7**. A representative scheme for the redox processes of the quinonediimine and polyaniline complexes is illustrated in Scheme 3.7.

Complexation of the *cis*-palladium(II) complex results in the quantitative formation of the cyclic 3 : 3 complex **13**, giving the cyclic conjugated system (Fig. 3.14). The single-crystal X-ray structure determination of **13** indicates the coordination of both quinonediimine nitrogens to palladium centers in *syn* configuration with the Pd-Pd separation 7.68 Å. Furthermore, a π-π interaction is observed between the dimethylaminobenzene rings of the neighboring quinonediimine units. The quinonediimine planes are inclined in the range of 28° to the plane defined by the six nitrogen atoms of the quinonediimine. An open cavity is created with different faces to provide a molecular recognition space. 1,2-Dimethoxybenzene is selectively recognized in this cavity as a guest molecule.

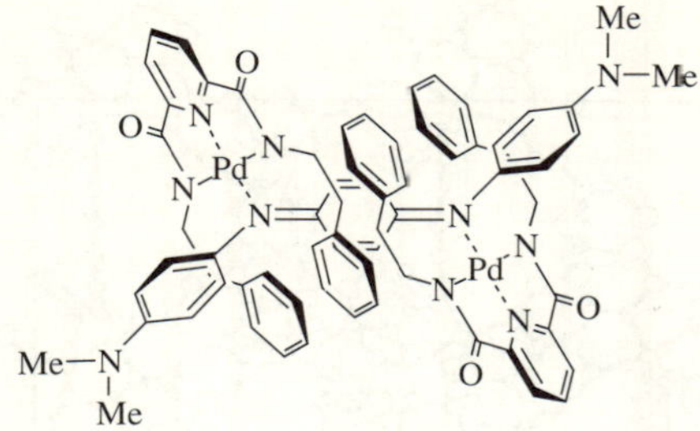


(a)

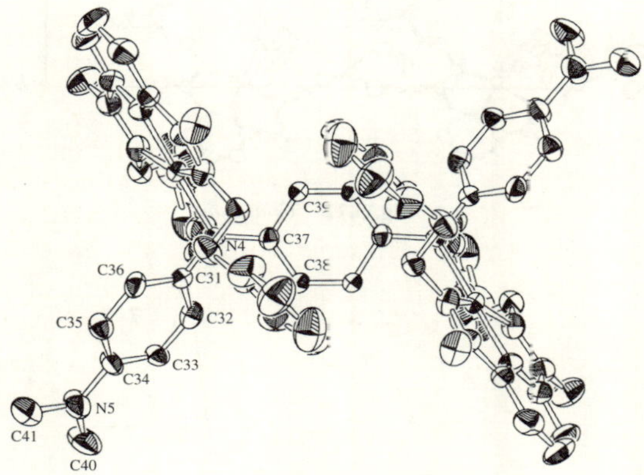


(b)

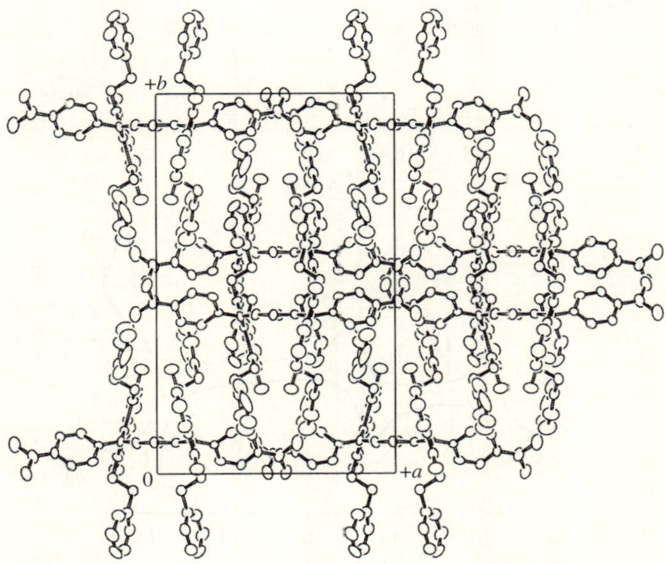


**Fig. 3.12** (a) The X-ray crystal structure of **12 *anti*** (40% probability ellipsoids. Hydrogen atoms are omitted for clarity), (b) A projection down the c axis and (c) a projection down the a axis of the molecular packing of **12 *anti***.

(c)

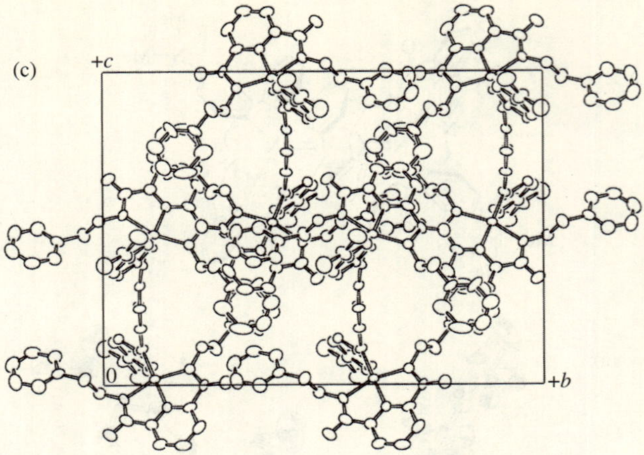


**Fig. 3.12** *Continued*

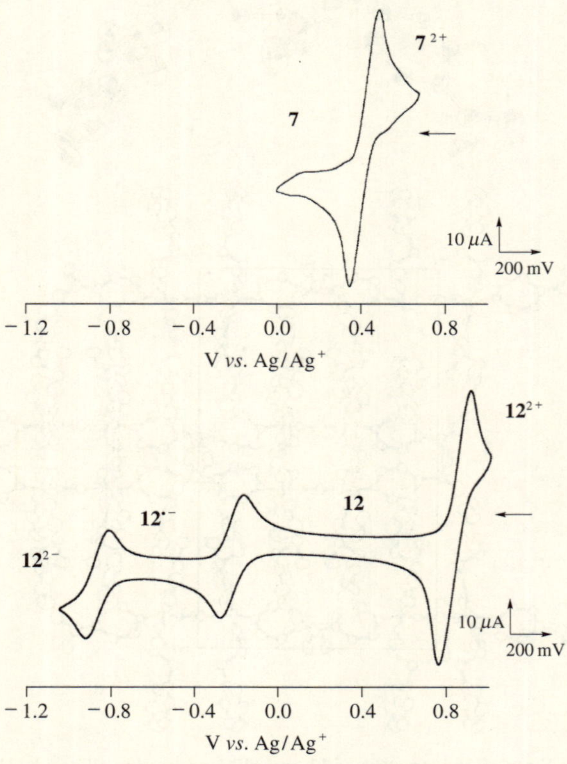


**Fig. 3.13** Cyclic voltammograms of **7** and **12** ($1.0 \times 10^{-3}$ M) in dichloromethane (0.1 M $Bu_4NClO_4$) as a glassy carbon working electrode with scan rate = 100 mV $s^{-1}$ under argon atmosphere.

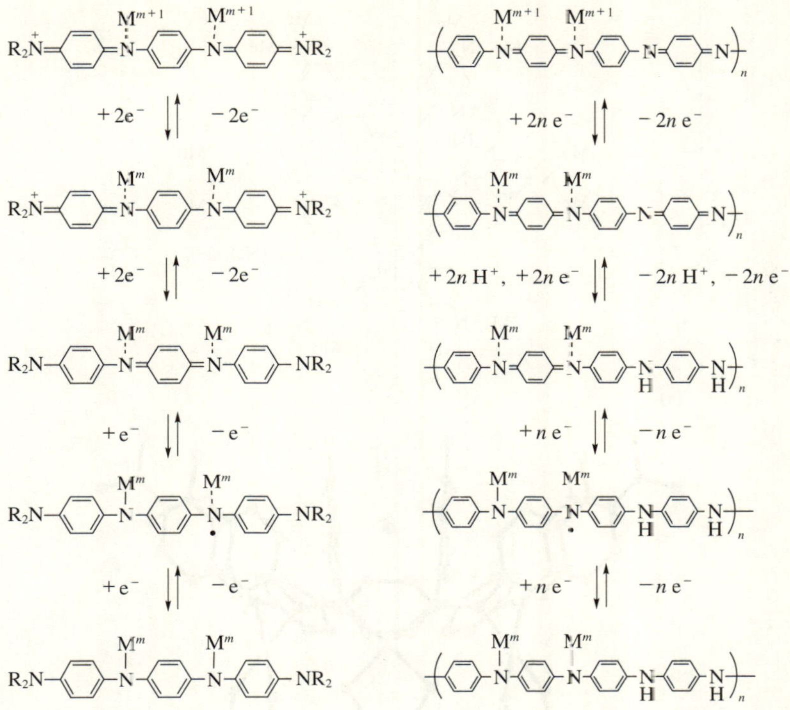

Representative redox processes of the quinonediimine and polyaniline complexes

**Scheme 3.7**

The present conjugated system appears to be of potential from the viewpoint of redox-active molecular recognition systems.

The linkage of the ruthenium π-complexes to the oligothiophene leads to the terminal π-conjugated bimetallic complex **14** in which the oligothiophene core is modulated via $[CpRu]^+$ moieties.[56] The cyclopentadithiophene bearing the pendant ferrocenyl group is electrochemically oxidized to the π-conjugated polymer.[57]

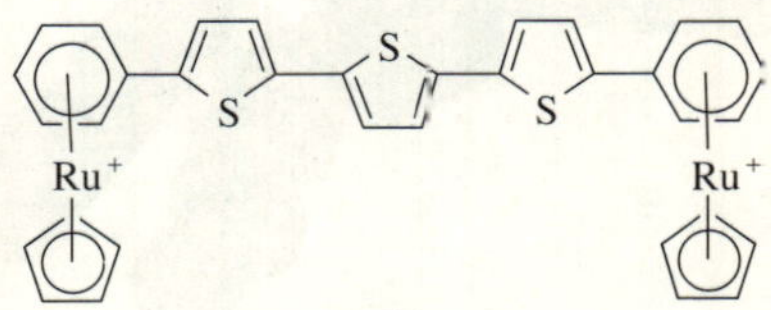

**14**

The terthienyl ligand of the ruthenium complex **15** is recognized as a redox-switchable hemilabile ligand (Scheme 3.8).[58]

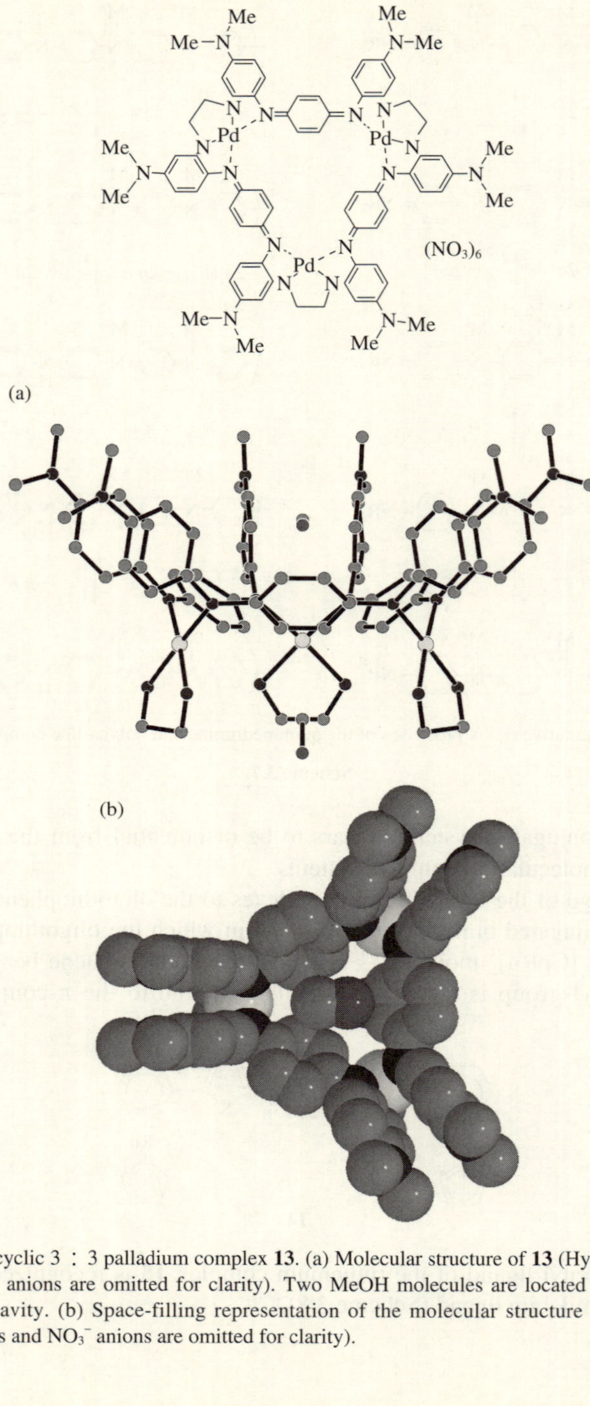


**Fig. 3.14** The cyclic 3 : 3 palladium complex **13**. (a) Molecular structure of **13** (Hydrogen atoms and $NO_3^-$ anions are omitted for clarity). Two MeOH molecules are located above and below the cavity. (b) Space-filling representation of the molecular structure of **13** (Hydrogen atoms and $NO_3^-$ anions are omitted for clarity).

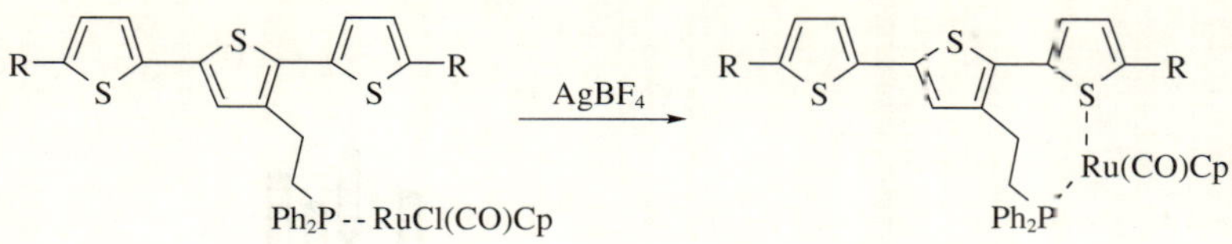

The ruthenium complex **15** with the terthienyl ligand

**Scheme 3.8**

### 3.2.4 Redox-active π-Conjugated Systems

A variety of hybrid redox-active systems are constructed as described above. The design of π-conjugated compounds is considered to widen their scope and utility. From this point of view, extended π-conjugated compounds have been prepared by using porphyrins and porphyrin complexes bearing oligothiophene or polypyrrole substituents at the *meso*-positions.[59] The architecturally ordered orientation and/or assembly of π-conjugated molecular chains are considered to afford three-dimensionally oriented π-electronic systems. Actually, such π-conjugated systems **16** (αααα: **16a** and αβαβ: **16b**, Fig. 3.15) are synthesized by the introduction of four phenylenediamine pendant groups to the *meso*-positions of the atropisomeric porphyrin scaffolds.[60] The αβαβ isomer appears to be in equilibrium with the conformers in which one or more strands lean towards the porphyrin ring. On the other hand, such a conformation is less accessible with the αααα isomer, probably due to the steric crowdedness. In the fluorescence emission spectra, the emission from the porphyrin moiety is almost completely quenched in both atropisomers. Less efficient quenching is observed with the porphyrin bearing only one

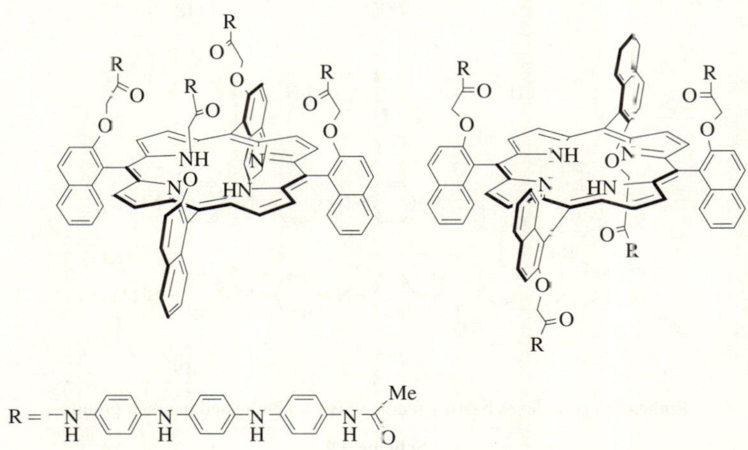

**16a** : αααα isomer **16b** : αβαβ isomer

**Fig. 3.15** Three-dimensionally oriented π-conjugated systems **16**.

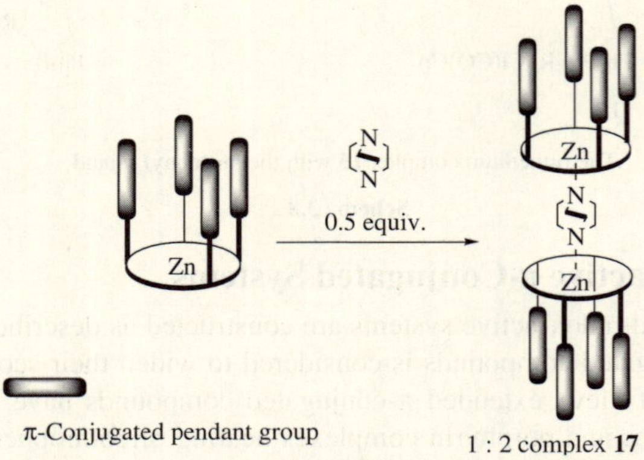


**Fig. 3.16** The sandwich-type zinc porphyrin **17** bearing phenylenediamine pendant groups.

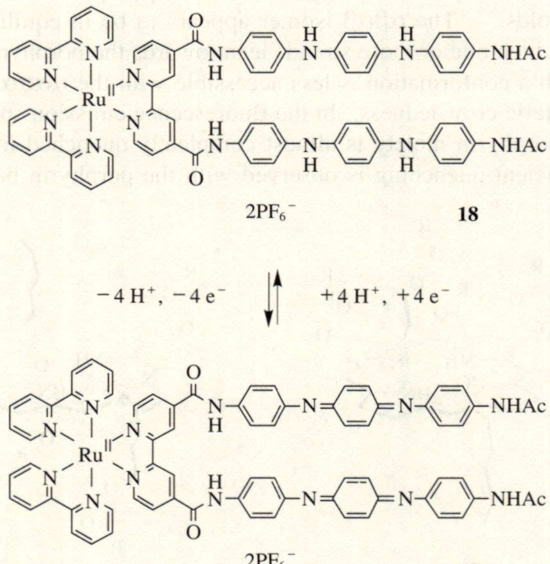


Ruthenium complexes bearing redox-active π-conjugated pendant groups

**Scheme 3.9**

phenylenediamine pendant group. This finding suggests an electronic interaction between the π-conjugated pendant groups.[61)]

Metal porphyrins provide various host-guest systems[62)] and electron- or energy-transfer systems.[63)] An additional functional moiety can be introduced to a vacant axial coordination site of metal porphyrins. Furthermore, the use of a bidentate ligand leads to a sandwich dimer-type metal porphyrin complex.[64)] This type of molecular design affords a dimensionally ordered π-conjugated system in which the central porphrin-bridging ligand-porphyrin moiety is surrounded by redox-active π-conjugated pendant groups. Treatment of the zinc porphyrins bearing four dimensionally oriented phenylenediamine pendant groups with a bidentate ligand, DABCO, results in the sandwich dimer complexes **17** (Fig. 3.16). Complexes thus obtained are considered to be a unique electronic system composed of the porphyrin surrounded by electron-donating phenylenediamine strands.

The ruthenium bipyridyl complex can be also used as a scaffold. The ruthenium(II) complex **18** bearing the phenylenediamine moieties has been designed and synthesized. Chemical oxidation leads to the complex **19** bearing the quinonediimine moieties (Scheme 3.9).[65)] An efficient photoinduced intramolecular electron transfer is observed with **18**.

### 3.2.5 Conclusions

The hybrid complexes composed of transition metals and redox-active π-conjugated polymers (synthetic metals) or molecules provide novel multi-redox systems, with much potential. The structural control in the complexation leads to structurally defined hybrid systems. These systems are promising for the development of efficient catalysts and functionalized materials.

## Acknowledgment

The present work was done by researchers in our laboratory whose names are cited in the references. This work is partly supported by a Grant-in-Aid for Scientific Research on Priority Areas from the Ministry of Education, Culture, Sports, Science and Technology, Japan.

# 3.3 Polymer-capped Bimetallic Nanoclusters as Active and Selective Catalysts

Naoki Toshima

## 3.3.1 Introduction

Recently much attention has been paid to metal nanoparticles, important building blocks of nanostructures to achieve the nanotechnology, which has been demonstrated by the U.S. government as one of the most important technologies of the 21st century.[66] The interest in metal nanoparticles arises from their unique large surface-to-volume ratio and quantum size effect (Kubo effect). The term "nanoparticle" is based on the technological background. In contrast, the term "cluster" is used for an aggregate of molecules or atoms with a precise structure. "Metal cluster" often means an aggregate of metal atoms that contains less than ten metallic atoms, for example, $[Ni_5(CO)_{12}]^{2-}$. They are the subjects of traditional organometallic chemistry. Recently, however, metal clusters that contain larger numbers of metal atoms than the conventional ones have attracted scientists' attention. For example, $Au_{55}$ is 1.4 nm in diameter. Some use the term "giant metal cluster" for such clusters in order to distinguish them from the conventional low-nuclearity organometallic clusters. Here the terminology "metal nanoclusters" is used to indicate fine metal particles with a definite size of less than 10 nm in diameter. If the fcc (face-centered cubic) structure is assumed, a metal particle composed of 13 atoms is the smallest metal nanocluster, the size of which is about 0.8 nm in diameter in the case of most metals.

Industrial catalysts are usually composed of inorganic supports and metal particles, the surface of which provides the active sites of catalyses. Thus, metal nanoclusters which have much larger surface area per unit volume or weight of metal than the bulk metal are considered to be promising materials for catalyses. In addition, since small amounts of additives can often improve the performance of industrial metal catalysts, bimetallic nanoclusters are much more interesting than monometallic ones from the viewpoint of sophisticated catalysts. Industrial metal catalysts are often prepared by heat treatment of metal ions on supports at high temperature often under hydrogen. They have very complex structures. For example, they are mixtures of metal particles of various sizes and shapes. The metal particles often interact strongly with the inorganic supports, resulting in the structure of half balls, for example.

Catalysts play a very important role not only in chemical processes in industry but also in chemical reactions (enzymatic reactions) in the human body.

Industry usually uses heterogeneous catalysts supported on inorganic supports under high pressure at high temperatures, and sometimes homogenous catalysts at lower temperatures. The former catalysts are much superior to the latter from the viewpoint of recycling of catalysts. In the human body or any other living organism, in contrast to industry, enzymes are used as catalysts in reactions at temperatures as low as 38 °C. They are usually not heterogeneous but homogeneous, and work as very active and selective catalysts. In both industry and living bodies, an acid or base is used as the most popular catalyst for simple reactions, and metals work as the active sites of catalysts for rather complicated reactions. In industry, transition metals are often used in the form of metal particles dispersed onto inorganic supports such as silica gel, alumina and activated charcoal.

In contrast, colloid chemistry has provided the colloidal dispersion of metal fine particles in water.[67) As early as the 1950's, colloidal dispersion of fine particles of precious metals was conducted and applied to catalyses. Although they contain fine metal particles, the size of the particles was not so uniform. In addition they were not very stable when used in solution. Thus, there still remained many problems in the reproducibility of the preparation and catalysis using metal nanoparticles. In 1976, the author's group prepared colloidal dispersions of rhodium particles protected by water-soluble polymers by reduction of rhodium(III) ions under mild conditions, i.e., reduction with refluxing alcohol in the presence of water-soluble polymers.[68) They were applied to the catalyst for hydrogenation of olefins. In 1989, we developed colloidal dispersions of Pd/Pt bimetallic nanoclusters by the simultaneous reduction of Pd and Pt ions in the presence of poly(*N*-vinyl-2-pyrrolidone) (PVP).[69)

Colloidal dispersions of metal nanoclusters usually work at rather low temperatures as homogeneous catalysts. On this point, metal nanoclusters are similar to enzymes and are often regarded as artificial enzymes. From the viewpoint of green chemistry (less energy, fewer by-products, more efficiency, more selectivity, etc.), enzymes are model industrial catalysts. Thus, metal nanoclusters may provide the means for industrial catalysts to step up from the present practical heterogeneous catalysts to more ideal and enzyme-like ones.

In this section the preparation and catalytic properties of the colloidal dispersion of polymer-capped bimetallic nanoclusters will be discussed, emphasizing the importance of polymers for controlling the structure and stabilizing the dispersions of bimetallic nanoclusters in preparation processes as well as for the improvement of the activity and selectivity in catalyses.[70) Although there are many reports on the preparation of industrial catalysts composed of bimetallic nanoclusters and inorganic supports, they are not included in this section.

### 3.3.2 Preparation and Structure of Polymer-capped Bimetallic Nanoclusters

A polymer-capped metal nanocluster catalyst has not only the advantages similar to those of particles dispersed on an inorganic support but also new benefits[71) in that

(a) the colloidal dispersions can form "homogeneous" solutions ;
(b) the capping polymer can shield the nanocluster catalyst from deactivation by catalytic poisons or air ;
(c) the capping polymer can interact attractively or repulsively with substrates, resulting in high selectivity and/or sometimes high activity ; and
(d) the colloidal dispersions of metal nanoclusters transmit light more readily than powders. As a result, colloidal dispersions of metal nanoclusters have frequently been used for investigations of various photocatalytic systems.

Metal nanoclusters can be stabilized not only by polymers but also by organic ligands of low molecular weight, organic surfactant molecules, and so on. Among them the polymer-capped metal nanoclusters may be the best as catalysts. Polymers that weakly coordinate to the surface of metal nanoclusters at many points are especially effective for stabilizing the metal nanocluster catalyst.[72] In the process of catalytic reaction, when the substrate is added to the reaction mixture, a weakly coordinating polymer can be easily replaced in part with the substrate molecule, resulting in the reaction of substrates on the surface of metal nanoclusters as shown in Fig. 3.17. Strongly coordinating organic ligands cannot be easily replaced by substrate molecules, making it difficult for the substrate to make the catalytic reaction proceed on the surface of metal nanoclusters.

In the case of bimetallic nanoclusters, their catalytic activities are usually much higher than those of the corresponding monometallic nanoclusters. Thus, the bimetallic nanoclusters are much more interesting not only for scientists in academia but also for researchers in industry. However, the bimetallic nanoclusters consisting of even the same elements can have various structures, on which the catalytic activities may depend.

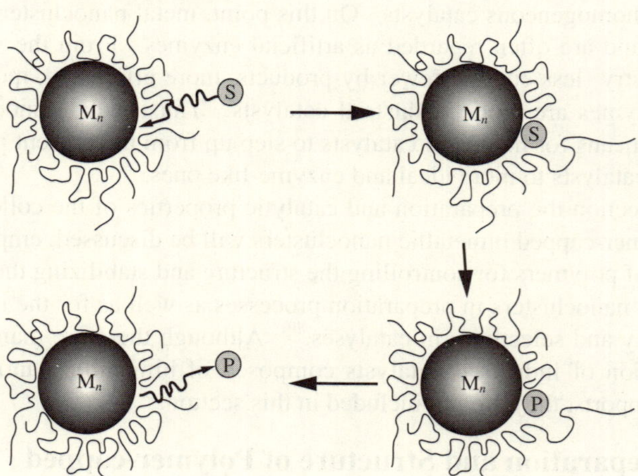


**Fig. 3.17** Catalysis process on the surface of polymer-protected metal nanoparticles. [Reprinted with permission from N. Toshima, *Nanoscale Materials* (L. M. Liz -Marzán and P. Kamat eds.), p.83, Fig. 3. Copyright 2003 Kluwer/Plenum Pub.]

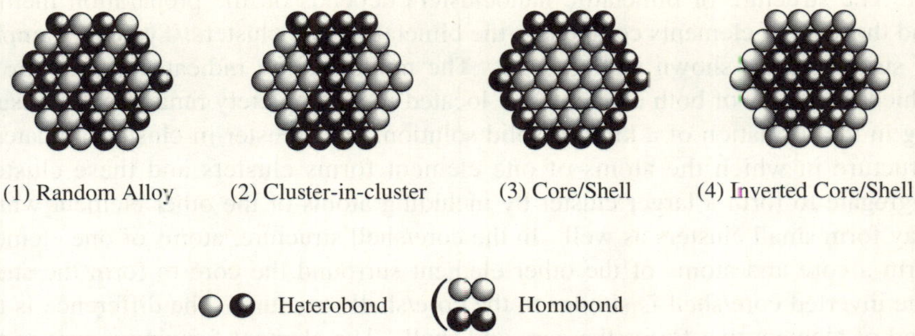


**Fig. 3.18** Various structures of bimetallic nanoparticles.

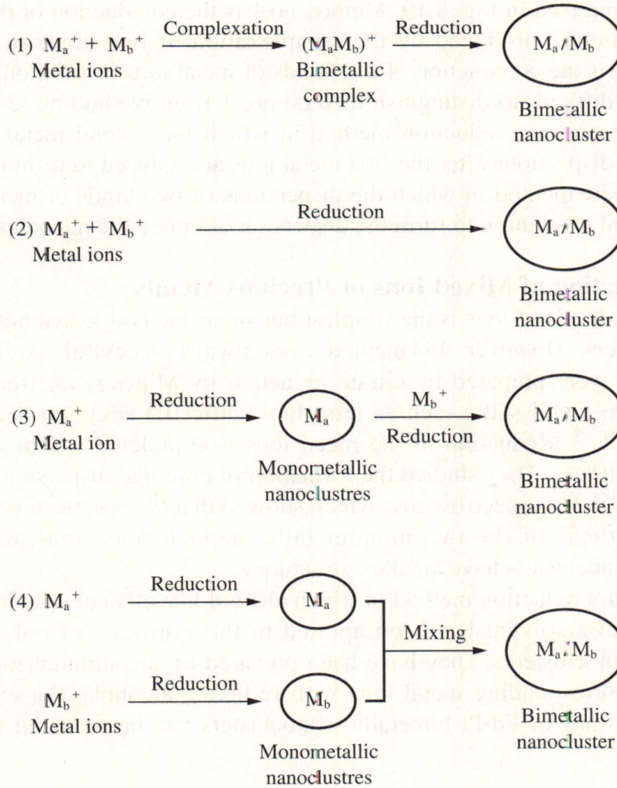


**Fig. 3.19** Various preparation schemes of bimetallic nanoclusters from the corresponding metal ions.

The structure of bimetallic nanoclusters depends on the preparation method and the kind of elements composing the bimetallic nanoclusters. Typical examples of structures are shown in Fig. 3.18. The random alloy indicates a structure in which the atoms of both elements are located in a completely random way, resulting in the formation of a kind of solid solution. The cluster-in-cluster indicates a structure in which the atoms of one element forms clusters and these clusters aggregate to form a larger cluster by including atoms of the other element which may form small clusters as well. In the core/shell structure, atoms of one element form a core and atoms of the other element surround the core to form the shell. The inverted core/shell is similar to the core/shell structure. The difference is the kind of element that forms the core and shell. The element forming a core in the core/shell structure now forms the shell in the inverted core/shell structure and vice versa. Thermodynamically, the core/shell structure is stable, while the inverted core/shell is metastable.

The preparation methods of the colloidal dispersions of bimetallic nanoclusters are summarized in Fig. 3.19. Method no.1 is the coreduction of the bimetallic complex which is produced by the complexation of two kinds of metal ions. Method no.2 is the coreduction of two kinds of metal ions in solution. It is sometimes rather difficult to distinguish method no. 1 from method no. 2. Method no. 3 shows a successive reduction method in which the second metal ions can be added to the dispersions after the first metal ions are reduced to form nanoclusters. No. 4 is a new method in which the dispersions of two kinds of metal nanoclusters are mixed in solution to form the dispersion of bimetallic nanoclusters.

## A. Coreduction of Mixed Ions of Precious Metals

Coreduction of mixed ions is the simplest but smart method to synthesize bimetallic nanoclusters. However, this method is not always successful. Au/Pt bimetallic nanoclusters were prepared by citrate reduction by Miner *et al.* from the corresponding two metal salts, such as tetrachloroauric(III) acid and hexachloroplatinic(IV) acid.[73] Reduction of the metal ions is completed within 4 h after the addition of citrate. They studied the formation of colloidal dispersion by ultraviolet-visible (UV-Vis) spectroscopy, which showed that the spectrum is not the simple sum of those of the two monometallic nanoclusters, indicating that the bimetallic nanoclusters have an alloy structure.

An alcohol reduction method in which alcohol like ethanol can work both as a reductant and a solvent has been applied to the synthesis of polymer-capped bimetallic nanoclusters. They have been prepared by the simultaneous reduction of the two corresponding metal ions with refluxing alcohol. For example, colloidal dispersions of Pd/Pt bimetallic nanoclusters can be prepared by refluxing

$$H_2PtCl_6 + 2\ C_2H_5OH \longrightarrow Pt + 6\ HCl + 2\ CH_3CHO$$

$$PdCl_2 + C_2H_5OH \longrightarrow Pd + 2\ HCl + CH_3CHO$$

**Scheme 3.10**

**Table 3.1** Coordination numbers of PVP-capped Pd/Pt (4/1) bimetallic nanoclusters and those calculated for the models

| Sample | Edge | Bond | Coordination number $N$ | | |
|---|---|---|---|---|---|
| | | | Observed | Core/shell | Random |
| Pd/Pt (4/1) | Pd-K | Pd-Pd | 4.4 ±1.0 | 4.6 | 6.0 |
| | | Pd-Pt | 2.3 ±1.3 | 2.0 | 1.9 |
| | Pt-$L_3$ | Pt-Pt | 5.5 ± 1.7 | 5.5 | 1.9 |
| | | Pt-Pd | 3.5 ±1.5 | 6.5 | 6.0 |

[Reprinted with permission from N. Toshima, M. Harada, T. Yomezawa, K. Kushihashi, K. Asakura, *J. Phys. Chem.*, 95(19), p.7451, Table II. Copyright 1991 American Chemical Society.]

the alcohol-water (1 : 1 v/v) solution of palladium(II) chloride and hexachloroplatinic(IV) acid in the presence of poly(*N*-vinyl-2-pyrrolidone) (PVP) at about 90–95 °C for 1 h (Scheme 3.10).[69] The resulting brownish colloidal dispersions are stable and neither precipitate nor flocculate over a period of several years. Interestingly, Pd/Pt bimetallic nanoclusters thus obtained have a core/shell structure, as proved by an EXAFS(Extended X-ray Absorption Fine Structure) technique (*vide infra*).[74] Au/Pd bimetallic nanoclusters, capped by PVP, were also prepared by the alcohol reduction method.[75]

The EXAFS technique was used to establish the structural model of several kinds of PVP-capped bimetallic nanoclusters.[74-76] PVP-capped Pd/Pt nanoclusters were chosen at the first stage because they are relatively small and uniform in size; in particular, the average diameter of the particles with a composition of Pd/ Pt = 4/1 (mol/mol) was as small as 1.5 nm (each particle should have only 55 atoms in it). Furthermore, the absorption edge of Pd and that of Pt are far enough apart. EXAFS measurements were thus performed on the Pd K-edge and Pt $L_3$-edge, providing the coordination number around Pd and Pt atoms, respectively.[74] From the analyses of reference samples, some fitting parameters, such as $\Delta E$ (the difference between the theoretical and experimental threshold energies) and $\sigma$ (the Debye-Waller factor of the coordination shell), which can be used for two-shell curve fitting of Pd/Pt bimetallic nanoclusters with higher accuracy, were obtained. In Fourier transforms of Pd K-edge EXAFS of PVP-capped Pd/Pt bimetallic nanoclusters at various Pd/Pt ratios, with decreasing Pd/Pt ratio, the main peak, assigned as a Pd-Pd bond, splits into two peaks, and is mainly attributed to Pd-Pt and Pd-Pd bonds. Coordination numbers of PVP-capped Pd/Pt (4/1) bimetallic nanoclusters determined by the EXAFS method are shown in Table 3.1. Each Pd/Pt (4/1) bimetallic nanocluster is estimated to contain 55 atoms in it and have a three-layered (two-shell type) fcc structure on the basis of the average diameter of 1.5 nm. Two structural models, random alloy and core/shell, were constructed for this particle. In the random model, Pd and Pt atoms are located at random in the particle, and the coordination numbers can be determined only by the composition of the two elements. In the other model, the Pt-core/Pd-shell model, the Pt atoms in the two-layered core number 13 (24%), and 42 atoms (76%) should be located in the surface layer. In this model, the calculated coordination numbers are mostly consistent with the experimentally observed values. The structural model of Pd/Pt (1/1) bimetallic nanoclusters can be constructed in a similar way. These considerations are the first report of a complete determination of the structure of metal nanoclusters using an EXAFS technique.

These core/shell structures were found to be very common in the PVP-capped bimetallic nanoclusters of precious metals prepared by the coreduction of precious metal ions with alcohol. For example, PVP-capped Au/Pd[75)] and Pt/Rh bimetallic nanoclusters[76)] prepared by alcohol coreduction in solution were also investigated in a similar way and have been proved to have a core/shell structure where Au and Pt form the core, and Pd and Rh form the shell, respectively. There appears to be a trend among the precious metals in ease of formation of the core or shell:

(core) Au < Pt < Pd < Rh (shell)

The metal on the left side more easily forms a core than that on the right side and vice versa for a shell. Thus, in the combination of Au and Pd, for example, Au forms the core and Pd the shell, and so on. The question that arises then is how the order of this trend is determined. The question was answered by the *in situ* UV-Vis spectra measured during the coreduction of Au and Pt ions in alcohol in the presence of PVP.[77)] According to precise analyses of spectra, the formation mechanism of core/shell-structured bimetallic nanoclusters has been proposed as shown in Fig. 3.20. Here the difference in the redox potential of both elements and that in the strength of coordination bonds between polymer and atom of both elements are the factors that determine the order of ease in forming the core or shell.

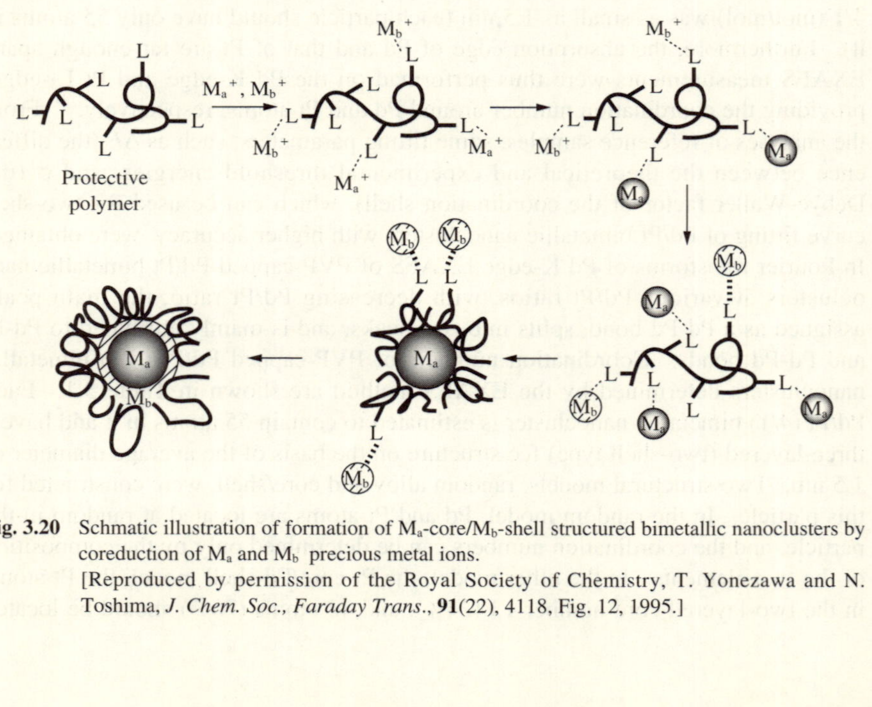


**Fig. 3.20** Schmatic illustration of formation of $M_a$-core/$M_b$-shell structured bimetallic nanoclusters by coreduction of $M_a$ and $M_b$ precious metal ions.
[Reproduced by permission of the Royal Society of Chemistry, T. Yonezawa and N. Toshima, *J. Chem. Soc., Faraday Trans.*, **91**(22), 4118, Fig. 12, 1995.]

## B. Coreduction of Mixed Ions of Light Transition Metals and Precious Metals

Nanoclusters of light transition metals (transition metals at the 4th period) are more difficult to prepare because of the lower redox potential of the metal ions than those of the precious metal ions. In other words, light transition metal ions are difficult to reduce to zero valence, and zero-valence light transition metals are easily oxidized to ions.

Polymer-capped bimetallic nanoclusters containing both light transition metal and precious metal can also be prepared by a modified alcohol reduction method. For example, Cu/Pd bimetallic nanoclusters were successfully prepared with various Cu:Pd ratios by refluxing a glycol solution of the hydroxides of Cu and Pd in the presence of PVP[78)] or by the thermal decomposition of metal acetates.[79,80)]

In the modified alcohol reduction procedure,[78)] metal hydroxides are prepared first by pH control of the ionic solutions (Scheme 3.11). Then, refluxing the solution at 198 °C with a nitrogen flow to remove water from the system gave a stable homogeneous dark-brown solution of PVP-capped Cu/Pd bimetallic nanoclusters. Auger and X-ray photoelectron spectroscopy (XPS) indicated that these nanoclusters contain $Cu^0$ species.

$$\mathrm{Pd(OCOCH_3)_3 + CuSO_4 + PVP \xrightarrow{NaOH} PVP\text{-}Cu/Pd(OH)}_n$$

$$\mathrm{PVP\text{-}Cu/Pd(OH)}_n \xrightarrow[\text{glycol}]{} \text{PVP-Cu/Pd (nanocluster)}$$

**Scheme 3.11**

Thermal decomposition of metal acetates in the presence of PVP was proposed by Bradley *et al.*,[80)] whereas the preparative procedure of Esumi *et al.*[79)] was modified. In the Bradley method the Cu/Pd bimethallic nanoclusters produced contain less than 50 mol% of Cu, while in the modified alcohol reduction method, bimetallic nanoclusters containing 95 mol% of Cu could be obtained.[81)]

The structure of the Cu/Pd bimetallic nanoclusters was investigated by XRD, XPS and EXAFS. The observed data showed that Cu/Pd bimetallic nanoclusters have a random alloy structure.[81)] However, recent precise analyses by EXAFS technique have indicated that these nanoclusters have a heterobond-philic structure in which the number of heterobonds (Cu-Pd bond) should be maximized.[82)] The concept of the heterobond-philic structure, where the heterobond is favored, is opposite to that of the homobond-philic structure, where the homobond (the bond between atoms of the same element) is favored. The core/shell and cluster-in-cluster are the examples of homobond-philic structures. PVP-capped Ni/Pd and Ni/Pt bimetallic nanoclusters could also be prepared by the modified alcohol reduction method at high temperatures.[83)] They may also have a structure similar to that of Cu/Pd nanoclusters.

## C. Successive Reduction of Metal Ions

Successive reduction of two metal salts can be considered as one of the most suitable methods to prepare core/shell structured bimetallic nanoclusters (Fig. 3.21).

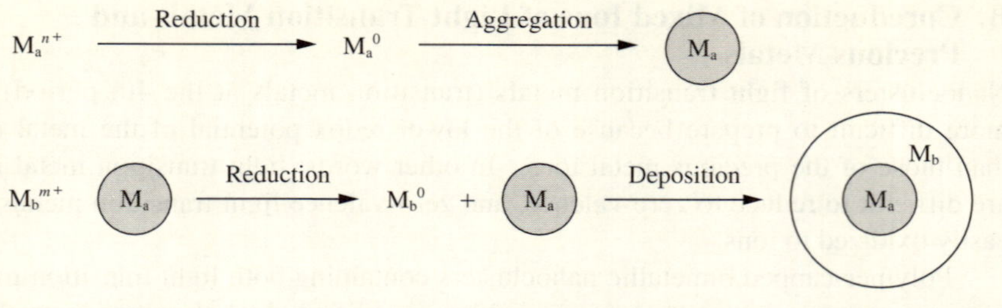


**Fig. 3.21** Schematic illustration of formation of $M_a$-core/$M_b$-shell structured bimetallic nanoclusters by successive reduction. [Reprinted from *Fine Particles* (T. Sugimoto ed.), p.438, Fig. 9.1.3, 2000 by courtesy of Marcel Dekker Inc.]

In 1970, Turkevich and Kim tried to grow gold on Pd nanoclusters to obtain gold-layered Pd nanoclusters.[84] The deposition of one metal element on preformed monometallic nanoclusters of another metal seems to be very effective. For this purpose, however, the second element must be deposited on the surface of preformed nanoclusters. If the second element cannot be deposited on the preformed nanoclusters, mixtures of two kinds of nanoclusters will be produced.

An attempt to prepare PVP-capped Au-core/Pd-shell bimetallic nanoclusters was made by the successive reduction procedure in ethanol/water.[85] When the preparation of Au nanoclusters preceded the reduction of Pd ion, only mixtures of Pd and Au monometallic nanoclusters were produced (Fig. 3.22(a)). On the other hand, when the preparation of Pd nanoclusters preceded the reduction of Au ions, some bimetallic nanoclusters were found. The Au/Pd bimetallic nanoclusters thus obtained did not have a core/shell structure but a cluster-in-cluster structure (Fig. 3.22(b)), although the bimetallic nanoclusters obtained by the coreduction have a core/shell structure. This strange phenomenon can be understood by considering

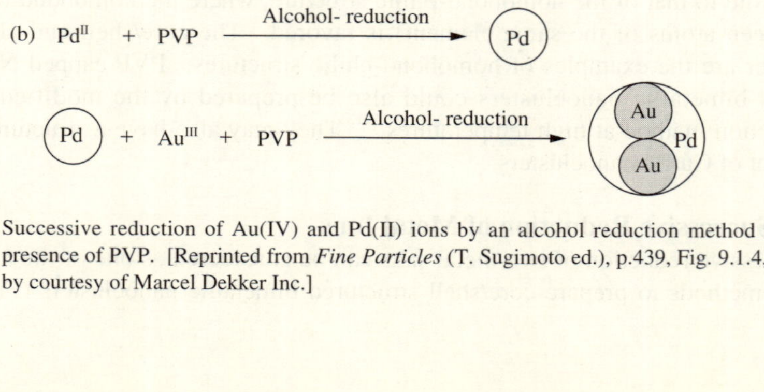


**Fig. 3.22** Successive reduction of Au(IV) and Pd(II) ions by an alcohol reduction method in the presence of PVP. [Reprinted from *Fine Particles* (T. Sugimoto ed.), p.439, Fig. 9.1.4, 2000 by courtesy of Marcel Dekker Inc.]

the reaction between preformed Pd nanoclusters and added Au(III) ions. Unlike the bulk metal of Pd, the Pd nanoclusters are very reactive and the redox reaction can occur between the Pd nanoclusters and Au(III) ions, resulting in the formation of Pd(II) ions and Au nanoclusters. Under reductive conditions complex mixtures of Pd and Au nanoclusters and Pd(II) and Au(III) ions can form Au/Pd bimetallic nanoclusters with a cluster-in-cluster structure.

Ligand-stabilized Au/Pd[86] and Au/Pt[37] bimetallic nanoclusters were prepared by Schmid *et al.* by successive reduction. Au colloids with a diameter of ~18 nm could be covered by Pt or Pd shells when the chloride precursors, $H_2PtCl_6$ or $PdCl_2$, were reduced. The color change from red to brown-black may suggest the covering of the Au core with Pt or Pd atoms. Addition of the water-soluble ligand *p*–$H_2NC_6H_4SO_3Na$ stabilized the bimetallic nanoclusters obtained. In the case of Au/Pt bimetallic nanoclusters, uniform heterogeneous agglomerates with a large spherical Au core surrounded by small (~5 nm diameter) spherical Pt clusters were obtained, as proved by EDX measurements.[87] In the Au/Pd bimetallic case, on the contrary, a rather continuous structure was formed. TEM investigations of Au/Pd bimetallic nanoclusters showed somewhat irregular structures, and no agglomeration of small clusters was observed. The EDX spectrum of a single particle showed only Pd on the surface. The reverse successive reduction also worked and Pd-core/Au-shell structured bimetallic nanoclusters were produced.

In an earlier study, Turkevich and Kim proposed gold-layered Pd nanoclusters.[84] Three types of Au/Pd bimetallic nanoclusters, such as Au-core/Pd-shell, Pd-core/Au-shell and random alloyed particles, were prepared by successive reduction. Two kinds of layered Pd/Pt bimetallic nanoclusters were also reported by successive reduction. However, detailed analyses of the structure of these bimetallic nanoclusters were not carried out at that time. Only the difference in UV-Vis spectra between the bimetallic nanoclusters and the physical mixtures of the corresponding monometallic nanoclusters was discussed.

## D. Sacrificial Hydrogen Reduction

As noted above, successive reduction of metal ions is sometimes successful in synthesizing core/shell structured bimetallic nanoclusters. However, sometimes it is not. For example, Schmid and coworkers were able to synthesize Pd-core/Au-shell bimetallic nanoclusters,[86] while we could not prepare the core/shell-structured bimetellic nanoclusters by the successive reduction.[85] The Au/Pd bimetallic nanoclusters prepared by successive reduction by us had a cluster-in-cluster stucture.[85] When considered the difference between two procedures, the size of the Pd core particles used by Schmid *et al.* is much larger than that used by ours. During the reduction of $AuCl_4^-$, the oxidation of $Pd^0$ atoms of the Pd core particles may also occur in the system of Schmid and coworkers. However, because of their large size, although small parts of Pd core particles may be oxidized by Au(III) ions, before the complete disappearance of Pd core particles by oxidation with Au(III) ions, the Au atoms generated may precipitate on them, thus also protecting Pd from further oxidation.

Recently, our group has invented a modified successive alcohol reduction process to prepare "inverted core/shell" structured bimetallic nanoclusters.[88] In practice, PVP-capped Pd-core/Pt-shell bimetallic nanoclusters ~2 nm in size were obtained. This method is based on the fact that precious metals like Pd have the ability to adsorb hydrogen and split it to form metal hydride on the surface of a precious metal. Hydride on precious metal has a very strong reducing ability, implying a quite low redox potential. As illustrated in Fig. 3.23, the Pt ions added into the dispersion of Pd nanoclusters are reduced by hydride on the surface of preformed Pd nanoclusters without oxidizing Pd atoms to Pd ions, and are deposited on the Pd nanoclusters to form Pd-core/Pt-shell bimetallic nanoclusters. Since the coreduction of Pd and Pt ions produces only Pt-core/Pd-shell bimetallic nanoclusters and this structure is controlled by nature, Pt-core/Pd-shell can be called a normal core/shell structure. The structure of Pd-core/Pt-shell is opposite to the normal structure. Thus, we call this an "inverted core/shell" structure.

In the modified successive reduction process to produce the inverted core/shell structure, hydrogen is additionally used for control of the structure. We call this the "sacrificial hydrogen reduction" method. This method can be very useful in controlling layered bimetallic and trimetallic structures in general.

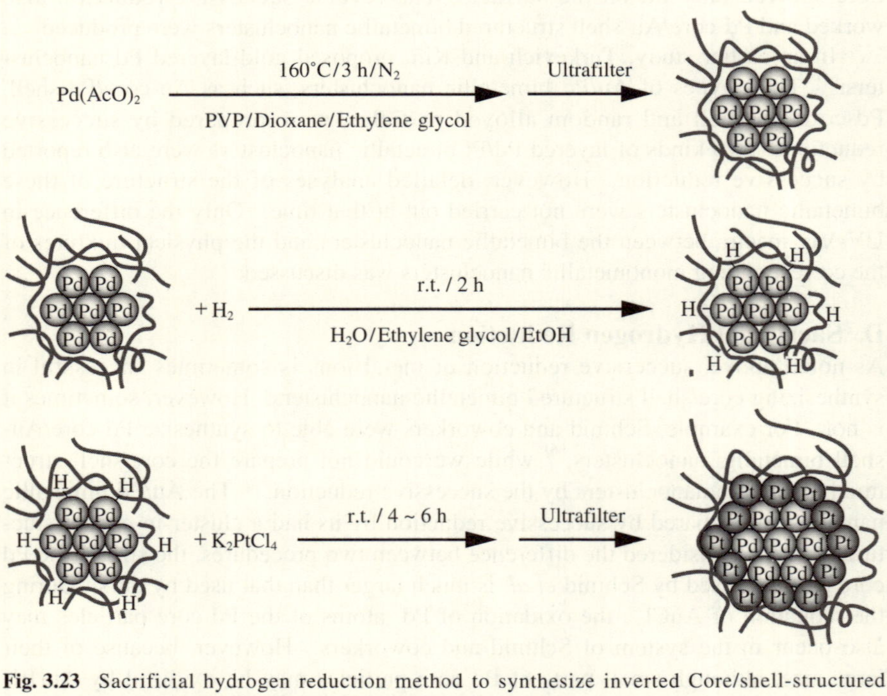


**Fig. 3.23** Sacrificial hydrogen reduction method to synthesize inverted Core/shell-structured bimetallic nanoclusters by successive reduction.
[Reprinted from N. Toshima, *Supramol. Sci.*, **5**(3−4), p.398, Fig.6. Copyright 1998, with permission from Elsevier Science.]

### E. Mixing of Two Kinds of Monometallic Nanoclusters

Recently we have found an interesting phenomenon to produce bimetallic nanoclusters.[89] Colloidal dispersions of PVP-capped Ag nanoclusters have a plasmon bond at about 420 nm. When the colloidal dispersions of Ag nanoclusters were mixed with those of PVP-capped Rh nanoclusters, the plasmon absorption band of PVP-capped Ag nanoclusters was extinguished within several minutes at room temperature. Transmission electron micrographs (TEM photographs), measured during the reaction, showed that Ag and Rh nanoclusters aggregate to form new bimetallic nanoclusters. X-ray diffraction (XRD) experiments of dry powders of the produced bimetallic nanoclusters suggest that these bimetallic nanoclusters tend to have the Ag-core/Rh-shell type structure. This is also supported by high-resolution TEM analyses. A similar aggregation occurs between the colloidal dispersion of Ag nanoclusters and those of Pd or Pt nanoclusters. Since this is a simple method just mixing dispersions it is expected to be applied to the preparation of a wide range of bimetallic nanoclusters.

## 3.3.3 Catalysis of Polymer-capped Bimetallic Nanoclusters

The catalytic activity of a metal is usually improved by the addition of another metal. Thus, the catalytic activity of polymer-capped bimetallic nanoclusters changes with their compositions. There are three types of changes in the catalytic activity of bimetallic nanoclusters depending on the composition of the metal elements. In the first case the additive can increase the catalytic activity of the original metal even though the additive itself has a low or no catalytic activity. In the second case there is no interaction between the additive and the original metal. Thus, the catalytic activity of bimetallic nanoclusters is just the average of those of two metals. In the third case the additive can decrease the catalytic activity of the original metal. This is the reverse of the first case.

The effect of the additive on the catalytic activity can be attributed to two kinds of effects. One is a ligand effect (an electronic effect) and the other is an ensemble effect (a steric effect). Let us consider the case in which the original metal works as the catalyst and provides a catalytic site(s). If the additive has the former effect on the catalysis of the original metal, the additive can influence the electronic density and/or electronic structure of the original metal. In this case the additive atom must be located near the catalyst metal. Since the catalytic site is located on the original metal, it is not necessary for the additive to be located on the surface. Nevertheless the additive must be located near the active site of the catalyst metal. The effect is the most effective when the additive atom is adjacent to the catalyst metal.

In the case of an ensemble effect the catalytic sites of the catalyst metal are located on the surface and the additive metal atom must be attacked by the substrate together with the catalyst metal. Thus, both catalyst and additive metals must be located on the surface of the metal particles. Although it is not necessary for the atom of the additive metal to be adjacent to the catalyst metal, it is better for the additive atom to be adjacent to the catalytic site. Then both additive atom and catalytic site can be easily attacked by the substrate at the same time, resulting in an ensemble effect.

When atoms of both the additive metal and the catalyst metal form a bimetallic nanocluster composed of the two elements, it is considered to have a variety of bimetallic structures. These have been discussed in the previous section and shown in Fig. 3.18. Here, some examples of the catalytic reactions in which bimetallic nanoclusters have higher activity than the corresponding monometallic nanoclusters are described.

### A. Hydrogenation

Hydrogenation reactions have been most extensively studied for evaluating the catalytic activity of nanoclusters. Precious monometallic (Pd, Pt, Rh) nanoclusters capped by PVP or polyvinylalcohol have high catalytic activities for hydrogenation of olefins. We applied PVP-capped Pd/Pt[90)] and Au/Pd[75)] bimetallic nanoclusters prepared by the coreduction method to the selective hydrogenation of 1,3-cyclooctadiene to cyclooctene. In both cases the bimetallic nanoclusters with a Pd content of 80 % showed the highest activity. As shown in the previous section, such bimetallic nanoclusters were found to have Pt- or Au-core/Pd-shell structures and Pt- or Au-cores are completely covered with Pd monoatomic layers. This improvement in catalytic activity can be interpreted only by a ligand effect of the core elements.

Taking the effect of particle size into consideration, the catalytic activity was normalized by the surface area calculated from the mean diameters of bimetallic nanoclusters. Since palladium has a high catalytic activity for hydrogenation of diene while platinum has a low activity, the selective existence of Pd atoms on the surface of bimetallic nanoclusters, confirmed by EXAFS and IR-CO, may account for the high catalytic activity. Fig. 3.24 shows the relationship between the Pd composition and the catalytic activity normalized for the surface Pd atom.[90)] The activity of surface Pd atoms is almost constant for Pd/Pt bimetallic nanoclusters containing between 50–95 mol% of Pd, and is larger than that of monometallic Pd nanoclusters. As the work function of Pd (5.55 eV) is slightly smaller than that of Pt (5.64 eV), the electronic interaction between the core Pt atom and the surface shell Pd atom results in the Pd atoms being deficient in electron density. This is a practical example of the ligand effect, since substrates with C=C bonds favor the electron-deficient catalytic sites. The selectivity of diene-to-monoene produced by Pd/Pt and Au/Pd bimetallic nanoclusters was almost 100%.

Ni/Pd alloy bimetallic nanoclusters prepared by the improved alcohol reduction method under high temperature were applied to the catalysis for hydrogenation of nitrobenzene[83)] and its derivatives.[91)] They were proved to be excellent catalysts with considerable air-resistance property for hydrogenation of nitrobenzene to aniline. Among the bimetallic nanoclusters with various Ni/Pd ratios, Ni/Pd(2/3) nanoclusters were found to be the most active catalyst for hydrogenation of nitrobenzene,[83)] whereas Ni/Pd(1/4) nanoclusters showed the highest activity for the cases of the derivatives of nitrobenzene.[91)] Two different kinds of

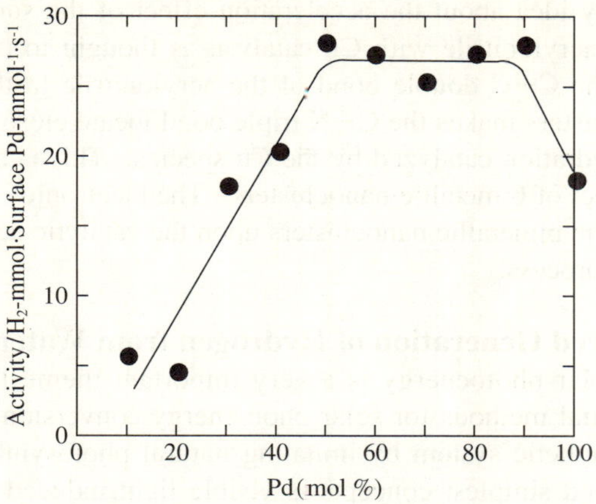


**Fig. 3.24** Relationship between the Pd composition and the catalytic activity of surface Pd atoms of core/shell-structured Pd/Pt bimetallic nanoclusters.
[Reproduced by permission of the Royal Society of Chemistry, N. Toshima, T. Yonezawa, K. Kushihashi, *J. Chem. Soc., Faraday Trans.*, **89**(14), 2542, Fig. 10, 1993.]

effects of bimetallization may be predicted for the hydrogenation catalyses: (1) The increase in Ni on the surface may decrease the amount of palladium hydride (Pd-H) on the surface and consequently decrease the activity. (2) Since a benzene ring more easily forms a charge-transfer complex with Ni than with Pd, the increase in Pd on the surface may decrease the interaction between the substrate and the particle surface, and consequently decrease the activity. Thus, coalescence of the above two effects may result in the present outcome.

## B. Hydration of Acrylonitrile

From a catalytic point of view, bimetallic nanoclusters composed of light transition metal and precious metal with specific structures will provide a great number of new candidates of catalysts for various chemical reactions, since the catalytic properties of bimetallic nanoclusters can be potentially tailored by both the ligand effect and the ensemble effect. Of particular interest are the excellent catalytic properties of Cu/Pd alloy bimetallic nanoclusters for both the selective hydration of acrylonitrile and the hydrogenation of carbon-carbon double bonds under mild conditions.[78] The rate of hydration of acrylonitrile over the PVP-capped Cu/Pd (2/1) bimetallic nanocluster catalyst is approximately seven times higher than that of the PVP-capped Cu monometallic nanoclusters. The activity increases with increasing Cu content within the range of Cu/Pd ratio from 1 to 3. The lack of cyanohydrine formation during the catalytic process suggests a nearly 100 % selectivity for the amide (Scheme 3.12).

$$CH_2=CH\text{-}CN + H_2O \longrightarrow CH_2=CH\text{-}CONH_2$$

**Scheme 3.12**

A preliminary idea about the acceleration effect of the surface Pd atoms on the hydration of acrylonitrile with Cu catalysts is thought to be as follows. The coordination of the C−C double bond of the acrylonitrile to the Pd atom in the bimetallic nanoclusters makes the C−N triple bond locate close to the Cu species, facilitating the hydration catalyzed by the Cu species. This is a good example of the ensemble effect of bimetallic nanoclusters. The electronic effect of the neighboring atoms in the bimetallic nanoclusters upon the catalytic activity may also be important in this process.

### C. Photo-induced Generation of Hydrogen from Water

Conversion of solar photoenergy is a very important theme for human beings. One of the potential methods for solar photoenergy conversion is to construct an artificial photosynthetic system by imitating natural photosynthesis. One of the most important and simplest concepts is visible light-induced decomposition of water to produce oxygen and hydrogen. Precious metal nanoclusters can work as a catalyst for visible light-induced hydrogen generation from water in a system like the EDTA/tris(bipyridine)ruthenium(II)/methyl viologen/metal nanocluster catalyst, where metal nanoclusters work as electron mediators to accept electrons from methyl viologen cation radicals and donate them to protons to produce hydrogen molecules. Until recently Pt nanoclusters was considered to be the best catalyst for this purpose. However, five monometallic (Au, Pd, Pt, Ru, and Rh) nanoclusters were investigated as electron mediators together with four core/shell structured bimetallic (Au/Pd, Au/Pt, Au/Rh, Pt/Ru) nanoclusters.[92] A linear relationship was observed between the electron transfer rate coefficients and the hydrogen generation rate coefficients. The bimetallic nanoclusters were generally more active than the corresponding monometallic nanoclusters. The highest catalytic activity was observed for Au/Rh and Pt/Ru bimetallic nanoclusters. In this reaction, the electron transfer from methyl viologen cation radicals to metal nanoclusters was found to be the rate-determining step. Electron deficiency of surface metal atoms of core/shell bimetallic nanoclusters caused by the difference in the work functions of the component metals may make it easier to transfer electrons from methyl viologen cation radicals to bimetallic nanoclusters.

## 3.3.4 Future Aspects

Polymer-capped bimetallic nanoclusters are very effective as catalysts. The combination of various metals can produce many kinds of bimetallic nanoclusters of various structures. We can now freely control the structure of bimetallic nanoclusters. Recently, we have succeeded in preparing triple core/shell structured trimetallic nanoclusters which have much higher catalytic activity than the corresponding bimetallic nanoclusters.[93] Thus, the present author believes that researches on polymer-capped bi- or tri-metallic nanoclusters are progressing rapidly and that the results will be applied to various practical chemical processes in the near future.

## 3.4 Organic-Inorganic Hybrid Materials Based on Silsesquioxanes

**Kyung-Min Kim and Yoshiki Chujo**

### 3.4.1 Introduction

Recently, nanocomposite materials combining organic and inorganic components have attracted great attention as potential candidates for use in optical devices, protective coatings, catalysis, sensors, membranes and biology from both industrial and academic groups.[94-99] A key aspect of the deep interest in hybrid nanocomposites is that they show improved mechanical, thermal, optical, or chemical properties relative to the organic or inorganic materials alone and that a numerous versatile and mild methods to prepare nanocomposites with new or enhanced properties are still being developed to satisfy their use for industrial application and new technologies. The sol-gel technology is the most popular, practical approach for preparing the diverse forms of nanocomposites such as particles, fibers, thin films and bulk materials because the reaction proceeds at ambient temperature and is performed in low viscosity solution that allows high uniformity of organic parts in the silica gel matrix.[100-103]

New preparative methods for organic-inorganic polymer hybrids and organic-inorganic polymer hybrids with responsive functions through processing by the sol-gel technique have been explored by our group.[104-114] That is, polymer nanocomposites made from building inorganic nanophases, capitalizing on the chemical functionalities of well-defined organic polymers help to control the final structures of the polymer hybrids. However, the silica gel matrix obtained has an irregular, amorphous structure and the morphology of the gel is dependent on the reaction conditions. Also, it is impossible to dissolve the silica gel in organic solvents for application. Through subsequent study for improving the weak points caused by the sol-gel reaction, we became interested in the smallest soluble particles of silica, so-called "polyhedral oligomeric silsesquioxanes (POSS)," with a well-defined hybrid structure consisting of a rigid inorganic silica-like core and eight functional organic groups bound to silicon. POSS contain various reactive or nonreactive functional groups bonded covalently to the inorganic $Si_8O_{12}$ core that make POSS easily compatible with common plastics or biological systems by means of copolymerization or blending and readily soluble in organic solvents.[115-119] The incorporation of POSS into polymeric materials can lead to dramatic improvements in polymer properties, including increases in oxygen permeability, thermoplastic or curable, oxygen resistance, surface hardening and improved thermal, mechanical properties, as well as reductions in heat evolution, flammability, viscosity.[120-123]

### 3.4.2 Organic-Inorganic Hybrid Nanocomposites Using Functionalized POSS as an Initiator

#### A. Synthesis of Hybrid Micelles Using Polyoxazolines Initiated by Monofunctional POSS[124]

New hybrid polyoxazolines were prepared by ring-opening polymerization of 2-methyl-2-oxazoline initiated by new monofunctional POSS having only one iodopropyl group as an initiator with various feed ratios. The hybrid micelles derived from hybrid polyoxazolines having both seven cyclopentyl groups of POSS as a hydrophobic segment and polyoxazoline as a hydrophilic segment were prepared in an aqueous phase (Scheme 3.13). Compared to polyoxazoline initiated by cyclohexyl iodide or methyl *p*-toluenesulfonate, increased hydrophobicity caused by the structure of POSS in hybrid polyoxazolines played a major role in the micellar formation. The micellar formation of hybrid polyoxazolines was dependent on the amount of POSS incorporated into POZO. This is a most interesting and unique method for preparing micelles using enhanced hydrophobicity caused by the structure of POSS even though POSS is very little incorporated into the hydrophilic polymer chain as an initiator. It is generally impossible to obtain homogeneous polymer hybrids using hydrophilic polyoxazolines and hydrophobic POSS in the silica gel matrix. On the other hand, hydrophobic POSS having seven cyclopentyl groups bonded covalently to POZO could be homogeneously dispersed in the silica gel matrix at the nanometer level through hydrogen bonding interaction between hybrid polyoxazolines and silica gel. The polymer hybrids obtained from hybrid polyoxazolines showed excellent solvent-resistant property, which can match that of polymer hybrids having IPN (interpenetrating polymer network) structure (Table 3.2). It was concluded that the solvent-resistant property came from the size of POSS and the hydrophobic interaction between POSS themselves in the silica gel matrix. In addition, thermal stability of the polymer hybrids from hybrid polyoxazolines was much increased compared to that of the polymer hybrids from polyoxazolines and POSS.

#### B. Synthesis of Star-shaped Polyoxazolines Using Octafunctional POSS Initiator[125]

Hybrid star-shaped polyoxazolines having a POSS core were prepared by ring-opening polymerization of 2-methyl-2-oxazoline using various octafunctional POSS as an initiator by changing the feed ratio of POSS to 2-methyl-2-oxazoline (Scheme 3.14). The core-first method, which uses an active multifunctional core to initiate growth of polymer chains, was applicable to making hybrid POSS-core star-shaped polyoxazolines. From the results of $^1$H NMR and GPC, it was concluded that star-shaped polyoxazolines having at most four or five different

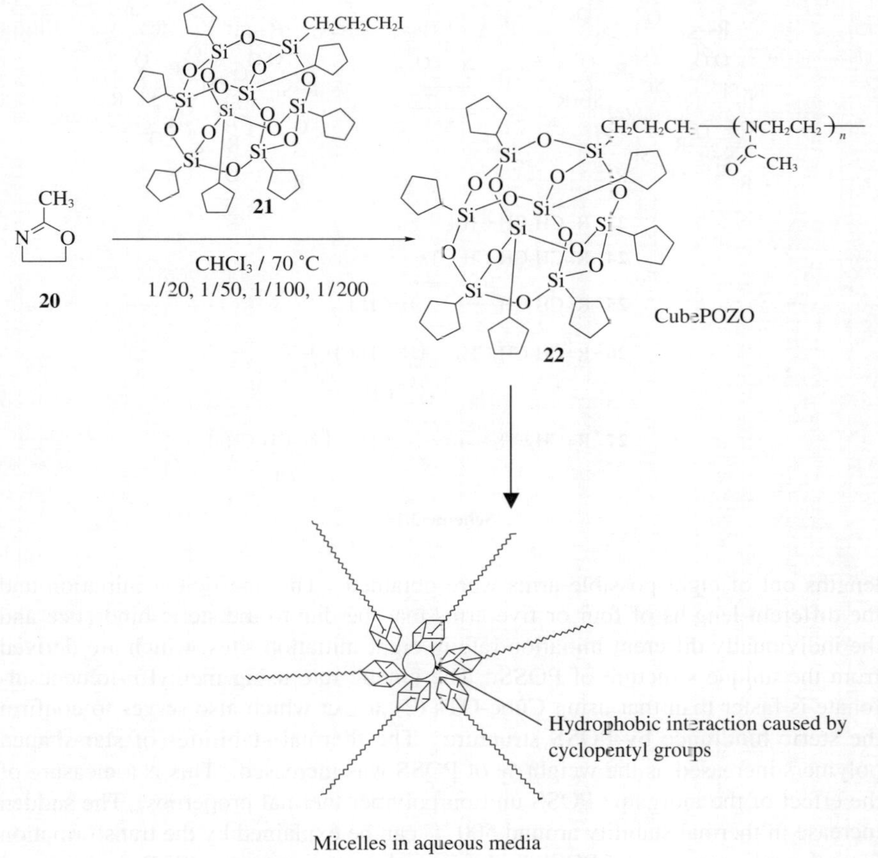

**Scheme 3.13**

**Table 3.2** Solvent-resistant Property of Polymer Hybrids from CubePOZO[a]

| run | solvent | solvent extraction efficiency (%) | polymer content (wt %)[c] $P_a$ | $P_b$ |
|---|---|---|---|---|
| 1 | methanol | 3.5 | 33.0 | 33.8 |
| 2 | chloroform | 5.7 | 32.5 | 33.8 |
| 3[b] (homoPOZO) | methanol | 70.7 | 17.4 | 41.8 |

[a] The polymer content in polymer hybrids before ($P_b$) and after ($P_a$), solvent extraction efficiency $E$ was calculated according to the equation $E = 100(P_b - P_a)/P_b(100 - P_a)$.

[b] Polymer hybrids from homoPOZO and TMOS.

[c] Measured by TGA.

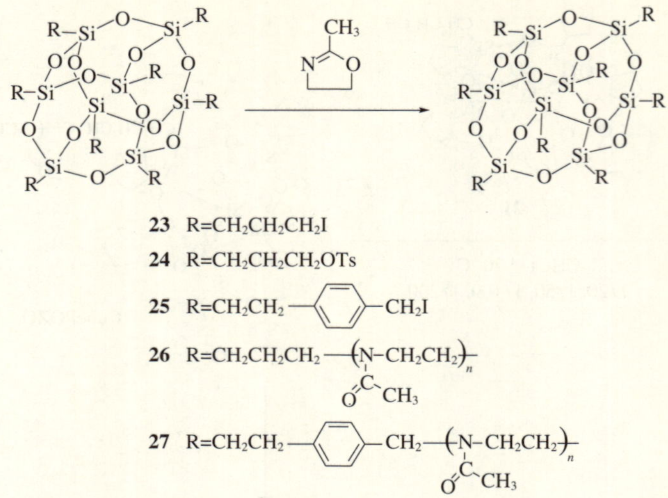


**Scheme 3.14**

lengths out of eight possible arms were obtained. This inefficient initiation and the different lengths of four or five arms may be due to the steric hindrance and the individually different initiation rate in eight initiation sites, which are derived from the unique structure of POSS. The kinetic rate using methyl *p*-toluenesulfonate is faster than that using Cube-OTs (2), a fact which also serves to confirm the steric hindrance by POSS structure. The thermal stabilities of star-shaped polymers increased as the weight % of POSS was increased. This is a measure of the effect of the inorganic POSS unit on polymer thermal properties. The sudden increase in thermal stability around 500 °C can be explained by the transformation from the cube structure of POSS to the amorphous silica matrix ($Si_8O_{16}$).

## 3.4.3 Organic-Inorganic Hybrid Polymers

### A. Synthesis of Hybrid Gels Having Functionalized POSS[126)]

Hybrid terpolymers consisting of POSS, *N,N*-dimethylacrylamide and bipyridine monomer were synthesized by common radical polymerization. The new hybrid gels having POSS were prepared through the coordination of various metal ions to 2,2'-bipyridine-modified hybrid terpolymers (Scheme 3.15). The highly concentrated solution gave the hybrid gel by iron (II) sulfate or ruthenium (III) chloride in good yield. On the other hand, no hybrid gel by nickel (II) chloride was formed even in much more concentrated solution due to the rapid ligand exchange reaction of the cross-linking points from intermolecular to intramolecular. The degree of swelling and the thermal stability of the gel in various solvents were dependent on the content of the POSS moiety in the hybrid gel, and the ruthenium gel was considerably stable compared with iron and nickel gels. The hybrid gel having

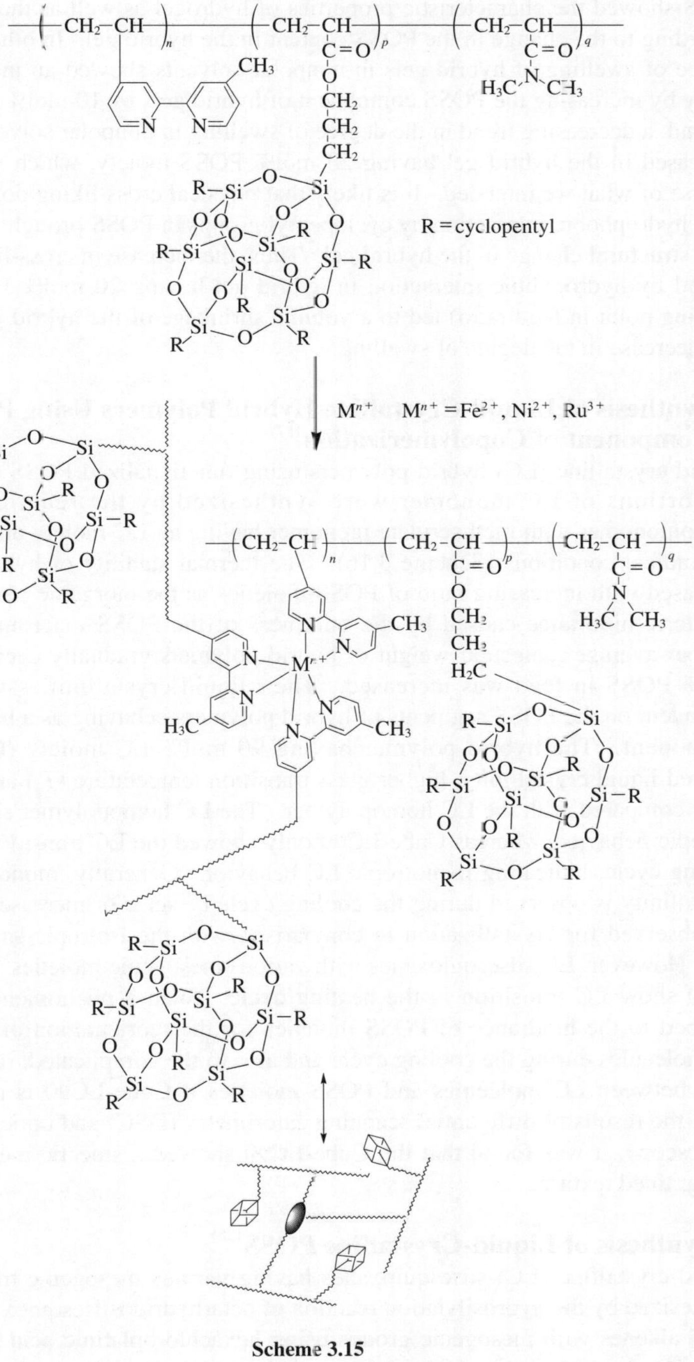


**Scheme 3.15**

POSS showed the characteristic properties of hydrogel as well as those of lipogel according to the change in the POSS content in the hybrid gel. In other words, the degree of swelling of hybrid gels in nonpolar solvents showed an increasing tendency by increasing the POSS component of hybrid gels by 10 mol%. On the other hand, a decreasing trend in the degree of swelling in nonpolar solvents suddenly decreased in the hybrid gel having 20 mol% POSS moiety, which was the very reverse of what we intended. It is likely that physical cross-liking points resulting from hydrophobic interaction by cyclopentyl groups in POSS brought about a possible structural change in the hybrid gel. Thus, the increase in cross-linking points caused by hydrophobic interaction in hybrid gel having 20 mol% POSS moiety (turning point in feed ratio) led to a volume shrinkage of the hybrid gel, resulting in a decrease in the degree of swelling.

### B. Synthesis of Liquid-Crystalline Hybrid Polymers Using POSS as a Component of Copolymerization[127)]

Liquid crystalline (LC) hybrid polymers using functionalized POSS with varying proportions of LC monomer were synthesized by the reaction of POSS macromonomer with methacrylate monomer having an LC moiety under common free-radical conditions (Scheme 3.16). The thermal stability of hybrid polymers increased with increasing ratio of POSS moieties as the inorganic part. Owing to the steric hindrance caused by the bulkiness of the POSS macromonomer, the number-average molecular weight of hybrid polymers gradually decreased as the mol% POSS in feed was increased. Their liquid crystallinities were closely dependent on the POSS segments of hybrid polymers behaving as a hard, compact component. The hybrid polymer having 90 mol% LC moiety (Cube-LC90) showed liquid crystallinity, higher glass transition temperature ($T_g$) and better stability compared with the LC homopolymer. The LC homopolymer showed enantiotropic behavior, whereas Cube-LC90 only showed the LC transition during the heating cycle, indicating monotropic LC behavior. Generally, monotropic liquid crystallinity is observed during the cooling cycle because of increased supercooling observed for crystallization in comparison with the isotropic-smectic transition. However, LC silsesquioxanes with various mesogenic moieties[128)] and Cube-LC90 show LC transition in the heating cycle. Such a phenomenon might be ascribed to the hindrance of POSS moieties for the reorientation of randomized LC molecules during the cooling cycle and also to the complicated, random structures between LC molecules and POSS moieties if Cube-LC90 is melted once. From the results of differential scanning calorimetry (DSC) and optical polarizing microscopy, it was found that the Cube-LC90 showed a smectic mesophase-like fine-grained texture.

### C. Synthesis of Liquid-Crystalline POSS[128)]

Liquid-crystalline (LC) silsesquioxanes having various mesogenic moieties were synthesized by the hydrosilylation reaction of octa(hydridosilsesquioxane) and terminal alkenes with mesogenic groups using hexachloroplatinic acid as a catalyst

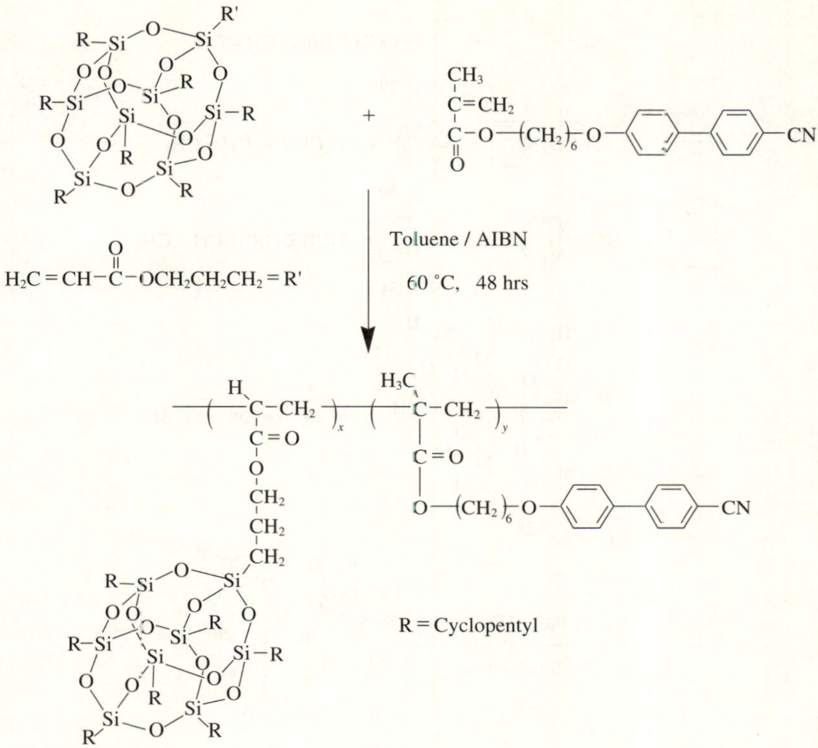


**Scheme 3.16**

(Scheme 3.17). Mesogenic compounds and LC-Cubes showed that the monotropic behaviors in LC phase appeared as a nematic phase on the cooling cycle and heating cycle, respectively. Their liquid crystallinities of nematic phase were found to be much more stable than those of the corresponding mesogenic compounds. In other words, the LC temperature ranges of LC-Cubes became wider than those of the corresponding mesogenic compounds (Table 3.3). This may be ascribed to the increase in local concentration of LC moieties as the mesogenic compounds were introduced to the eight hydro groups of silsesquioxane. The increase in local concentration caused by compact, fixed directionality of silsesquioxane makes it easier to reorient mesogenic moieties. It was also observed that the thermal stabilities of LC silsesquioxanes improved after the hydrosilylation reaction because of the rigid, hard inorganic core of silsesquioxane.

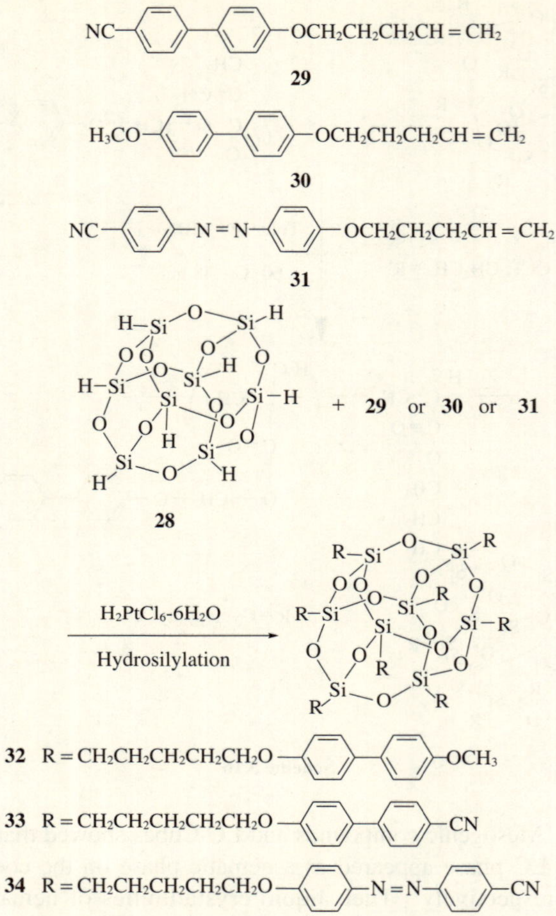

**Scheme 3.17**

**Table 3.3** Phase Transition Temperatures of Mesogenic Compounds and LC-Cubes

| Samples | phase transition temperature(°C) | |
|---|---|---|
| | thermal transition[a] | $T_m$[b] |
| Vinyl-$OCH_3$ | i 109 n 102 k | 123 |
| Vinyl-CN | i 65 n 48 k | 87 |
| Vinyl-AZO | i 113 n 101 k | 124 |
| Cube-$OCH_3$ | k 166 n 215 i | 215 |
| Cube-CN | k 70 n 139 i | 139 |
| Cube-AZO | k 140 n 173 i | 173 |

[a] Observed by optical polarizing microscope.

[b] Melting temperature was detected by DSC thermogram.

## 3.4.4 Transparent Organic-Inorganic Polymer Hybrids with Functionalized POSS

### A. Synthesis of Polymer Hybrids Using POSS with Hydroxyl Groups[129,130)]

Polymer hybrids were prepared using three kinds of POSS having hydroxyl groups (Cube-3DOH, Cube-3OH, or Cube-6OH) and organic polymers such as poly(2-methyl-2-oxazoline) (POZO), poly(*N*-vinylpyrrolidone) (PVP) and poly(*N*,*N*-dimethylacrylamide) (PDMAAm) using hydrogen bonding interaction (Scheme 3.18). In the case of polymer hybrids from Cube-3OH and Cube-6OH, the transparent and homogeneous polymer hybrids could be obtained only with a higher polymer ratio (10/1), and a lower polymer ratio (1/1) brought about phase separation. High homogeneity was achieved in a wide range of the ratio of polymer-to-POSS (10/1–1/1) with various organic polymers, whose result was different from that of polymer hybrids obtained with Cube-3OH or Cube-6OH. This result may be explained by the flexibility of the eight arms of POSS derived from dimethylsiloxy groups enhancing the hydrogen bonding ability of POSS toward organic polymers for the homogeneity of the polymer hybrids. The polymer hybrids obtained from polystyrene and Cube-3DOH were turbid, demonstrating the effectiveness of hydrogen bonding interaction between polymer and POSS. Also, FT-IR measurement of these polymer hybrids was conducted to confirm the

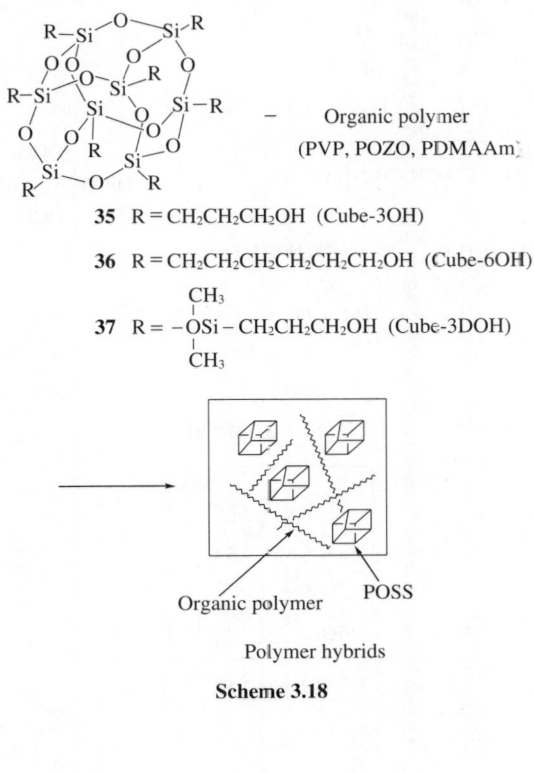


**Scheme 3.18**

shift of amide carbonyl peaks caused by hydrogen bonding interaction between polymer and POSS. The homogeneous and transparent hybrid films obtained could be dissolved in solvents and cast again without any separation.

### B. Synthesis of Ternary Polymer Hybrids of POSS and Organic Polymers Utilizing the Sol-Gel Reaction[131)]

Ternary polymer hybrids were prepared by organic polymers such as poly(2-methyl-2-oxazoline) (POZO) or poly(*N*-vinylpyrrolidone) (PVP) and aminopropylsilsesquioxane (Cube-aminopropyl) utilizing the sol-gel reaction of tetramethoxysilane (TMOS) (Scheme 3.19). Polymer hybrids using POZO showed high transparency and homogeneity without any influence of Cube-aminopropyl ratio to POZO. On the other hand, in the case of polymer hybrids from PVP, higher Cube-aminopropyl ratio to PVP brought about phase separation. From FT-IR measurement, the intensity of the hydrogen bonding interaction between PVP and residual silanol groups of silica gel was much stronger than that between PVP and Cube-aminopropyl. This unbalanced hydrogen bonding interactions resuled in single-phase polymer hybrids of PVP and silica gel causing the aggregation of Cube-aminopropyl having weak hydrogen bonding strength in ternary systems. Thus, the homogeneity of ternary polymer hybrids was found to be closely depen-

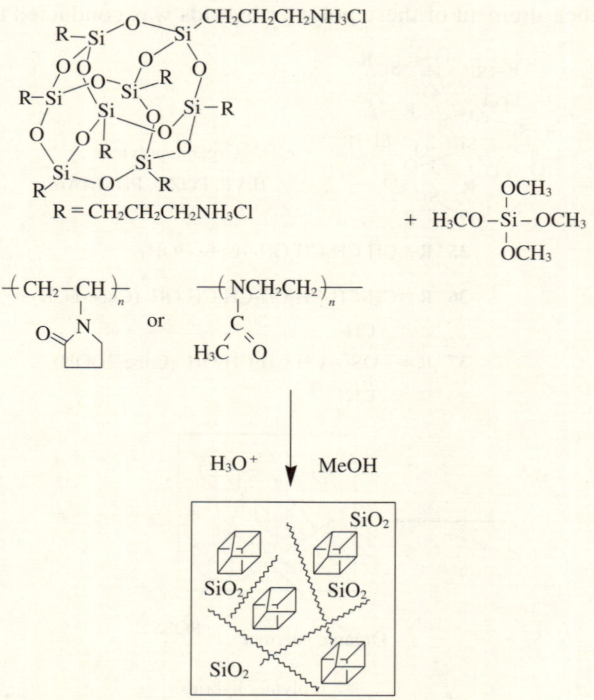


**Scheme 3.19**

dent on the difference between the strength of the hydrogen bonding interaction of polymer and residual silanol groups of silica gel and the strength of the hydrogen bonding interaction of polymer and Cube-aminopropyl. The ternary polymer hybrids from POZO obtained by balanced hydrogen bonding interactions have higher solvent-resistance property compared to those from PVP. The ternary polymer hybrids were also found to have high thermal stability due to the rigid silica matrix as well as the hard and compact inorganic silica-like core of POSS.

### C. Synthesis of Polymer Hybrids Using Two Physical Bonding Interactions in One System[132)]

Homogeneous organic-inorganic polymer hybrids could be prepared by using physical interactions such as hydrogen bonding, aromatic (π-π) and ionic interactions between organic polymer and silica gel. On the other hand, it is difficult to make transparent polymer hybrids using two physical interactions such as hydro-

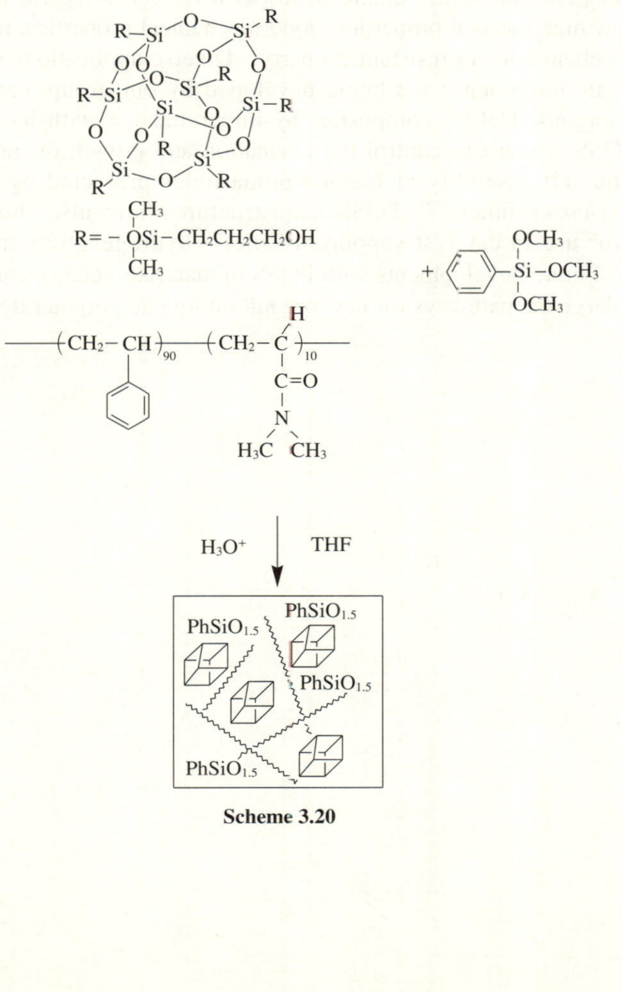

**Scheme 3.20**

gen bonding and aromatic (π-π) interactions at the same time in one system. Transparent polymer hybrids were obtained from the copolymer having an amide group for hydrogen bonding interaction with POSS and a phenyl group for aromatic (π-π) interaction with silica gel (Scheme 3.20). When homopolystyrene and homopolyacrylamide were used instead of a copolymer, phase separation was turbid. FT-IR measurement of these polymer hybrids was conducted to confirm the shift of amide carbonyl peaks caused by the hydrogen bonding interaction between copolymer and POSS. Evidence of aromatic (π-π) interaction in this system was made clear by making turbid polymer hybrids without PhTMOS. In the powder X-ray diffraction (XRD) analysis, the polymer hybrids utilizing two physical interactions in one system showed the halo peak ($SiO_2$ matrix), indicating that POSS and copolymer were homogeneously dispersed in silica matrix.

### 3.4.5 Conclusions and Future Perspectives

The use of POSS compounds provides many new synthetic possibilities for various hybrid materials with enhanced properties. These hybrid materials using POSS show high thermal properties, good mechanical properties, high transparency and excellent solvent-resistant property. Diverse applications of POSS to the field of materials science are being developed by our group, e.g., hydrophobic organic-inorganic $CaCO_3$ composites by mineralization with incompletely condensed POSS which can control the nucleation and growth of inorganic crystallites[133)] and self-assembly of metal nanoparticles protected by functionalized POSS as a cross-linker.[134)] POSS nanostructures have also shown significant promise for use in catalyst supports of olefin hydrogenation and polymerization.[135-137)] Future developments with POSS in materials science can potentially be used to enlarge the pathways for new organic-inorganic polymer hybrids.

# 3.5 Protonation-induced Intramolecular Electron Transfer in the Ferrocene-Quinone Conjugated System

**Hiroshi Nishihara and Masaki Murata**

## 3.5.1 Introduction

Intelligent molecules, the structures and properties of which are facilely and reversibly changeable by application of external physical and chemical stimuli, have attracted much recent attention relative to the development of new molecule-based devices.[138)] One category of such molecules comprises transition metal complexes with π-conjugated spacers; the combination of the flexibility of d- and π-orbitals can yield unique optical, magnetic and electronic properties. Because the ferrocenyl moiety works as a good donor, conjugated ferrocene systems constructed by combining ferrocene with electron-accepting molecular units have attracted attention for their fluctuating electronic structures,[139-142)] which are sensitive to outer fields, and exhibit novel behavior resulting from their second-order nonlinear optical (NLO) properties[143,144)] and facile intramolecular electron transfer, which causes dramatic changes in their structure and physical properties. Quinonoid compounds are one of the most versatile electron-accepting and proton-responsive organic molecules.[145)] They are employed in covalently and non-covalently linked donor-acceptor systems of particular interest in the study of proton-coupled electron transfer[146-148)] and photo-induced electron transfer.[149-152)] Thus, the quinonoid compound linked to ferrocene by a π-conjugated spacer model provides insight into proton-coupled intramolecular electron transfer, because the redox potentials of the two molecules are similar and the redox potential of the quinonoid molecule can be readily controlled by external perturbation such as pro-

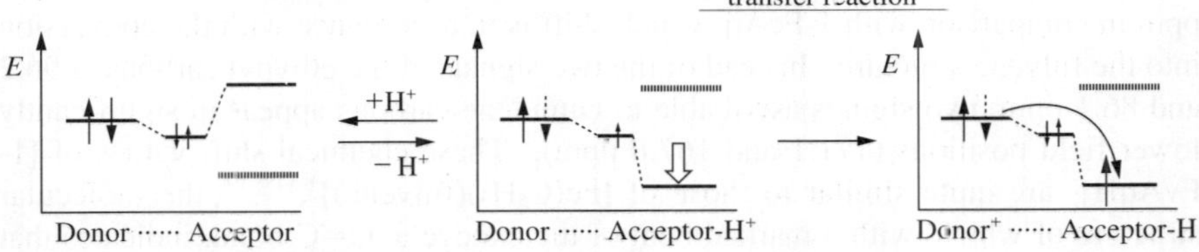


**Fig. 3.25** Concepts of protonation-induced intramolecular electron transfer.

tonation and solvation (Fig. 3.25).[153,154] In this section, ethynylene-bridged anthraquinone-ferrocene complexes from monomer to polymer are described. The complexes exhibit novel structural changes induced by proton-coupled intramolecular reactions.

### 3.5.2 1 : 1-D/A Complexes

We start here to describe the protonation behavior of ferrocenyethynylanthraquinones from the simplest D/A 1 : 1 complexes, 1-ferrocenylethynylanthraquinone (1-FcAq)[155] and 2-ferrocenylethynylanthraquinone (2-FcAq).[156] There are two carbonyl groups in each compound and only one of them is conjugated to the ferrocenyl group as shown in Scheme 3.21. This conjugation affords a specific absorption band at 510 nm attributable to the MLCT transition from the Fe(II) to the $\pi^*$-orbital of the ethynylanthraquinone moiety, the molar extinction coefficient $\varepsilon$ of which is larger for 2-FcAq ($2.5 \times 10^3$ $mol^{-1}$ $dm^3$ $cm^{-1}$) than 1-FcAq ($1.7 \times 10^3$ $mol^{-1}$ $dm^3$ $cm^{-1}$). The difference between the two isomers also appears in the X-ray crystallographic structures of Fig. 3.26, showing that the cyclopentadienyl rings of the ferrocenyl moiety are perpendicular and parallel to the plane of the anthraquinone moiety for 1-FcAq and 2-FcAq, respectively.

The protonation behavior of 1-FcAq and 2-FcAq in a benzonitrile solution supplemented with $CF_3SO_3H$ is also entirely different. The color of the 1-FcAq solution changed immediately from red to deep reddish-pink after the addition of the stoichiometric amount of $CF_3SO_3H$. This corresponds to a spectral change in which the n-$\pi^*$ band decreased and the band at 510 nm increased in intensity, and in which a new broad band with the half-width $\Delta\nu_{1/2} = 5.2 \times 10^3$ $cm^{-1}$ appeared over the visible and near-IR regions ($\lambda_{max}$ = 939 nm) (Fig. 3.27). Judging from the NMR spectra as described below, the protonated complex of 1-FcAq is a diamagnetic complex in benzonitrile with a unique structure involving a fulvene complex (Fv) moiety, $[1\text{-FvAqH}]^+CF_3SO_3^-$. The band at $\lambda_{max}$ = 939 nm can be ascribed to the valence tautomerization of $[1\text{-FvAqH}]^+$ into a spin-separated form containing a ferrocenium ($Fc^+$) moiety, $[1\text{-Fc}^{\bullet}\text{AqH}^{\bullet}]^+$ (see Scheme 3.21), given that such a broad low-energy band has been recognized in other valence tautomerization complexes.[157,158] This consideration is supported by the fact that the solvent effect on $\lambda_{max}$ is large; $\lambda_{max}$ = 1051 nm in dichloromethane, which is less polar than benzonitrile.

Notable features in the $^1H$ NMR spectrum of $[1\text{-FvAqH}]^+CF_3SO_3^-$ in $CD_3CN$ include a new singlet due to the hydroxyl group appearing at $\delta$ 8.23 ppm, and two double-doublets, one at 4.67 and one at 4.33 ppm, due to the cyclopentadienyl (Cp) ring connecting to the ethynyl group of 1-FcAq, shift to the lower field by 1 ppm in comparison with 1-FcAq, which shift is in accordance with the conversion into the fulvene structure. Instead of the two signals of the ethynyl carbons at 96.2 and 86.1 ppm, two signals ascribable to cumulene carbons appear in significantly lower field positions (171.1 and 167.8 ppm). These chemical shift values of $[1\text{-FvAqH}]^+$ are quite similar to those of $[Fe(C_5H_5)(\text{fulvene})]^+$,[159-162] the molecular structure of which, with a marked bent of the exocyclic C=C bond, indicates that the fulvene unit of $[Fe(C_5H_5)(\text{fulvene})]^+$ is characterized by both $\eta^4$- and $\eta^6$-coor-

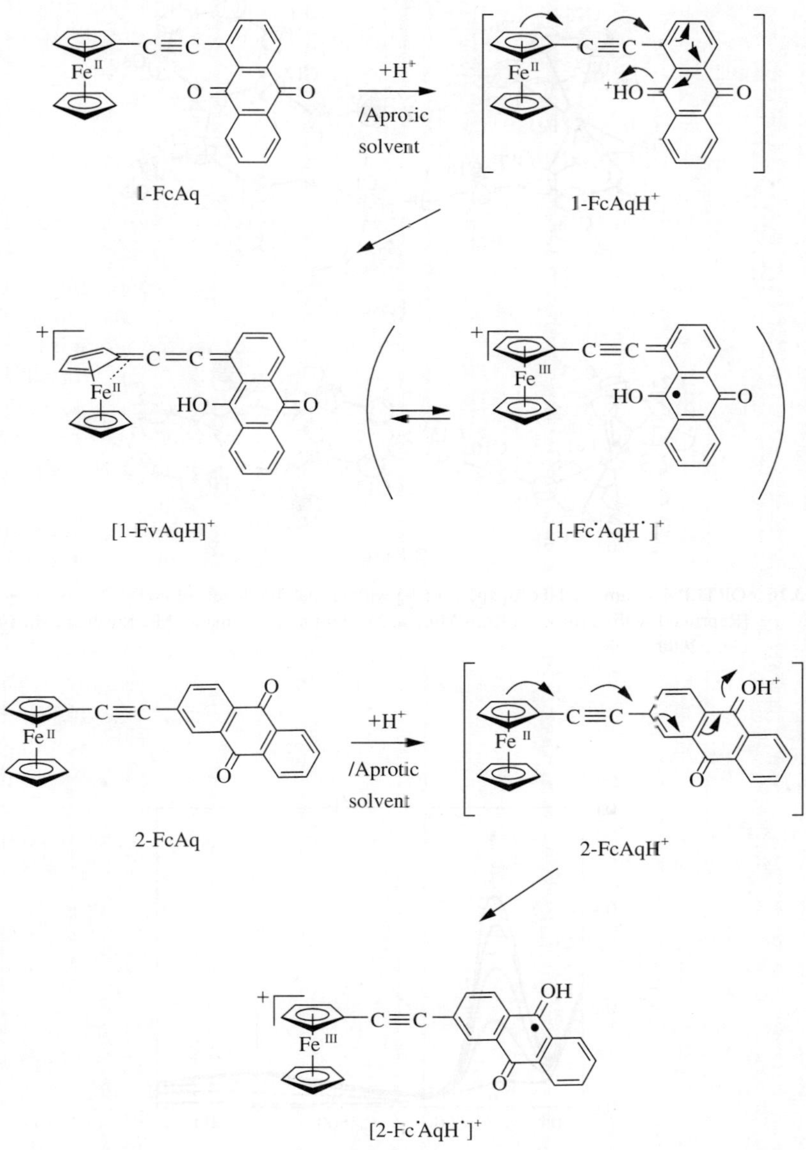


Protonation reaction pathway of 1-FcAq and 2-FcAq.

**Scheme 3.21**

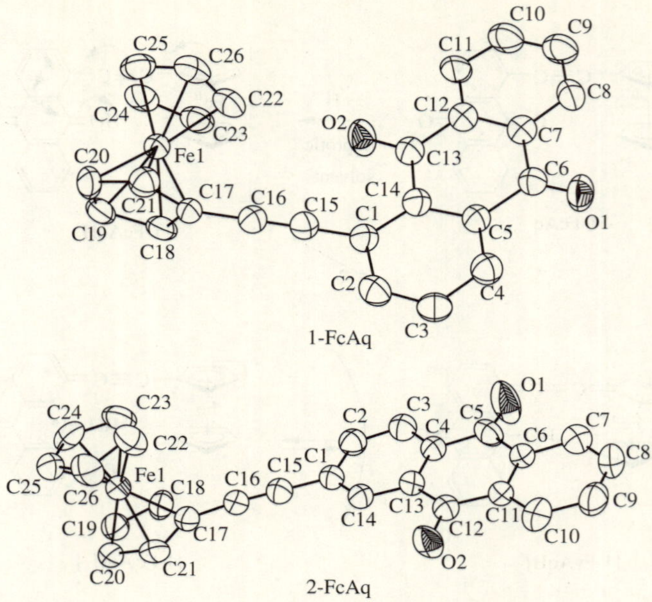


**Fig. 3.26** ORTEP diagrams of 1-FcAq and 2-FcAq with probability levels of 50 %.
[Reprinted with permission from Murata, M., Fujita, T., Yamada, M., Kurihara, H., *Chem. Lett.*, **2000**, 1328.]

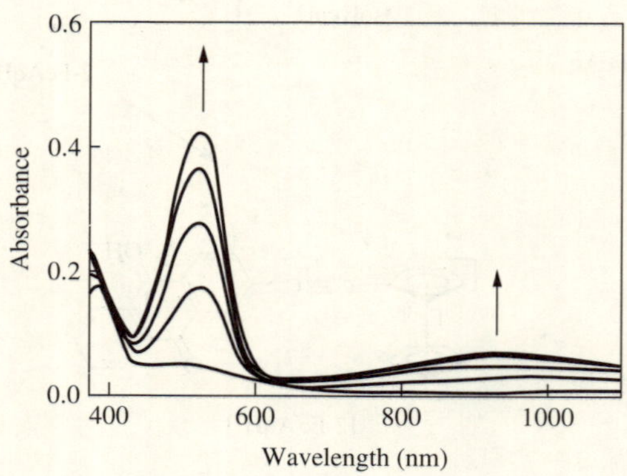


**Fig. 3.27** UV-vis-near-IR absorption spectral change of 1-FcAq ($3.0 \times 10^{-4}$ mol $dm^{-3}$) in benzonitrile upon stepwise addition of 0 – 2 equivalents of $CF_3SO_3H$.
[Reprinted with permission from Murata, M., Fujita, T., Yamada, M., Kurihara, H., *Chem. Lett.*, **2000**, 1328.]

**Table 3.4** Redox Potentials of Ferrocene-Anthraquinone Complexes and Their Protonated Species in 0.1 mol $dm^{-3}$ $Bu_4NClO_4$-$CH_2Cl_2$[a]

| | $E^{0'}_{red}$ / V *vs* $Fc^+/Fc$ | | | | $E^{0'}_{ox}$ / V *vs* $Fc^+/Fc$ | | | |
|---|---|---|---|---|---|---|---|---|
| 1-FcAq | −1.33 (Aq) | −1.39 (Aq) | | | | 0.11 (Fc) | | |
| $[1\text{-FvAqH}]^+$ | | −1.05 (AqH) | −0.31 (AqH) | | | | 0.30 (Fv) | |
| 1,8-$Fc_2Aq$ | −1.75 (Aq) | −1.33 (Aq) | | | 0.07 (Fc) | 0.14 (Fc) | | |
| $[1,8\text{-FcFvAqH}]^+$ | | −1.01 (AqH) | −0.31 (AqH) | | | 0.19 (Fc) | 0.30 (Fv) | |
| 1,5-$Fc_2Aq$ | −1.78 (Aq) | −1.39 (Aq) | | | | 0.11 ($2e^-$, Fc) | | |
| $[1,5\text{-FcFvAqH}]^+$ | | −1.07 (AqH) | −0.32 (AqH) | | | 0.11 (Fc) | 0.27 (Fv) | |
| $[1,5\text{-FcFvAqH}_2]^{2+}$ | | | −0.16 ($AqH_2$) | 0.11 ($Fc^+$) | | | 0.27(Fv) | 0.42($AqH_2$) |
| 1,1'-$FcAq_2$ | −1.85 ($2e^-$, Aq) | −1.39 ($2e^-$, Aq) | | | | 0.21 (Fc) | | |
| $[1,1'\text{-FvAq}_2\text{H}]^+$ | −1.35 (Aq) | −1.39 ($2e^-$, Aq& AqH) | −0.32 (AqH) | | | | 0.43 (Fv) | |
| $[1,1'\text{-FvAq}_2\text{H}_2]^{2+}$ | | −1.39 ($2e^-$, AqH×2) | −0.14 (AqH) | −0.06 (AqH) | | | 0.43 (Fv) | |

[a] The number of electrons for the reaction is one except otherwise stated in parentheses.

dination.[163] This implies a significant contribution of the 18 $e^-$ form rather than the 16 $e^-$ form. The IR spectrum of $[1\text{-FvAqH}]^+$ shows the disappearance of the ν(C≡C) band of 1-FcAq, and the molecular ion peak of $[1\text{-FvAqH}]^+$(*m/z* 417) detected in the ESI-mass spectrum indicates the addition of one hydrogen atom to the composition of 1-FcAq (*m/z* 416).

The deprotonated form, 1-FcAq, shows reversible two-step 1 $e^-$ reduction at $E^{0'} = -1.26$ and $-1.71$ V versus ferrocenium/ferrocene ($Fc^+$/Fc) derived from the anthraquinone moiety, and reversible 1 $e^-$ oxidation at $E^{0'} = 0.22$ V due to the ferrocenyl moiety in $Bu_4NClO_4$-$CH_2Cl_2$ (Table 3.4). The first reduction potential shifts dramatically in the positive direction to $E^{0'} = -0.06$ V, and the oxidation potential shifts moderately in the positive direction to $E^{0'} = 0.33$ V in the protonation product, $[1\text{-FvAqH}]^+$, whereas the second reduction potential is little changed. These results correspond to the structural changes in both ferrocenyl and anthraquinone moieties by protonation.

Contrary to the facile change of 1-FcAq, the deep-red solution of 2-FcAq did not change significantly after the addition of the stoichiometric amount of $CF_3SO_3H$. The color of the solution changed gradually to deep-green by the addition of an excess amount (50 equivalents) of $CF_3SO_3H$. The n-$\pi^*$ and the MLCT bands decreased in intensity, and two new bands appeared at 414 and 615 nm. The intense band at 414 nm is appreciably consistent with that of the semiquinone form of anthraquinone derivatives.[164] The $^1$H NMR spectrum of a $CD_2Cl_2$ solution of 2-FcAq indicated that its complex lost its diamagnetic nature by the protonation. The EPR spectrum of 2-FcAq in benzonitrile on addition of 6 equiv of $CF_3SO_3H$ at 6.5 K showed well-resolved signals interpreted as the superposition of a ferrocenium cation including paramagnetic low-spin Fe(III) nuclei ($g_{//} = 4.26$, $g_\perp = 1.49$) and a semiquinone radical of the anthraquinone moiety ($g = 2.00$). The latter signal did not exhibit a hyperfine structure due to the coupling between the electron spin and the protons of the anthrasemiquinone moiety, indicating a slight fluctuation of the electron spin state. From the results of the UV-Vis absorption and the EPR spectra, it is postulated that 2-FcAq is changed into a "spin-separated" ferrocenium-semiquinone combination linked by an ethynyl group [2-Fc$^{\bullet}$AqH$^{\bullet}$]$^+$ through protonation (Scheme 3.21).

### 3.5.3 2 : 1-D/A Complexes

The next simplest complexes are 2 : 1-D/A compounds, 1,8-bis(ferrocenylethynyl)-anthraquinone, 1,8-$Fc_2$Aq, and 1,5-bis(ferrocenylethynyl) anthraquinone, 1,5-$Fc_2$Aq.[165] Comparison of protonation behavior between them is intriguing because the quinone moiety can undergo double protonation, drastically changing the acceptor level and because there is a significant difference in conjugation of the ferrocenyl groups with the carbonyl groups of the anthraquinone moieties between the isomers. Two ferrocenyl groups in 1,5-$Fc_2$Aq conjugate with different carbonyl moieties, whereas those in 1,8-$Fc_2$Aq conjugate with the same carbonyl moiety (see Scheme 3.22). This actually influences the

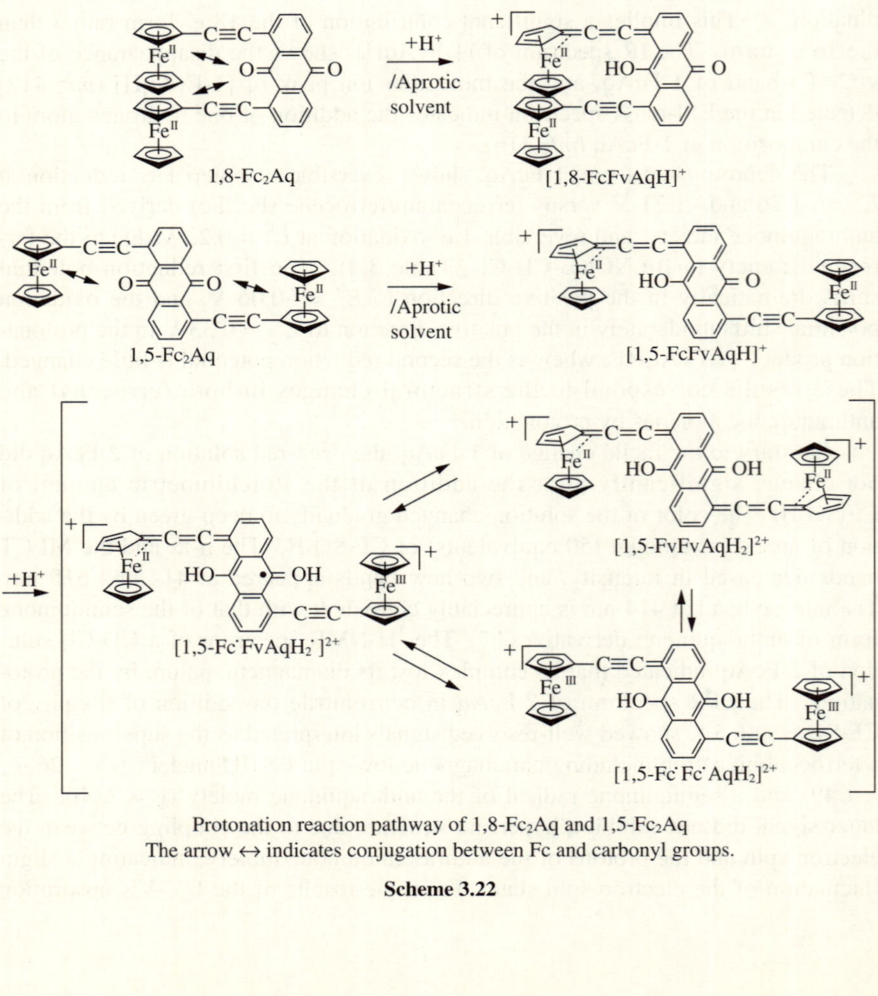


Protonation reaction pathway of 1,8-$Fc_2$Aq and 1,5-$Fc_2$Aq
The arrow ↔ indicates conjugation between Fc and carbonyl groups.

**Scheme 3.22**

protonation behavior of the isomers, i.e., only 1,5-$Fc_2Aq$ undergoes a facile two-step proton response, leading to a doubly protonated species with unique physical properties due to the existence of three possible valence tautomers, whereas 1,8-$Fc_2Aq$ does only single protonation as described below.

The X-ray crystallography of both 1,8-$Fc_2Aq$ and 1,5-$Fc_2Aq$ indicates that the Cp rings and the anthraquinone plane are nearly perpendicular to each other as displayed in Fig. 3.28. The spectroscopic responses of 1,8-$Fc_2Aq$ and 1,5-$Fc_2Aq$ to $CF_3SO_3H$ in benzonitrile are significantly different. The solution of 1,8-$Fc_2Aq$ changed immediately from deep red to deep reddish-purple after the addition of the acid, corresponding to the UV-vis-near-IR absorption spectral change. The n-$\pi^*$ band of the anthraquinone moiety decreases, two visible bands increase in intensity at 491 and 546 nm, and a new broad band with the half-width $\Delta\nu_{1/2}$ = 5.4

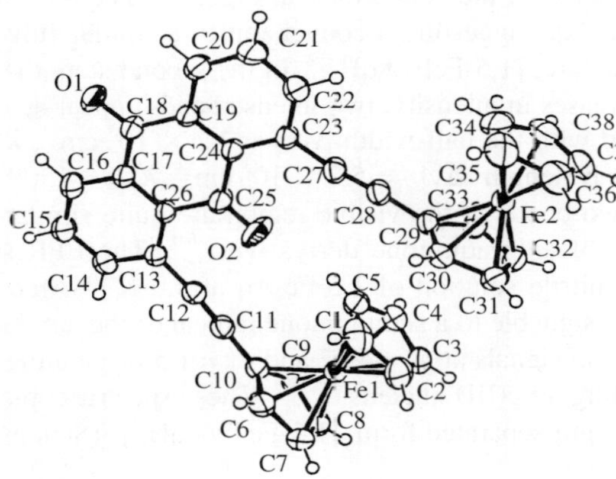


1,8-$Fc_2Aq$

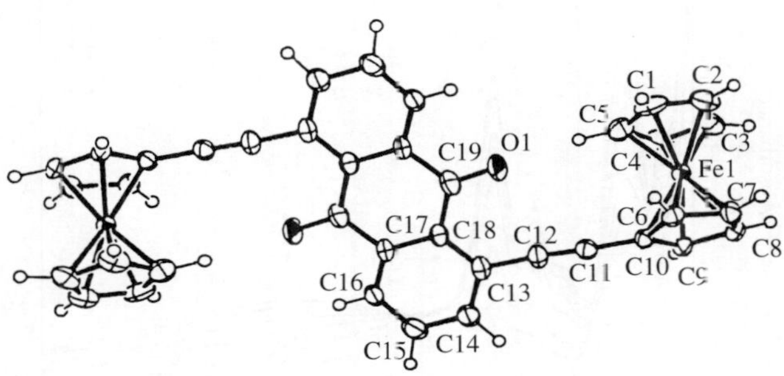


1,5-$Fc_2Aq$

**Fig. 3.28** ORTEP diagrams of 1,8-$Fc_2Aq$ and 1,5-$Fc_2Aq$ with probability levels of 50 %. [Reprinted with permission from Murata, M., Yamada, M., Fujita, T., Kojima, K., Kurihara, M., Kudo, K., Kobayashi, Y., Nishihara H., *J. Am. Chem. Soc.*, 2001, **123**, 12903.]

$\times 10^3$ $cm^{-1}$ appears over the visible and near-IR regions ($\lambda_{max}$ = 934 nm). This spectral change is quite similar to that of 1-FcAq.[155] The protonated species of 1,8-$Fc_2$Aq was isolated from the acidic solution, and its ESI-mass, $^1$H NMR, $^{13}$C NMR and IR spectra have revealed that 1,8-$Fc_2$Aq undergoes a single protonation of the carbonyl oxygen adjacent to the ethynylene bond, causing a structural change to create a compound involving both ferrocenyl-ethynyl and $\eta^6$-fulvene-cumulene moieties, $[1,8\text{-FcFvAqH}]^+$ (Scheme 3.22).

In contrast, the addition of several equivalents of $CF_3SO_3H$ into a benzonitrile solution of 1,5-$Fc_2$Aq resulted in two-step color changes, initially to deep reddish-purple, and then to deep green. ESI-mass spectra have indicated that deep reddish-purple and deep green solutions are derived from singly and doubly protonated species, respectively. The first spectral change in the UV-Vis-near-IR absorption spectra, as depicted by the dotted lines in Fig. 3.29, is quite similar to those of 1-FcAq and 1,8-$Fc_2$Aq, suggesting a conversion to a similar fulvene-Fe(II) complex-cumulene structure, $[1,5\text{-FcFvAqH}]^+$. In the second step, the MLCT band at $\lambda_{max}$ = 522 nm decreases in intensity, two intense bands appear at 748 and 850 nm, and the broad band with the half-width $\Delta\nu_{1/2} = 7.2 \times 10^3$ $cm^{-1}$, $\lambda_{max}$ = 909 nm is shifted to longer wavelength ($\Delta\nu_{1/2} = 5.2 \times 10^3$ $cm^{-1}$, $\lambda_{max}$ = 1309 nm) (Fig. 3.29, solid line). The two bands in the visible region are quite similar to those of the semiquinone form of anthraquinone derivatives.[164] The EPR spectrum of the frozen acidic acetonitrile solution of 1,5-$Fc_2$Aq at 5.0 K showed a strong, sharp signal at $g$ = 2.02 assignable to a semiquinone radical of the anthraquinone moiety as well as weak broad signals at $g_{//}$ = 4.12 and $g_{\perp}$ = 1.5 originating from a ferrocenium cation including a Fe(III) nucleus.[166,167] These spectroscopic results suggest the generation of a spin-separated form, $[1,5\text{-Fc}^{\bullet}\text{FvAqH}_2^{\bullet}]^{2+}$(Scheme 3.22).

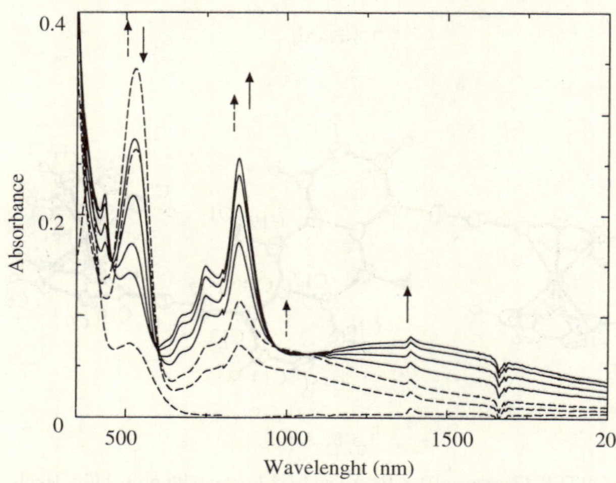


**Fig. 3.29** UV-vis-near-IR absorption spectral change of 1,5-$Fc_2$Aq ($3.0 \times 10^{-4}$ mol $dm^{-3}$) in benzonitrile upon stepwise addition of 0–2 equivalents (dotted lines and arrows) and 2–4 equivalents (solid lines and arrows) of $CF_3SO_3H$.

Electrochemistry of the protonated compounds supported the reactions given in Scheme 3.22. The series of monoprotonated complexes exhibits a reversible two-step one-electron reduction of the protonated anthraquinone moiety (AqH) in the cyclic voltammograms, whose potentials are largely shifted to more positive than those of nonprotonated forms (Table 3.4). The reversible oxidation waves of non- and monoprotonated complexes are derived from the metal-centered oxidation of the ferrocenyl and fulvene complex moieties. In $[1,5\text{-Fc}^{\bullet}\text{FvAqH}_2^{\bullet}]^{2+}$, the redox reaction occurs in a different pattern from that of non- and monoprotonated complexes, with the rest potential being positioned between the reduction wave of the ferrocenium moiety and the oxidation waves of Fv. Therefore, based on the above results, the spin-separated form, $[1,5\text{-Fc}^{\bullet}\text{FvAqH}_2^{\bullet}]^{2+}$, among the three possible canonical structures, $[1,5\text{-Fc}^{\bullet}\text{FvAqH}_2^{\bullet}]^{2+}$, $[1,5\text{-FvFvAqH}_2]^{2+}$, and $[1,5\text{-Fc}^{\bullet}\text{Fc}^{\bullet}\text{AqH}_2]^{2+}$, is considered to be the most thermodynamically favorable in solution as the doubly protonated species of 1,5-$Fc_2Aq$.

Protonation of 1,5-$Fc_2Aq$ with $CF_3SO_3H$ in dichloromethane afforded a deep green precipitate (species A), the ESI-mass spectrum of which indicated the formation of a doubly protonated structure. Its UV-Vis-near-IR absorption spectrum in Nujol mull, showing the $^2E_{2g}\rightarrow{}^2E_{1u}$ transition characteristic of a ferrocenium ion ($\lambda_{max}$ = 766 nm)[167] and significantly broad absorption (1456 nm), is different from that of $[1,5\text{-Fc}^{\bullet}\text{FvAqH}_2^{\bullet}]^{2+}$. The spectrum was gradually changed into that of $[1,5\text{-Fc}^{\bullet}\text{FvAqH}_2^{\bullet}]^{2+}$, with isosbestic points appearing after dissolution in acetonitrile. These results suggest that species A is a valence tautomer of $[1,5\text{-Fc}^{\bullet}\text{FvAqH}_2^{\bullet}]^{2+}$.

The $^{57}$Fe Mössbauer spectra of species A designate strong temperature dependence, as shown in Fig. 3.30. The spectrum at 10 K (Fig. 3.30) shows the superposition of two components of Fe(II) ($QS$ = 2.10 mms$^{-1}$, $IS$ = 0.40 mms$^{-1}$) and Fe(III) nuclei ($QS$ = 0.30 mms$^{-1}$, $IS$ = 0.40 mms$^{-1}$), for which the areal intensity ratio ($\phi$ = Fe(III)/Fe(II)+Fe(III)) is 0.44 (± 0.01). The $QS$ value of the Fe(II) component is smaller than the value of the Fe(II) nucleus of the non-protonated 1,5-$Fc_2Aq$ ($QS$ = 2.39 mms$^{-1}$, $IS$ = 0.55 mms$^{-1}$, at 9.2 K), and this smaller $QS$ value indicates a conversion to the fulvene-Fe(II) complex structure through protonation. The absorption of the ferrocenium-Fe(III) nucleus increases compared to that of Fe(II), when the temperature is raised, as shown in the spectrum at 293 K (Fig. 3.30). The $\phi$ values at 50, 150, 220 and 293 K were estimated to be 0.51 (± 0.04), 0.70 (± 0.08), 0.70 (± 0.08) and 1.00 (± 0.09), respectively.

The EPR spectrum of the solid state of the doubly protonated 1,5-$Fc_2Aq$ at 6.6 K showed the superposition of two components of Fe(III) nuclei from a ferrocenium cation ($g_{//}$ = 3.93, $g_{\perp}$ = 1.59) and a semiquinone radical of the anthraquinone moiety ($g$ = 2.00). The ratio in signal intensity of the semiquinone radical to the ferrocenium cation is considerably smaller in the solid state than in the frozen solution state (*vide supra*). The $\chi_M T$-$T$ plot of this doubly protonated 1,5-$Fc_2Aq$ is shown in Fig. 3.31. The $\chi_M T$ value decreases with decrease in temperature.

The temperature-dependent change of the $^{57}$Fe Mössbauer spectra can be ascribed to the thermal equilibrium between the three valence tautomers, $[1,5\text{-FvFvAqH}_2]^{2+}$, $[1,5\text{-Fc}^{\bullet}\text{FvAqH}_2^{\bullet}]^{2+}$ and $[1,5\text{-Fc}^{\bullet}\text{Fc}^{\bullet}\text{AqH}_2]^{2+}$ in the solid state (Scheme

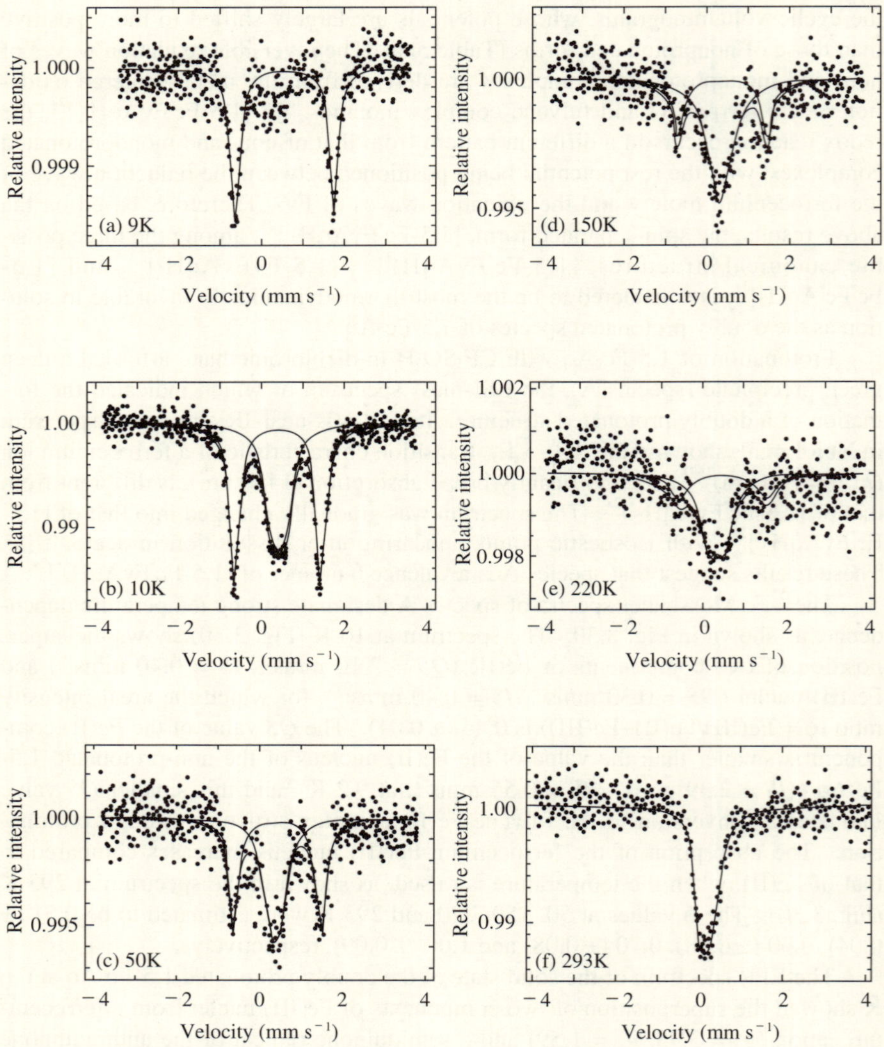


**Fig. 3.30** $^{57}$Fe Mössbauer spectra of 1,5-$Fc_2Aq$ at 9 K (a), and doubly protonated 1,5-$Fc_2Aq$ at 10 K (b), 50 K (c), 150 K (d), 220 K (e), and 293 K (f).
[Reprinted with permission from Murata, M., Yamada, M., Fujita, T., Kojima, K., Kurihara, M., Kudo, K., Kobayashi, Y., Nishihara, H., *J. Am. Chem. Soc.*, 2001, **123**, 12903.]

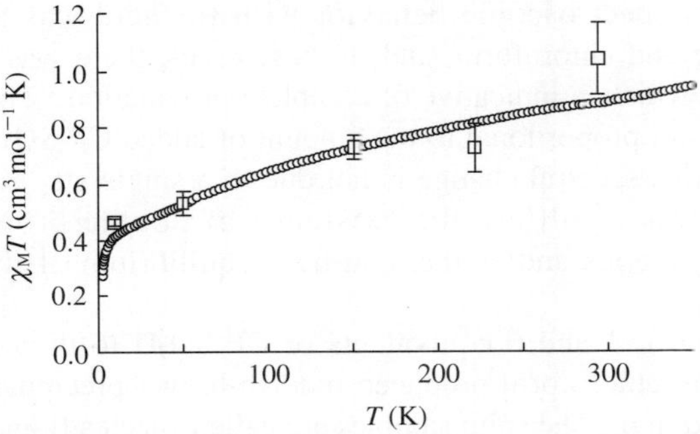


**Fig. 3.31** The $\chi_M T$-$T$ plot of doubly protonated 1,5-$Fc_2Aq$ (○) and $\chi_M T$ values estimated from the EPR $g$ values and $^{57}$Fe Mössbauer φ values (□) (b).
[Reprinted with permission from Murata, M., Yamada, M., Fujita, T., Kojima, K., Kurihara, M., Kudo, K., Kobayashi, Y., Nishihara, H., *J. Am. Chem. Soc.*, 2001, **123**, 12903.]

3.22). At higher temperatures, the paramagnetic $[1,5\text{-}Fc^{\bullet}\ Fc^{\bullet}AqH_2]^{2+}$, with its two ferrocenium Fe(III) nuclei, is dominantly formed. The $\chi_M T$ value at 348 K is 0.95 $cm^3\ mol^{-1}$ K, which is close to the calculated $\chi_M T$ value of 1.05 for $[1,5\text{-}Fc^{\bullet}Fc^{\bullet}AqH_2]^{2+}$, as estimated from the EPR $g$ values. With decreases in temperature, $[1,5\text{-}Fc^{\bullet}Fc^{\bullet}AqH_2]^{2+}$ might primarily be transformed into the diamagnetic $[1,5\text{-}FvFvAqH_2]^{2+}$, which has two fulvene-Fe(II) and doubly protonated anthrahydroquinone moieties, since the calculated $\chi_M T$ values in typical temperatures estimated from the EPR $g$ values and $^{57}$Fe Mössbauer φ values correspond well with the experimental $\chi_M T$ values (Fig. 3.31). A small distribution of $[1,5\text{-}Fc^{\bullet}FvAqH_2^{\bullet}]^{2+}$, which is dominant in the solution, cannot be ruled out at low temperature in the solid state because of the existence of a weak EPR signal for the radical.

### 3.5.4 A 1 : 2-D/A Complex

The other simplest combination of the non-equivalent D-A members is a 1 : 2-donor/acceptor compound, 1,1'-bis(1-(anthraquinonyl)ethynyl)ferrocene (1,1'-$FcAq_2$). To this complex, two-step protonation occurs, while only the first protonation causes the intramolecular electron transfer reaction.[168)]

When 8 equivalents of $CF_3SO_3H$ was added to a benzonitrile solution of 1,1'-$FcAq_2$, the color of the solution changed immediately from red to deep reddish-pink. This color change corresponds to the spectral change in which the band at $\lambda_{max}$ = 510 nm with ε = 2.8 × $10^3$ $mol^{-1}dm^3\ cm^{-1}$ increases four times in intensity with a slight blue-shift to $\lambda_{max}$ = 504 nm, together with an appearance of a new broad band with the half-width over visible and near-IR regions ($\lambda_{max}$ = 936 nm, ε = 1.3 × $10^3$ $mol^{-1}dm^3\ cm^{-1}$). These spectral changes are exceedingly similar to those of other Fc-Aq complexes as noted above,indicating that the ferrocenyl moiety is converted into butatriene-fulvene-Fe(II)-Cp.

The spectroscopic change due to the proton-coupled phenomenon of 1,1'-$FcAq_2$ in benzonitrile continued even after the addition of 8 equivalents of

$CF_3SO_3H$. This spectroscopic behavior with the acid was also observed in dichloromethane and chloroform, and in these cases, there was saturation of the spectral change, which is indicative of complete protonation. The degree of spectral change was not proportional to the amount of added $CF_3SO_3H$ in any solvent, suggesting that this spectral change is not due to a single step reaction. Possible rationales for this result are the existence of double protonation to both anthraquinone moieties and/or the acid-base equilibrium of the anthraquinone moiety.

The addition of 1 and 6 equivalents of $CF_3SO_3H$ to the concentrated 1,1'-$FcAq_2$ solution in chloroform produced reddish-brown precipitates that could be separated by filtration. The solid samples are called species B and C, respectively. The molecular ion peak of species B was detected at $m/z$ = 647 MH+ (solvent: acetonitrile), and that of species C at $m/z$ = 648 $MH_2$+ (solvent: acetonitrile) in their spectra, indicating the addition of one and two hydrogen atoms to 1,1'-$FcAq_2$ (FAB-mass ($m/z$): 646 M+), respectively. The resulting isotope patterns are quite in agreement with that of singly and doubly protonated 1,1'-$FcAq_2$, respectively. These results suggest that 1,1'-$FcAq_2$ undergoes a two-step protonation; the first step occurring with the addition of 1 equivalent of $CF_3SO_3H$ involves the monoprotonation of 1,1'-$FcAq_2$ causing a structure conversion identical to that of 1-FcAq, producing a unique structure involving an AqH-butatriene-fulvene-Fe(II)-Cp-ethynylene-Aq moiety, $[1,1'\text{-}FvAq_2H]^+$ (Scheme 3.23), while the second step achieved by the addition of 6 equivalents of $CF_3SO_3H$ involves the double protonation of 1,1'-$FcAq_2$ to give $[1,1'\text{-}FvAq_2H_2]^{2+}$, with the UV-vis-near-IR absorption not changing significantly. This process results in no additional intramolecular electron transfer, which is reasonable since the butatriene-fulvene-Fe(II)-fulvene-butatriene structure should be unstable.

The results of cyclic voltammetry of 1,1'-$FcAq_2$ with the stepwise addition of $CF_3SO_3H$ in $Bu_4NClO_4$-$CH_2Cl_2$ are shown in Table 3.4. The nonprotonated com-

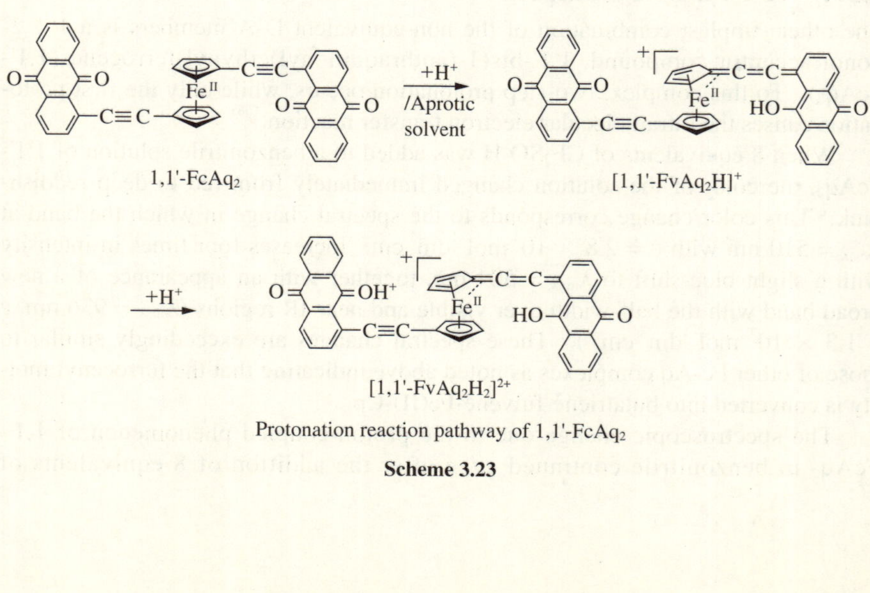


Protonation reaction pathway of 1,1'-$FcAq_2$

**Scheme 3.23**

plex, 1,1'-$FcAq_2$ shows reversible two-step 2 $e^-$ reduction derived from the two anthraquinone moieties (–1.85 and –1.39 V *vs*. $Fc^+$/Fc), with reversible 1 $e^-$ oxidation originating from the ferrocene moiety (0.21 V), respectively. The reduction of two neutral anthraquinone units occurs simultaneously, due to the relatively little interaction between the two units through the linked ethynylene and ferrocene units. The addition of 1 equivalent of $CF_3SO_3H$ for 1,1'-$FcAq_2$ shows the redox behavior of [1,1'-$FvAq_2H]^+$. This monoprotonated complex exhibits three reversible reduction waves at –1.85 (1 $e^-$), –1.39 (2 $e^-$), and –0.32 V (1 $e^-$) and an oxidation wave at 0.43 V (1 $e^-$), which can be interpreted as the overlapping of redox waves of anthraquinone (Aq; –1.85 and –1.39 V), protonated anthraquinone (AqH; –1.39 and –0.32 V) and fulvene-Fe(II) (Fv; 0.43 V). Further addition of the acid to cause one more protonation to [1,1'-$FvAq_2H]^+$ produces a serious change in the redox wave of the anthraquinone moiety, and this behavior appears identical to that occurring with the formation of [1,1'-$FvAq_2H_2]^{2+}$. The reduction waves at –1.85 (1 $e^-$), –1.39 (2 $e^-$) and –0.32 V (1 $e^-$) are changed to –1.39 (2 $e^-$), –0.14 (1 $e^-$) and –0.06 V (1 $e^-$), whereas the oxidation wave derived from the fulvene-Fe(II) unit is not influenced by the further addition of $CF_3SO_3H$ (0.43 V), indicating that the fulvene-Fe(II) structure is not changed. Disappearance of the most negative wave at –1.85 V (1 $e^-$) derived from the nonprotonated anthraquinone unit of [1,1'-$FvAq_2H]^+$ indicates that the second protonation for [1,1'-$FvAq_2H]^+$ occurs at the nonprotonated anthraquinone moiety. Consequently, two monoprotonated anthraquinone moieties are formed in the double protonation of 1,1'-$FcAq_2$.

### 3.5.5 Polymeric 1 : 1-D/A Complexes

The results noted above present peculiar structure-conversion and spin-separation behavior of a series of ferrocenylethynylanthraquinones caused by protonation that depends on the substituent position, the number of donor and acceptor units. This predicts the intriguing properties of polymer complexes of this series. From this point of view, 1 : 1-donor/acceptor polymer, (1,1'-Fc-1,8-Aq)$_n$, comprising ferrocenylethynyl-anthraquinone units in the main chain, was synthesized and its protonation-induced electron transfer behavior investigated.[169)]

The polymer (1,1'-Fc-1,8-Aq)$_n$ was synthesized by the reaction of this bis(trimethyl stanyletnylferrocene) and this dibromoanthraquinone. This polymer was obtained as deep brown powder, containing soluble and insoluble components in regular organic solvents. The soluble part gives a film when it is dried, and its average molecular weight determined by GPC based on the polystylene standard was about 2000. As for the insoluble component of the polymer, the addition of $CF_3SO_3H$ changes the polymer soluble in common organic solvents such as acetonitrile or dichloromethane, and the color changes from deep red to deep purple-pink.

The UV-vis-near-IR absorption spectrum of the solution of the protonated polymer exhibits MLCT bands and valence tautomerization similar to those of 1,8-$Fc_2Aq$ as shown in Fig. 3.32, indicating the formation of fulvene complexes in the main chain in solution. Infrared spectrum of the protonated polymer also shows the similarity with that of 1,8-$Fc_2Aq$, namely, half the CC triple bond remains.

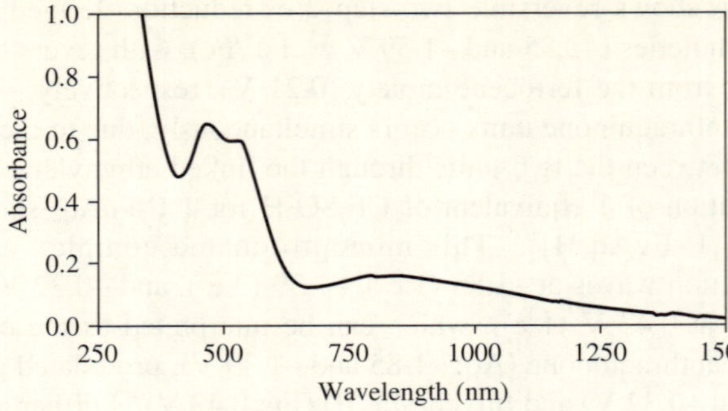


**Fig. 3.32** UV-vis-near-IR absorption spectrum of protonated (1,1'-Fc-1,8-Aq)$_n$ (1.3 × $10^{-3}$ mol $dm^{-3}$ as Fc-Aq monomer unit) in acetonitrile.

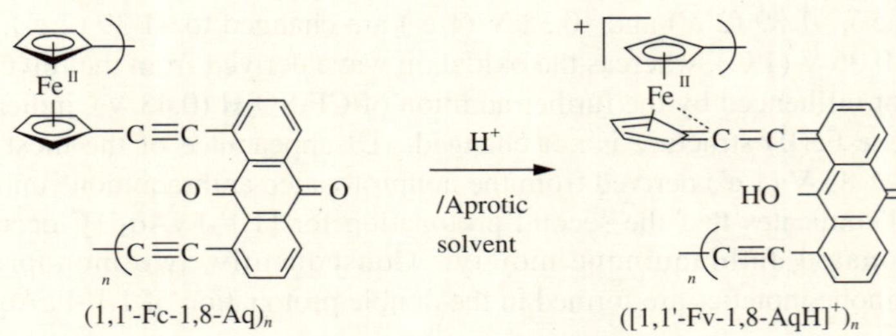


Protonation reaction pathway of (1,1'-Fc-1,8-Aq)$_n$

**Scheme 3.24**

These results indicate the formation of ([1,1'-Fv-1,8-AqH]$^+$)$_n$ in Scheme 3.24.

## 3.5.6 Concluding Remarks

In this section, novel protonation-induced intramolecular electron transfer causing a drastic change in structure and physical properties in the ethynylene-bridged ferrocene-anthraquinone donor-acceptor conjugated complexes, 1-FcAq, 2-FcAq, 1,8-$Fc_2$Aq, 1,5-$Fc_2$Aq, 1,1'-Fc$Aq_2$, and (1,1'-Fc-1,8-Aq)$_n$ have been described. The singly and doubly protonated forms of them reach an equilibrium involving two and three valence tautomers, depending on the matrix and temperature. It should be noted that other donor-acceptor combinations provide other unique behaviors as we demonstrated for the ferroccenylvinylhydroquinone system.[170] These findings should be useful in designing novel functional molecular systems based on strong intramolecular donor-acceptor interaction triggered by chemical perturbation.

## Acknowledgments

The authors thank the following people who contributed to the present study: Dr. Kenya Kubo (presently at ICU), Dr. Masato Kurihara (presently at Yamagata University), Dr. Mami Yamada, Takako Fujita and Kohei Kojima at the University of Tokyo, and Dr. Yoshio Kobayashi at RIKEN. This work was supported in part by Grants-in-Aid from the Ministry of Culture, Education, Science, Sports, and Technology of Japan.

# 3.6 Oligomers of Non-natural Metal Complex Amino Acids

**Taka-aki Okamura and Norikazu Ueyama**

## 3.6.1 Introduction

Polymers containing metal complexes in the main chain are one of the most attractive materials having electronic, optical and magnetic properties.[171,172] Ferrocene and bipyridine derivatives especially, have been used for their stability, synthetic facility and redox properties. Polyferrocene[173,174] or poly(vinylferrocene) have been reported since the 1960's and before.[175-183] Recently, Manners *et al.* have reported ferrocene polymers connected by various linkages, e.g., silicon,[184-190] germanium,[188,191] hydrocarbon,[192-195] sulfur[196,197] and phosphrous.[198,199] Many π-conjugated bipyridine derivatives have been synthesized and are known to exhibit electric conductivity or photoconductivity when ruthenium ions coordinate to bipyridine units.[200-203] In these polymers, bipyridine units are symmetrically linked by relatively rigid spacers, e.g., unsaturated hydrocarbons. The introduction of electric potential orientation to the polymers will lead to rectifier or finely designed electric devices. For this purpose, unsymmetrical units and synergetic effects are necessary.

On the other hand, peptides have been known to form unique secondary conformations, helices or sheets. The overall structure is determined by the content and sequence of amino acids. If appropriate amino acids are selected, the peptide readily forms well-defined secondary structures. The conformation is determined by the orientation of the amide plane and direction of elongation. Fig. 3.33 shows a schematic drawing of L-amino acid. The white arrows show the orientation of the dipole of the amide plane and the gray arrows show the direction of elongation from the N- to C-terminus. If the side group, R is preferentially selected, the desired conformation is realized. For example, L-leucine containing an isopropyl side chain trends to form an α-helical structure.[204] Different spacing lengths between amino acids, e.g., β- and δ-amino acid are effective in controlling the secondary structures.[205,206] Proper design of the spacer with suitable bulkiness, length and flexibility is necessary to construct the required secondary structure. Our purpose is to achieve various combinations of such peptides with metal compounds, i.e., non-natural peptides containing metal complexes in the main chain. For this purpose, various amino acids are necessary. The metal complex opens the way to numerous possibilities and capabilities. However, selection is limited by their stability and reactivity.

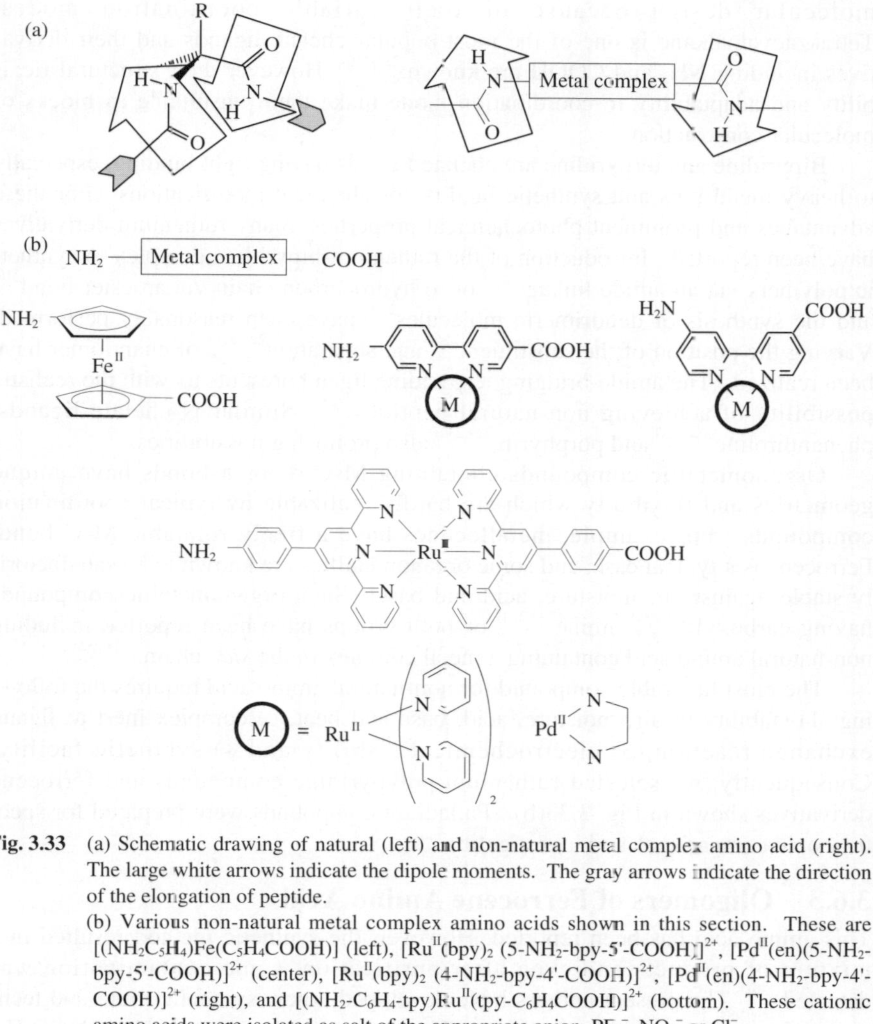


**Fig. 3.33** (a) Schematic drawing of natural (left) and non-natural metal complex amino acid (right). The large white arrows indicate the dipole moments. The gray arrows indicate the direction of the elongation of peptide.
(b) Various non-natural metal complex amino acids shown in this section. These are $[(NH_2C_5H_4)Fe(C_5H_4COOH)]$ (left), $[Ru^{II}(bpy)_2$ (5-$NH_2$-bpy-5'-COOH)$]^{2+}$, $[Pd^{II}(en)$(5-$NH_2$-bpy-5'-COOH)$]^{2+}$ (center), $[Ru^{II}(bpy)_2$ (4-$NH_2$-bpy-4'-COOH)$]^{2+}$, $[Pd^{II}(en)$(4-$NH_2$-bpy-4'-COOH)$]^{2+}$ (right), and $[(NH_2\text{-}C_6H_4\text{-tpy})Ru^{II}(\text{tpy-}C_6H_4COOH)]^{2+}$ (bottom). These cationic amino acids were isolated as salt of the appropriate anion, $PF_6^-$, $NO_3^-$ or $Cl^-$.

## 3.6.2 Design of Non-natural Amino Acid

Introduction of amino and carboxyl groups to metal compounds has been tried and some compounds have been reported. For stability (against oxidation, heat, moisture, acid, base, etc.) and synthetic facility, the selection of the ligand is confined to chelates with N or O donor sets and sufficiently stable organometals like ferrocene. Among monodentate ligands, pyridine is very useful for chemical modification but insufficiently inert.[207-209] Pyrazole[210] and uracil[211-213] derivatives having a $NH_2$ group have been reported but these ligands are not appropriate for strategic molecular design because of their variable coordination modes. Tetraazacycloalkane is one of the most popular chelate ligands and their derivatives including $NH_2$ and COOH are known.[214-216] However their structural flexibility and adaptability to coordination mode make them unsuitable as blocks of molecular construction.

Bipyridine and terpyridine are chelate ligands having tight binding, especially to heavy metal ions and synthetic facility for chemical modifications. For these advantages and prominent photochemical properties, many ruthenium derivatives have been reported. Introduction of the ruthenium-bipyridine complex to synthetic polymers via an amide linkage[217] or to hydrocarbon chain via an ester bond[218] and the synthesis of dendrimeric molecules[219] have been reasonably performed. Varying the position of the substituent[220] and separation[221-224] of enantiomer have been realized. The amide-bridging terpyridine ligand presents us with the realistic possibility of achieving non-natural peptides.[225] Similar N-chelate ligands, phenanthroline[226-228] and porphyrin,[229,230] also promising possibilities.

Organometallic compounds containing M-C $\sigma$- or $\pi$-bonds have unique geometries and flexibility which are hardly realizable by typical coordination compounds. For example, metallocenes have a freely rotatable M-C bond. Ferrocene is a typical case, and some organometallics are known to be satisfactorily stable against air, moisture, acid and base. Such organometallic compounds having carboxyl,[231,232] amine[233,234] or both groups have been reported including non-natural amino acid containing a metal complex in the *side* chain.[234,235]

The most favorable compound for non-natural amino acid requires the following: 1) stability to air, moisture, acid, base and heat, 2) complex inert to ligand exchange reaction, 3) electrochemical stability and 4) synthetic facility. Consequently, we selected ruthenium-polypyridine compounds and ferrocene derivatives shown in Fig. 3.33(b). Palladium compounds were prepared for spectroscopic comparisons and investigation.

### 3.6.3 Oligomers of Ferrocene Amino Acid

This amino acid has been reported. However, the synthetic method resulted in a mixture of products and even after purification, some contamination was detected.[236] The present authors have developed a new synthetic route and techniques for stepwise polymerization.[237] The monomer obtained, $[(CH_3CONHC_5H_4)Fe(C_5H_4CONHCH_3)]$ (**38**), and the dimer, $[(CH_3CONHC_5H_4)Fe(C_5H_4CONHC_5H_4)Fe(C_5H_4CONHCH_3)]$ (**39**), were characterized by spectroscopic measurements and single crystal X-ray analysis. These compounds form inter- or intramolecular hydrogen bonds which are illustrated in Fig. 3.34. The trimer, $[(CH_3CONHC_5H_4)Fe(C_5H_4CONHC_5H_4)Fe(C_5H_4CONHC_5H_4)Fe(C_5H_4CONHCH_3)]$ (**40**), was also successfully isolated. Polymerization of the amino acid, $[(NH_2C_5H_4)Fe(C_5H_4COOH)]$ (**41**), using active ester method gave oligometric products up to the hexamer, which was detected by ESI-MS spectroscopy. The limited solubility restricted the propagation step. The insoluble product was not characterized by MALDI-TOF because of decomposition by laser excitation.

The formation of hydrogen bonds in [($CH_3CONHC_5H_4$)Fe($C_5H_4CONHCH_3$)] (**38**), and [($CH_3CONHC_5H_4$)Fe($C_5H_4CONHC_5H_4$)Fe($C_5H_4CONHCH_3$)] (**39**) were investigated using X-ray analysis, $^1$H NMR and IR. The amide planes of **38** and **39** are stacked on each other to cancel the amide dipole moments in the crystal (Fig. 3.34(a)). Significantly short intermolecular N···O (2.74 – 2.90 Å) and NH···O=C distances (1.84 – 1.95 Å) clearly indicate that all amide groups are intermolecularly NH···O=C hydrogen bonded in the crystal, forming an effective hydrogen bond network. Crystallographically distinguishable hydrogen-bonded amides, two kinds for **38** and three for **39**, were also detected by the IR spectra. The N- and C-terminal and bridging amide NHs are assignable based on the ν(NH) bands of the analogous compounds, [($C_5H_5$)Fe($C_5H_4NHCOCH_3$)] and

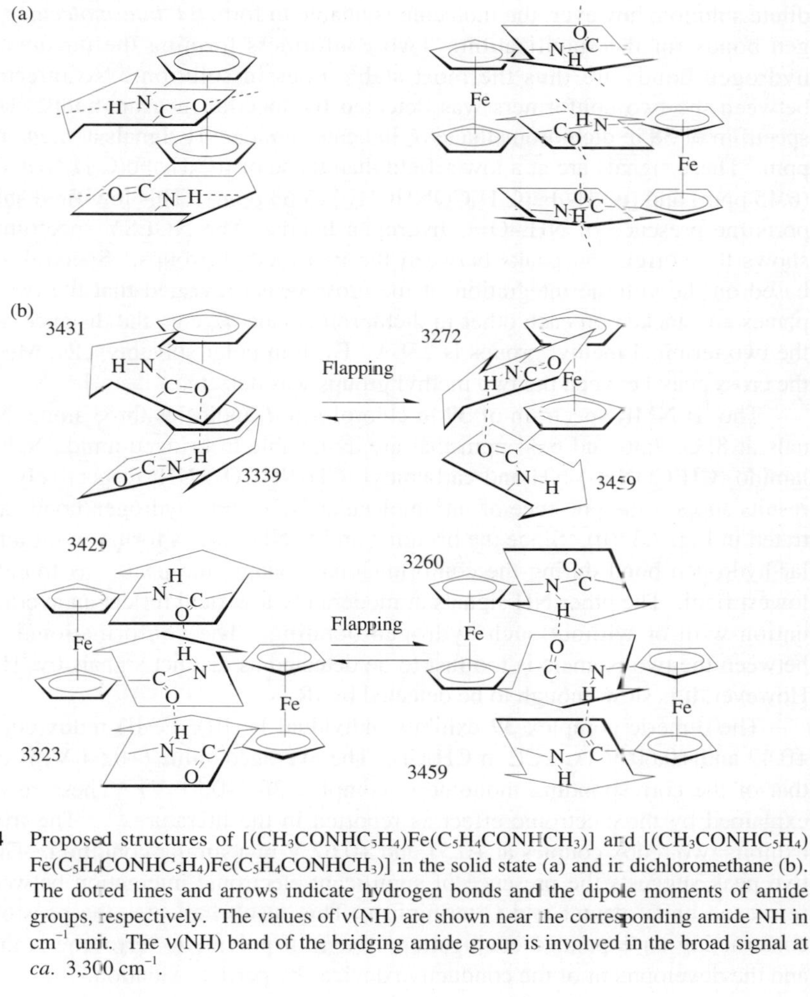


**Fig. 3.34** Proposed structures of [($CH_3CONHC_5H_4$)Fe($C_5H_4CONHCH_3$)] and [($CH_3CONHC_5H_4$)Fe($C_5H_4CONHC_5H_4$)Fe($C_5H_4CONHCH_3$)] in the solid state (a) and in dichloromethane (b). The dotted lines and arrows indicate hydrogen bonds and the dipole moments of amide groups, respectively. The values of ν(NH) are shown near the corresponding amide NH in $cm^{-1}$ unit. The ν(NH) band of the bridging amide group is involved in the broad signal at *ca.* 3,300 $cm^{-1}$.

$[(C_5H_5)Fe(C_5H_4CONHCH_3)]$. The value of ν(NH), i.e., the strength of the NH bond is considered to reflect the acidity of amide NH. The adjacent ferrocenyl group contributes to the increase in the acidity of NH resulting in the strong hydrogen bond formation.

When complexes **38** and **39** were dissolved in nonpolar solvents such as dichloromethane and chloroform, they formed *intramolecular* hydrogen bonds as detected by IR and $^1$H NMR spectroscopic analysis. The spectral analysis reveals the presence of the flapping equilibrium in solution shown in Fig. 3.34(b). The observed ν(NH) in $cm^{-1}$ units are shown in the figure. The intensity of individual stretching bands indicates the validity of our assignments. In the crystal, two amide planes are in antiparallel orientation to minimize total dipole moments and form *intermolecular* hydrogen bonds to realize the most stable conformation. In a dilute solution, however, the molecule is unable to form the *intermolecular* hydrogen bonds for the stabilization. Two conformers forming the *intramolecular* hydrogen bonds are thus the most stable ones in solution. No intermediate between the two conformers was detected by spectral measurements. $^1$H NMR spectrum of **38** in dichloromethane-$d_2$ indicates amide NH signals at 6.24 and 7.30 ppm. These signals are at a lower field than those of $[(C_5H_5)Fe(C_5H_4NHCOCH_3)]$ (6.45 ppm) and $[(C_5H_5)Fe(C_5H_4CONHCH_3)]$ (5.63 ppm). This low-field shift supports the presence of NH···O=C hydrogen bonds. The NOESY spectrum of **38** shows the correlation peaks between the two methyl groups. Spectral analysis based on the volume integration of the cross peaks revealed that the two amide planes are stacked on each other in dichloromethane-$d_2$, e.g., the distance between the two terminal methyl groups is 2.9 Å. Even in polar solvents, e.g., $Me_2SO$-$d_6$, the cross peak between the two methyl groups was detected.

The $^1$H NMR spectrum of **39** in chloroform-$d_1$ indicates three amide NH signals at 8.33, 7.36 and 6.34 ppm that are assignable to bridged amide NH, acetylamino ($CH_3CON$**H**-) NH and carbamoyl ($CH_3N$**H**CO-) NH, respectively. These results support the presence of intramolecular NH···O=C hydrogen bonds as illustrated in Fig. 3.34(b). Since the bridging amide NH always forms an intramolecular hydrogen bond during the conformational change, the peak was found at the lowest field. The other NH signals at moderately low field reflect an averaged situation with or without such hydrogen bonding. The conformational change between the two isomers is too fast to be detected as distinct signals by $^1$H NMR. However, it is slow enough to be detected by IR.

The dimeric complex **39** exhibits individual Fe(III)/ Fe(II) redox couples at +0.47 and +0.60 V *vs.* SCE in $CH_2Cl_2$. The averaged value (+0.54 V) is close to that of the corresponding monomeric complex **38** (+0.51 V). These results are explained by the electronic effect as reported in the literature.[238)] The trimer **40** exhibits two redox couples at +0.51 and +0.67 V in a similar condition. The positive peak suggests the presence of significant electronic interaction between two ferrocene units (center and C-terminus). This result indicates the possibility of constructing an expanded conjugation system of polymeric ferrocene amino acid and the development of the conductive devices by partial oxidation.

Unfortunately, **39** and **40** show poor solubility in less polar solvents. Thus we could not determine their solution structure by $^1$H NMR. In a polar solvent that is $(CD_3)_2SO$, the hydrogen bonds are broken or competing with solvation. The conformation was found to be a turn-like structure.

The unique and appealing conformations led us to propose a unique structure for a novel organometallic non-natural polypeptide with skeletal ferrocenes as shown in Fig. 3.35(a). When the polymer is dissolved in polar solvent, the secondary structure is probably changed to a turn or random structure (Fig. 3.35(b)). The folding structure reminds us of secondary structures such as the α-helix and β-sheet of polypeptides, and the conformational change calls to mind the helix-coil transition of natural peptides.

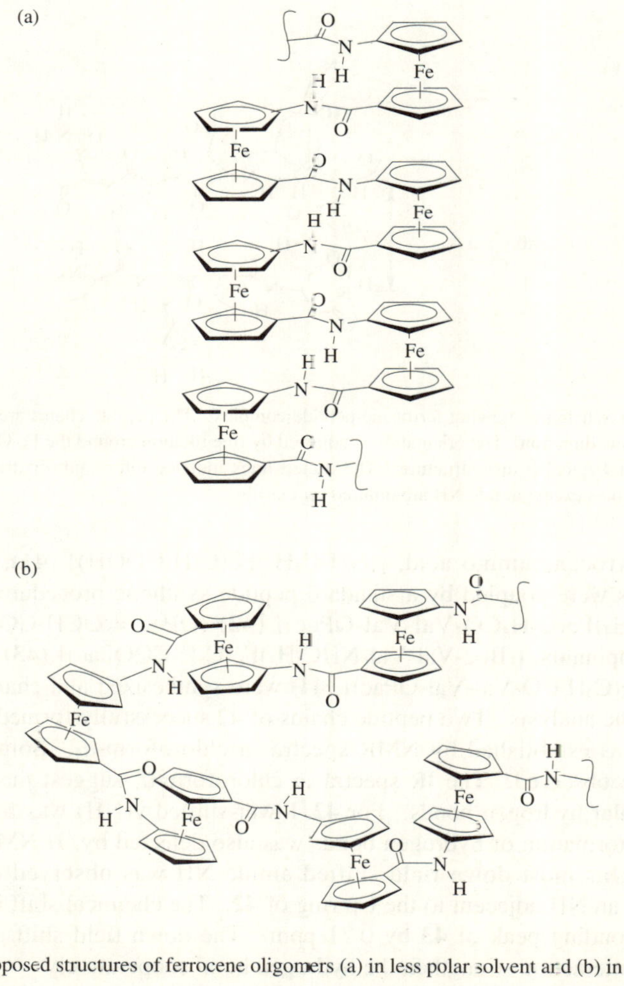

**Fig. 3.35** Proposed structures of ferrocene oligomers (a) in less polar solvent and (b) in polar solvent.

### 3.6.4 Ferrocene-Peptide Complexes

Ferrocene has a suitable distance between Cp (cyclopentadienyl) ligands for the spacer of anti-parallel β-sheet (Fig. 3.36(a)). The freely rotatable bond is appropriate for adjusting the distance between two peptide chains In this section, the combination of non-natural amino acid with natural peptide is described.

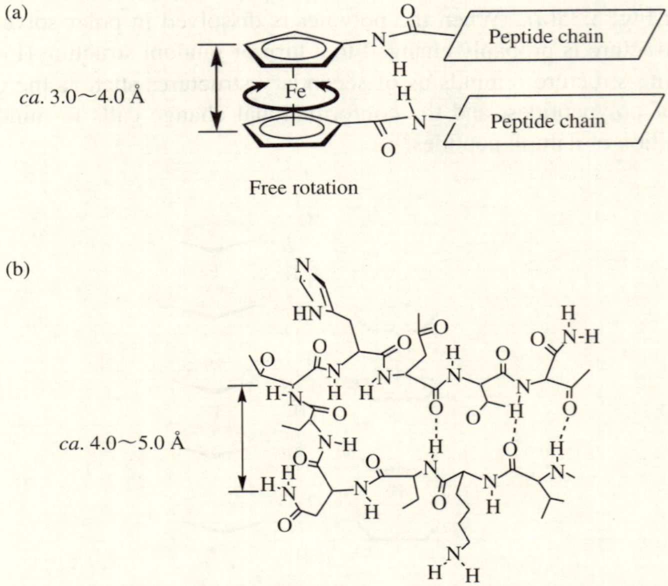


**Fig. 3.36** (a) Schematic drawing ferrocene-peptide complex. The peptide chains are oriented in the same direction. The orientation is adjusted by free rotation around the Fe-Cp bond.
(b) Typical β-turn structure. The dotted lines indicate inter-chain hydrogen bonds. H atoms except amide NH are omitted for clarity.

The ferrocene amino acid, [($NH_2C_5H_4$)Fe($C_5H_4COOH$)] (**41**), and natural amino acids were coupled by a standard peptide synthetic procedure. [(Boc-Val-Val-$NHC_5H_4$)Fe($C_5H_4$CO-Val-Val-OPac)] (**42**) (OPac = $OCH_2COPh$) and the related compounds, [(Boc-Val-Val-$NHC_5H_4$)Fe($C_5H_4COOPac$)] (**43**) and [(Boc - $NHC_5H_4$)Fe($C_5H_4$CO-Val-Val-OPac)] (**44**) were synthesized and characterized by spectroscopic analysis. Two peptide chains of **42** successfully formed an anti-parallel sheet as established by NMR spectra in chloroform-$d_1$. Some interchain NOEs were observed. The IR spectra in chloroform-$d_1$ suggest the presence of intramolecular hydrogen bonds. For **42**, lower-shifted ν(NH) was found at 3249 $cm^{-1}$. The formation of hydrogen bonds was also detected by $^1H$ NMR in chloroform-$d_1$. This most down-field shifted amide NH was observed at 9.50 ppm assigned to an NH adjacent to the Cp ring of **42**. The chemical shift is lower than the corresponding peak of **43** by 0.71 ppm. The down-field shift suggests that interchain interaction strengthen the hydrogen bond cooperatively.

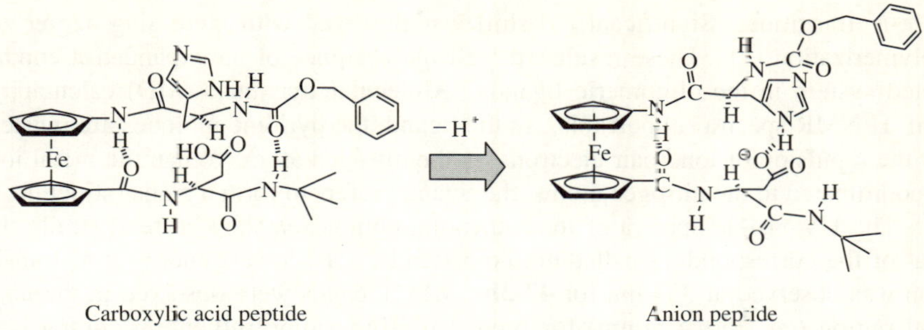

Fig. 3.37 Proposed conformational change of [Z-His-NHC$_5$H$_4$]Fe(C$_5$H$_4$CO-Asp-NH-*t*-Bu)] by deprotonation. In the sheet structure, two inter-chain NH···O=C hydrogen bonds exist (left). When the carboxyl group releases a proton, ImNH···O=C is formed instead of terminal NH···O=C hydrogen bond to form a turn structure (right).

When ionic side chains were introduced, interesting behavior was found. The compound containing aspartic acid and histidine residues, [Z-His-NHC$_5$H$_4$]Fe(C$_5$H$_4$CO-Asp-NH-*t*-Bu)] (**45**), was synthesized. The peptide forms strong inter-chain NH···O=C hydrogen bonds. When one proton was removed, the imidazole of histidine formed a hydrogen bond with the carboxylate anion, as established by $^1$H NMR and IR spectra. Solution structure analysis and spectral data suggest a conformational change induced by deprotonation as shown in Fig. 3.37. When the deprotonation occurs, the hydrogen bonds are rearranged to N···O=C and ImNH···O$^-$. The structure is changed from sheet to turn, resulting in significant change. A negative shift of the redox potential was also found.

### 3.6.5 Bipyridine Complex Amino Acid

The oligomer of bipyridine amino acid,[239] *t*-BuCO-(4-NH-bpy-4'-CO)$_n$-NH-*t*-Bu (**46-na**, $n$ = 1,2) and *t*-BuCO-(5-NH-bpy-5'-CO)$_n$-NH-*t*-Bu (**46-nb**, $n$ = 1−3) were prepared by a modified method described in the literature[240] and general procedures for liquid-phase peptide synthesis. The ruthenium and palladium compounds, [{Ru$^{II}$(bpy)$_2$}$_n${*t*-BuCO-(4-NH-bpy-4'-CO)$_n$-NH-*t*-Bu}](PF$_6$)$_{2n}$ (**47-na**), [{Ru$^{II}$(bpy)$_2$}$_n${*t*-BuCO-(5-NH-bpy-5'-CO)$_n$-NH-*t*-Bu}](PF$_6$)$_{2n}$(**47-nb**), [{Pd$^{II}$(en)}$_n${*t*-BuCO-(4-NH-bpy-4'-CO)$_n$-NH-*t*-Bu}](NO$_3$)$_{2n}$(**48-na**) and [{Pd$^{II}$(en)}$_n${*t*-BuCO-(5-NH-bpy-5'-CO)$_n$-NH-*t*-Bu}](NO$_3$)$_{2n}$ (**48-nb**) were obtained by the reaction of the ligand with [Ru(bpy)$_2$Cl$_2$],[241] [Ru(bpy)$_2$(OCMe$_2$)$_2$](BF$_4$)$_2$[242] and [Pd(en)(H$_2$O)$_2$](NO$_3$)$_2$[243] (bpy = 2,2'-bipyridine, en = ethylenediamine). The ruthenium derivatives were obtained as PF$_6^-$ salts by the anion-exchange reaction using an aqueous NH$_4$PF$_6$ solution. The products were characterized by $^1$H NMR, ESI-MS and elemental analysis.

For the palladium compounds, the 4,4'-disubstituted derivatives show an MLCT band at *ca.* 310–320 nm in water as found for usual palladium-bipyridine complexes.[244] On the other hand, the 5,5'-disubstituted derivatives exhibit characteristic absorption at 344 (for **48-1b**), 361 (for **48-2b**) and 367 nm (for **48-3b**) due to π-π* transition. Significant red shift was observed with increasing degree of polymerization (*n*). These results suggest the presence of an expanded π-conjugated system in the oligomeric ligands. Molecular dynamics (MD) calculation and $^1$H NMR spectra support this. In the ligand, the pyridine moieties are twisted by the repulsion of lone-pair electrons on the nitrogen atom. When the metal ion is coordinated to the nitrogen atoms, the ligand prefers to form a planar structure.

The UV-visible spectra of the ruthenium complexes show a trend similar to that of the corresponding palladium compounds. The lowest energy π-π* transition was observed at 357 nm for **47-3b**. MLCT bands were observed in the normal region (*ca.* 450–470 nm) for typical tris(bipyridine)ruthenium complexes. The proposed structure of **47-3b** is shown in Fig. 3.38.

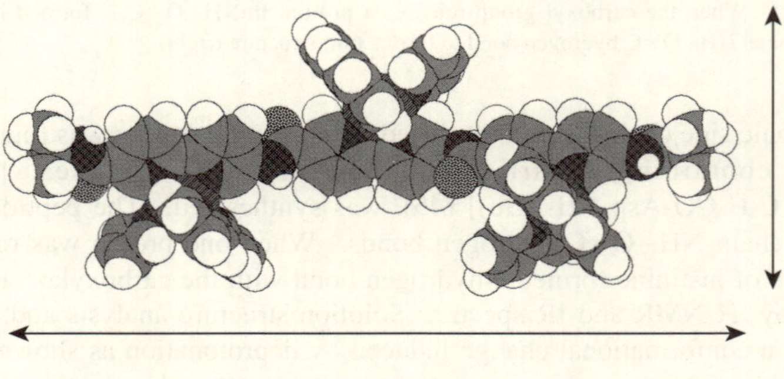


**Fig. 3.38** Proposed structure of $[\{Ru^{II}(bpy)_2\}_3\{t\text{-BuCO-(5-NH-bpy-5'-CO)}_3\text{NH-}t\text{-Bu}\}]^{6+}$ using molecular dynamics calculation.

The ruthenium compounds show characteristic emission from $^3$MLCT state at *ca.* 600–700 nm (Fig. 3.39). In comparison with $[Ru(bpy)_3](PF_6)_2$, the relatively intense peaks with lower energy show stabilization of the excited state by the conjugation of bipyridine with amide groups. Emission spectrum (excited at 488 nm in $CH_3CN$) of each dimeric Ru(II) compound, **47-2a** and **47-2b**, show red-shifted maximum by *ca.* 30 nm compared with the corresponding monomeric compounds. These results suggest the presence of low-lying π* orbital in oligomeric compounds. The low intensity of **47-2b** and **47-3b** indicates the presence of another quenching pathway that is probably an intersystem crossing to the $^3$π-π* state with lower energy.[245)]

The amide-substituted bipyridine ligand is reduced more easily than the normal bipyridine. $E_{red}^{I}$ separates into two peaks in the dimeric derivatives, **47-2a** and **47-2b**, and into three in the trimeric **47-3b**, whereas the Ru(III/II) redox couple is found as only one peak. This indicates that the LUMO (lowest unoccupied molecular orbital) of the current Ru-complexes is localized on the bipyridine ligands interacting with each other through the amide bridges, without detectable delocalization onto the metal atoms.

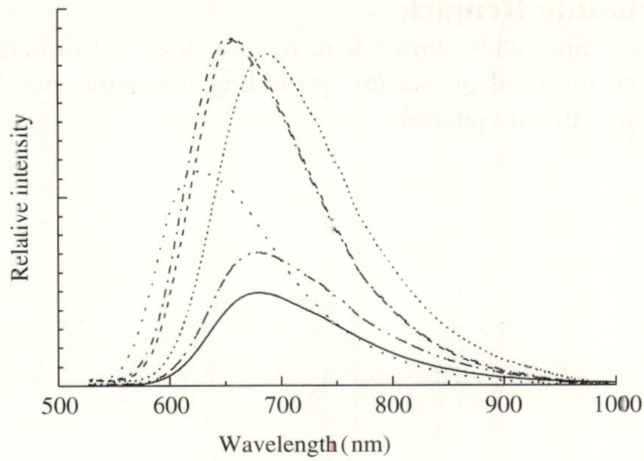


**Fig. 3.39** Emission spectra of [Ru$^{II}$(bpy)$_2$(*t*-BuCO-4-NH-bpy-4'-CONH-*t*-Bu)](PF$_6$)$_2$ (·······), [Ru$^{II}$(bpy)$_2$ (*t*-BuCO-5-NH-bpy-5'-CONH-*t*-Bu)](PF$_6$)$_2$ (-·-·-·) [{Ru$^{I}$(bpy)$_2$}$_2${*t*-BuCO-(4-NH-bpy-4'-CO)$_2$-NH-*t*-Bu}](PF$_6$)$_4$ (------), [{Ru$^{II}$(bpy)$_2$}$_2${*t*-BuCO-(5-NH-bpy-5'-CO)$_2$-NH-*t*-Bu}](PF$_6$)$_4$ (--···--), [{Ru$^{II}$(bpy)$_2$}$_3${*t*-BuCO-(5-NH-bpy-5'-CO)$_3$-NH-*t*-Bu}](PF$_6$)$_6$ (——) and [Ru(bpy)$_3$](PF$_6$)$_2$ (· · · · ·) excited at 488 nm in acetonitrile.

### 3.6.6 Terpyridine Complex Amino Acid

The ruthenium-terpyridine complex, [(*t*-BuCONH-C$_6$H$_4$-tpy)Ru$^{II}$(tpy-C$_6$H$_4$CONHC$_6$H$_4$-tpy)Ru$^{II}$(tpy-C$_6$H$_4$CONH-*t*-Bu)](PF$_6$)$_4$ (**49**), was synthesized by stepwise N-terminal elongation including coordination and coupling reaction. The reaction was started from the preparation of [Cl$_3$Ru$^{III}$(tpy-C$_6$H$_4$CONH-*t*-Bu)] and then [NH$_2$C$_6$H$_4$-tpy)Ru$^{II}$(tpy-C$_6$H$_4$CONH-*t*-Bu)](PF$_6$)$_2$, [(tpy-C$_6$H$_4$CONHC$_6$H$_4$-tpy)Ru$^{II}$(tpy-C$_5$H$_4$CONH-*t*-Bu)](PF$_6$)$_2$, and the products were successively obtained. The proposed structure of **49** is shown in Fig. 3.40. The each ruthenium ion in **49** is electrochemically independent.

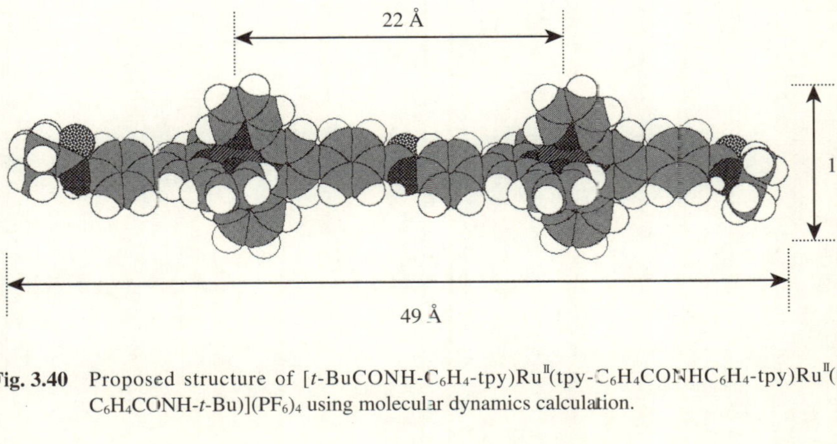


**Fig. 3.40** Proposed structure of [*t*-BuCONH-C$_6$H$_4$-tpy)Ru$^{II}$(tpy-C$_6$H$_4$CONHC$_6$H$_4$-tpy)Ru$^{II}$(tpy-C$_6$H$_4$CONH-*t*-Bu)](PF$_6$)$_4$ using molecular dynamics calculation.

### 3.6.7 Concluding Remark

The non-natural amino acids shown here have a unique 3-dimensional structure and chemical and physical properties, providing numerous possibilities for the development of functional materials.

# References


< Section 3.1 >

1. Dickerson, R.E., Geis, I., *Hemoglobin*, Benjamin, Menlo Park, 1983.
2. Metzler, D.E., *Biochemistry*, Academic Press, 1977.
3. Klingelhöfer, S., Heitz, W., Greiner, A., Oestreich, S., Förster, S.K., Antonietti, M., *J. Am. Chem. Soc.*, 1997, **119**, 10116.
4. Zhang, L., Eisenberg, A., *J. Am. Chem. Soc.*, 1996, **118**, 3168.
5. Hawker, C.J., Wooley, K.L., Fréchet, J.M.J., *J. Chem. Soc., Perkin Trans. 1*, **1993**, 1287.
6. Fuhrhop, J.-H., Köning, J., Molecular Assemblies and Membranes, Monographs in *Supramolecular Chemistry* (Stoddart, J.F. ed), Royal Soc. Chem., London, I-xiii, 1994.
7. Fuhrhop, J.-H., Endisch, C., *Molecular and Supramolecular Chemistry of Natural Products and Model Compounds*, Marcel Dekker, New York, 2000.
8. Draeger, C., Boettcher, C., Messerschmidt, C., Ruhlmann, L., Fuhrhop, J.-H., Siggel, U., Hammerström, L., *Langmuir*, 2000, **16**, 2068.
9. Gouzy, M. F.,Lauer, M., Gonzaga, F., Schulz, A., Fuhrhop, J.-H., *Langmuir*, 2002, **41**, 2730.
10. Li, G., Fudickar, W., Skupin, M., Klyszcz, A., Draeger, C., Lauer, M., Fuhrhop, J.-H., *Angew. Chem. Int. Ed.*, 2002, **41**, 1828.
11. Komatsu, T., Tsuchida, E., Boettcher, C., Donner, D., Messerschmidt, C., Siggel, U., Stocker, W., Rabe, J.P., Fuhrhop, J.-H., *J. Am. Chem. Soc.*, 1997, **119**, 11660.
12. a) Kelch, S., Rehahn, M., *Macromolecules*. 1997, **30**, 6185.
    b) Kelch, S., Rehahn, M., *Macromolecules*. 1999, **32**, 5818.
13. Kobuke, Y., Miyaji, H., *J. Am. Chem. Soc.*, 1994, **116**, 4111.
14. a) Fuhrhop, J.-H., Schnieder, P., Rosenberg, J., Boekema, E., *J. Am. Chem. Soc.*, 1987, **109**, 3387.
    b) Köning, J., Boettcher, C., Winkler, H., Zeitler, E., Talmon, Y., Fuhrhop, J.-H., *J. Am. Chem. Soc.*, 1993, **115**, 693.
15. Fuhrhop, J.-H., Boettcher, C., *J. Am. Chem. Soc.*, 1990, **112**, 1768.
16. Fuhrhop, J.-H., Spiroski, D., Boettcher, C., *J. Am. Chem. Soc.*, 1993, **115**, 1600.
17. Tachibana, T., Yoshizumi, T., Hori, K., *Bull. Chem. Soc. Jpn.*, 1979, **52**, 34.
18. Fuhrhop, J.-H., Demoulin, C., Boettcher, C., Köning, J., Siggel, U., *J. Am. Chem. Soc.*, 1992, **114**, 4159.
19. Fuhrhop, J.-H., Bindig, U., Siggel, U., *J. Am. Chem. Soc.*, 1993, **115**, 11036.
20. Rosengarten, B., Boettcher, C., Schulz, A., Fuhrhop, J.-H., Siggel, U., *J. Porphyrins Phthalocyanines*, 1998, **2**, 273.
21. Messerschmidt, C., Svenson, S., Stocker, W., Fuhrhop, J.-H., *Langmuir*, 2000, **16**(19), 7445.
22. Schneider, J., Messerschmidt, C., Schulz, A., Gnade, M., Schade, B., Luger, P., Bombicz, P., Hubert, V., Fuhrhop, J.-H., *Langmuir*, 2000, **16**(23), 8575.
23. Ungashe, S. B., Wilson, W. L., Katz, H.E., Scheller, G. R., Putvinski, T. M., *J. Am. Chem. Soc.*, 1992, **114**, 8717.
24. Král, V., Rusin, O., Charvátová, J., Anzenbacher, P., Fogl, J., *Tetrahedron Lett.*, 2000, **41**, 10147.
25. Alberti, G., *Comprehensive Supramolecular Chemistry*, Vol 7 (Lehn, J.-M. ed.), Elsevier Science, p.151, 1996.
26. Hong, H. G., Sackelt, D. D., Mallouk, T. E., *Chem. Mater.*, 1991, **3**, 521.
27. Neuhaus, A., *Fortschr. Mineral.*, 1950/51, **29/30**, 138.
28. van der Boom, T., hayes, R. T., Zhao, Y., Bushard, P. J., Weiss, E. A., Wasielewsi, M. R., *J. Am. Chem. Soc.*, 2002, **124**, 9582.
29. Fudickar, W., Zimmermann, J., Ruhlmann, L., Roeder, B., Siggel, U., *J. Am. Chem. Soc.*, 1999, **121**, 9539.
30. Skupin, M., Li, G., Fudickar, W., Zimmermann, J., Röder, B., Fuhrhop, J.-H., *J. Am. Chem. Soc.*, 2001, **123**, 3454.
31. a) Li, G., Fuhrhop, J.-H., *Langmuir*, 2002, **18**, 7740.
    b) Li, G., Lauer, M., Schulz, A., Boettcher, C., Li, F., Fuhrhop, J.-H., *Langmuir*, 2003, **19**, 6483.

32. Endisch, C., Fuhrhop, J.-H., Buschmann, J., Luger, P., Siggel, U., *J. Am. Chem. Soc.*, 1996, **118**, 6671.

**< Section 3.2 >**

33. a) Hiramatsu, M., Nakano, H., Fujinami, T., Sakai, S., *J. Organomet. Chem.*, 1982, **236**, 131.
  b) Hiramatsu, M., Shiozaki, K., Fujinami, T., Sakai, S., *J. Organomet. Chem.*, 1983, **246**, 203.
  c) Massoud, S. S., Jordan, R. B., *Inorg. Chem.*, 1991, **30**, 4851.
  d) DelMedico, A., Auburn, P. R., Dodsworth, E. S., Lever, A. B. P., Pietro, W. J., *Inorg. Chem.* 1994, **33**, 1583.
  e) Klein, R. A., Elsevier, C. J., Hartl, F., *Organometallics*, 1997, **16**, 1284.
34. a) Hirao, T., Higuchi, M., Ikeda, I., Ohshiro, Y., *J. Chem. Soc., Chem. Commun.*, **1993**, 194.
  b) Hirao, T., Higuchi, M., Ohshiro, Y., Ikeda, I., *Chem. Lett.*, **1993**, 1889.
  c) Higuchi, M., Ikeda, I., Hirao, T., *J. Org. Chem.*, 1997, **62**, 1072.
35. Hirao, T., Fukuhara, S., *J. Org. Chem.*, 1998, **63**, 7534.
36. Higuchi, M., Imoda, D., Hirao, T., *Macromolecules*, 1996, **29**, 8277.
37. a) Allgeier, A. M., Mirkin, C. A., *Angew. Chem. Int. Ed.*, 1998, **37**, 894.
  b) Kingsborough, R. P., Swager, T. M., *Prog. Inorg. Chem.*, 1999, **48**, 123.
  c) Sauvage, J.-P. ed., *Transition Metals in Supramolecular Chemistry*, John Wiley & Sons, Chichester, 1999.
38. a) Haga, M., Ano, T., Kano, K., Yamabe, S., *Inorg. Chem.*, 1991, **30**, 3843.
  b) Sauvage, J.-P., Collin, J.-P., Chambron, J.-C., Guillerez, S., Coudret, C., Balzani, V., Barigelletti, F., De Cola L., Flamigni, L., *Chem. Rev.*, 1994, **94**, 993.
  c) Weng, W., Bartik, T., Gladysz, J. A., *Angew. Chem. Int. Ed. Engl.*, 1994, **33**, 2199.
  d) Ward, M. D., *Chem. Soc. Rev.*, **1995**, 121.
  e) Le Narvor, N., Toupet, L., Lapinte, C., *J. Am. Chem. Soc.*, 1995, **117**, 7129.
  f) Harriman, A., Ziessel, R., *Chem. Commun.*, **1996**, 1707.
  g) Lavastre, O., Plass, J., Bachmann, P., Guesmi, S., Moinet, C., Dixneuf, P. H., *Organometallics*, 1997, **16**, 184.
  h) Jones, N. D., Wolf, M. O., Giaquinta, D. M., *Organometallics*, 1997, **16**, 1352.
  i) Richter, M. M., Bard, A. J., Kim, W., Scmehl, R. H., *Anal. Chem.*, 1998, **70**, 310.
  j) Mccleverty, J. A., Ward, M. D., *Acc. Chem. Res.*, 1998, **31**, 842.
  k) Hissler, M., El-ghayoury, A., Harriman, A., Ziessel, R., *Angew. Chem. Int. Ed.*, 1998, **37**, 1717.
  l) Bartik, T., Weng, W., Ramsden, J. A., Szafert, S., Falloon, S. B., Arif, A. M., Gladysz, J. A., *J. Am. Chem. Soc.*, 1998, **120**, 11071.
  m) Constable, E. C., Housecroft, C. E., Schofield, E. R., Encinas, S., Armaroli, N., Barigelletti, F., Flamigni, L., Figgemeier, E., Vos, J. G., *Chem. Commun.*, **1999**, 869.
39. Venugopal, G., Quan, X., Johnson, G. E., Houlihan, F. M., Chin, E., Nalamasu, O., *Chem. Mater.*, 1995, **7**, 271.
40. Hirao, T., Higuchi, M., Hatano, B., Ikeda, I., *Tetrahedron Lett.*, 1995, **36**, 5925.
41. Higuchi, M., Yamaguchi, S., Hirao, T., *Synlett,* **1996**, 1213.
42. Hirao, T., Higuchi, M., Yamaguchi, S., *Macromol. Symp.*, **1998**, 131.
43. Hirao, T., Yamaguchi, S., Fukuhara, S., *Tetrahedron Lett.*, 1999, **40**, 3009.
44. a) Hirao, T., Moriuchi, T., Mikami, S., Ikeda, I., Ohshiro, Y., *Tetrahedron Lett.*, 1993, **34**, 1031.
  b) Hirao, T., Moriuchi, T., Ishikawa, T., Nishimura, K., Mikami, S., Ohshiro, Y., Ikeda, I., *J. Mol. Catal. A: Chem.*, 1996, **113**, 117.
45. Hirao, T., Yamaguchi, S., Fukuhara, S., *Synthetic Met.*, 1999, **106**, 67.
46. Hirao, T., Fukuhara, S., Otomaru, Y., Moriuchi, T., *Synthetic Met.*, 2001, **123**, 373.
47. Wei, Y., Yang, C., Ding, T., *Tetrahedron Lett.*, 1996, **37**, 731.
48. Hirao, T., Shiori, J. unpublished results.
49. a) Yamamoto, T., Maruyama, T., Zhou, Z., Ito, T., Fukuda, T., Yoneda, Y., Begum, F., Ikeda, T., Sasaki, S., Takazoe, H., Fukuda, A., Kubota, K., *J. Am. Chem. Soc.*, 1994, **116**, 4832.
  b) Shen, X., Moriuchi, T., Hirao, T., *Tetrahedron Lett.*, 2003, **44**, 7711.
50. a) Maruyama, T., Yamamoto, T., *Inorg. Chim. Acta*, 1995, **238**, 9.
  b) Zhu, S. S., Carroll, P. J., Swager, T. M., *J. Am. Chem. Soc.*, 1996, **118**, 8713.
51. a) Rasmussen, S. C., Thompson, D. W., Singh, V., Petersen, J. D., *Inorg. Chem.*, 1996, **35**, 3449.
  b) Peng, Z., Yu, L., *J. Am. Chem. Soc.*, 1996, **118**, 3777.
  c) Zhu, S. S., Swager, T. M., *Adv. Mater.*, 1996, **8**, 497.
  d) Cameron, C. G., Pickup, P. G., *Chem. Commun.*, **1997**, 303.

52. Reddinger, J. L., Reynolds, J. R., *Macromolecules*, 1997, **30**, 673.
53. Moriuchi, T., Bandoh, S., Miyaishi, M., Hirao, T., *Angew. Chem. Int. Ed.*, 2002, **40**, 3042.
54. Moriuchi, T., Bandoh, S., Miyaishi, M., Hirao, T., *Eur. J. Inorg. Chem.*, **2001**, 651.
55. Moriuchi, T., Watanabe, T., Ikeda, I., Ogawa, A., Hirao, T., *Eur. J. Inorg. Chem.*, **2001**, 277.
56. Graf, D. C., Mann, K. R., *Inorg. Chem.*, 1997, **36**, 141.
57. Zotti, G., Zecchin, S., Schiavon, G., Berlin, A., Pagani, G., Canavesi, A., *Chem. Mater.*, 1995, **7**, 2309.
58. Weiberger, D. A., Higgins, T. B., Mirkin, C. A., Liable-Sands, L. M., Rheingold, A. L., *J. Am. Chem. Soc.* 2001, **123**, 2503.
59. a) Bettelheim, A., White, B. A., Raybuck, S. A., Murray, R. W., *Inorg. Chem.*, 1987, **26**, 1009.
    b) Bedioui, F., Merino, A., Devynck, J., Mestres, C., Bied-Charreton, C., *J Electroanal. Chem.*, 1988, **239**, 433.
    c) Segawa, H., Wu, F.-P., Nakayama, N., Maruyama, H., Sagisaka, S., Higuchi, N., Fujitsuka, M., Shimidzu, T., *Synthetic Met.*, 1995, **71**, 2151.
    d) Vollmer, M. S., Würthner, F., Effenberger, F., Emele, P., Meyer, D. U., Stümpfig, T., Port, H., Wolf, H. C. *Chem. Eur. J.*, 1998, **4**, 260.
60. Hirao, T., Saito, K., *Tetrahedron Lett.*, 2000, **41**, 1413.
61. Saito, K., Hirao, T., *Tetrahedron*, 2002, **58,** 7491.
62. For a recent review, Weiss, J., *J. Inclu. Phenom. Macrocylic Chem.*, 2001, **40**, 1.
63. Examples for electron- and energy-transfer systems with axially coordinated zinc porphyrin,
    a) Hunter, C. A., Hyde, R. K., *Angew. Chem. Int. Ed. Engl.*, 1996, **35**, 1936.
    b) Otsuki, J., Harada, K., Araki, K., *Chem. Lett.*, **1999**, 269.
    c) Armaroli, N., Diederich, F., Echegoyen, L., Habicher, T., Flamigni, L., Marconi, G., Nierengarten, J. F., *New J. Chem.*, 1999, **23** 77.
    d) Guldi, D. M., Luo, C., Ros, T. D., Prato, M., Dietel, E., Hirsch, A., *Chem. Commun.*, **2000**, 375.
    e) Ros, T. D., Prato, M., Guldi, D. M., Ruzzi, M., Pasimeni, L., *Chem. Eur. J.*, 2001, **7**, 816.
    f) D'Souza, F., Deviprasad, G. R., El-Khouly, M. E., Fujitsuka, M., Ito, C., *J. Am. Chem. Soc.*, 2001, **123**, 5277.
64. a) Anderson, H. L., Hunter, C., Sanders, J. K. M., *J. Chem. Soc., Chem. Commun.*, **1989**, 226.
    b) Hunter, C. A., Meath, M. N., Sanders, J. K. M., *J. Am. Chem. Soc.*, 1990 **112**, 5773.
    c) Anderson, H. L., *Inorg. Chem.*, 1994, **33**, 972.
    d) Yamada, K., Imahori, H., Yoshizawa, E., Gosztola, D., Wasielewski, M. R., Sakata, Y., *Chem. Lett.*, **1999**, 235.
    e) Guldi, D. M., Luo, C., Swartz, A., Scheloske, M., Hirsch, A., *Chem. Commun.*, **2001**, 1066.
65. Hirao, T., Iida, K., *Chem. Commun.*, **2001**, 431.

**< Section 3.3 >**

66. Schmid, G. ed., *Clusters and Colloids. From Theory to Applications*, VCH, Wienheim, 1994.
67. Toshima, N., Shiraishi,Y., *Encyclopedia of Surface and Colloid Science* (Hubbard, A. T. ed.), p.879, Marcel Dekker, New York, 2002.
68. Hirai, H., Nakao, Y., Toshima, N., Adachi, K., *Chem. Lett.*, **1976**(9), 905.
69. Toshima, N., Kushihashi, K., Yonezawa, T., Hirai, H., *Chem. Lett.*, **1989**(10), 1769.
70. Toshima N., Yonezawa, T., *New J. Chem.*, 1998, **22**(11), 1179.
71. Hirai H., Toshima, N., *Tailored Metal Catalysts* (Iwasawa, Y. ed.), D. Reidel Pub., Dordrecht, p.87, 1985.
72. Toshima, N., *Nanoscale Materials* (Liz-Marzán, L. M. and Kamat, P. eds.), Kluwer Pub, New York, p.79, 2003.
73. Miner, R. S., Namba, S., Turkevich, J., *Proceedings of the 7th International Congress on Catalysis*, p.160, Kodansha, Tokyo, 1981.
74. Toshima, N., Harada, M., Yonezawa, T., Kushihashi, K., Asakura, K., *J. Phys. Chem.*, 1991, **95**(19), 7448.
75. Toshima, N., Harada, M., Yamazaki, Y., Asakura, K., *J. Phys. Chem.*, 1992, **96**(24), 9927.
76. Harada, M., Asakura, K., Toshima, N., *J. Phys. Chem.*, 1994, **98**(10), 2653
77. Yonezawa, T., Toshima, N., *J. Chem. Soc., Faraday Trans.*, 1995, **91**(22), 4111.
78. Toshima, N., Wang, Y., *Langmuir*, 1994, **10**(12),4574.
79. Esumi, K., Tano, T., Meguro, K., *Chem. Mater.*, 1990, **2**(5), 554-567.
80. Bradley, J. S., Hill, E. W., Klein, C., Chandret, B., Duteil, A., *Chem. Mater.*, 1993, **5**, 254.
81. Lu, P., Dong, J., Toshima, N., *Langmuir*, 1999, **15**(23), 7980.

82. Bian, C.-R., Suzuki, S., Asakura, K., Ping, L., Toshima, N., *J. Phys. Chem., B*, 2002, **106**(34), 8587.
83. Lu, P., Teranishi, T., Asakura, K., Miyake, M., Toshima, N., *J. Phys. Chem., B*, 1999, **103**(44), 9673.
84. Turkevich J., Kim, G., *Science*, 1970, **169**, 873.
85. Harada, M., Asakura, K., Toshima, N., *J. Phys. Chem.*, 1993, **97**(17), 5103.
86. Schimid, G., Lehnert, A., Malm, J.-O., Bovin, J.-O., *Angew. Chem. Int. Ed. Engl.*, 1991, **30**, 874.
87. Schmid, G., West, H., Malm, J.-O., Bovin, J.-O, Grenthe, C., *Chem. Eur. J.*, 1996, **2**, 1099.
88. Wang, Y., Toshima, N., *J. Phys. Chem. B*, 1997, **101**(27), 5301.
89. Hirakawa, K., Toshima, N., *Chem. Lett.*, 2003, **32**(1), 78.
90. Toshima, N., Yonezawa, T., Kushihashi, K., *J. Chem. Soc., Faraday Trans.*,1993, **89**(14), 2537.
91. Lu, P., Toshima, N., *Bull. Chem. Soc. Jpn.*, 2000, **73**(3), 751.
92. Toshima, N., Hirakawa, K., *Polym. J.*, 1999, **31**(11-2), 1127.
93. Toshima, N., Shiraishi, Y., Matsushita, T., Mukai, H., Hirakawa, K., *Int. J. Nanosci.*, *2003*, **1**(5-6), 397.

**< Section 3.4 >**

94. Giannelis, E. P., *Appl. Organomet. Chem.*, 1998, **12**, 675.
95. Innocenzi, P., Brusatin, G., *Chem. Mater.*, 2001, **13**, 3126.
96. Wen, J., Wilkes, G. L., *Chem. Mater.*, 1996, **8**, 1667.
97. Loy, D. A., Shea, K. J., *Chem. Rev.*, 1995, **95**, 1431.
98. Livage, J., *Bull. Mater. Sci.*, 1999, **22**, 201.
99. Beecroft, L. L., Ober, K., *Chem. Mater.*, 1997, **9**, 1302.
100. Novak, B. M., *Adv. Mater.*, 1993, **5**, 422.
101. Schmidt, H., Scholze, H., Kaiser, A., *J. Non-Cryst. Solids*, 1984, **63**, 1.
102. Hench, L. L., West, J. K., *Chem. Rev.*, 1990, **90**, 33.
103. Schubert, U., Husing, N., Lorenz, A., *Chem. Mater.*, 1995, **7**, 2010.
104. Chujo, Y., Saegusa, T., *Adv. Polym. Sci.*, 1992, **100**, 11.
105. Chujo, Y., *Curr. Opin. Solid-State Mater. Sci.*, 1996, **1**, 806.
106. Tamaki, R., Chujo, Y., *Chem. Mater.*, 1999, **7**, 1719.
107. Tamaki, R., Samura, K., Chujo, Y., *Chem. Commun.*, **1998**, 1131.
108. Tamaki, R., Naka, K., Chujo, Y., *Polym. Bull.*, 1997, **39**, 303.
109. Tamaki, R., Chujo, Y., *J. Mater. Chem.*, 1998, **5**, 1113.
110. Imai, Y., Naka, K., Chujo, Y., *Polym. J.*, 1998, **12**, 990.
111. Imai, Y., Naka, K., Chujo, Y., *Macromolecules*, 1999, **32**, 1013.
112. Imai, Y., Itoh, H., Naka, K., Chujo, Y., *Macromolecules*, 2000, **33**, 4343.
113. Imai, Y., Yoshida, N., Naka, K., Chujo, Y., *Polym. J.*, 1999, **3**, 258.
114. Ogoshi, T., Itoh, H., Kim, K. M., Chujo, Y., *Macromolecules*, 2002, **35**, 334.
115. Lichtenhan, J. D., Schwab, J. J., Reinerth, W. A., *Chem. Innov.*, 2001, **1**, 3.
116. Haddad, T. S., Stapleton, R., Jeon, H. G., Mather, P. T., Lichtenhan, J. D., Phillips, S., *Polym. Prepr.*, 1999, **40**, 496.
117. Ellsworth, M. W., Gin, D. L., *Polym. News*, 1999, **24**, 331.
118. Lee, A., Lichtenhan, J. D., *J. Appl. Polym. Sci.*, 1999, **73**, 1993.
119. Schwab, J. J., Lichtenhan, J. D., *Appl. Organomet. Chem.*, 1998, **12**, 707.
120. Feher, F. J., Soulivong, D., Eklud, A. G., Wyndham, K. D., *Chem. Commun.*, **1997**, 1185.
121. Feher, F. J., Budzichowski, T. A., *J. Organomet. Chem.*, 1989, **379**, 33.
122. Zhang, C., Laine, R. M., *J. Am. Chem. Soc.*, 2000, **122**, 6979.
123. Zhang, C., Babonneau, F., Bonhomme, C., Laine, R. M., Soles, C. L., Hristov, H. A., Yee, A. F., *J. Am. Chem. Soc.*, 1998, **120**, 8380.
124. Kim, K. M., Keum, D. K., Chujo, Y., *Macromolecules*, 2003, **36**, 867.
125. Kim, K. M., Ouchi, Y., Chujo, Y., *Polym. Bull.*, 2003, **49**, 341.
126. Kim, K. M., Chujo, Y., *J. Mater. Chem.*, 2003, **13**, 1384.
127. Kim, K. M., Chujo, Y., *J. Polym. Sci., Part A: Polym. Chem.*, 2001, **39**, 4035.
128. Kim, K. M., Chujo, Y., *Polym. Bull.*, 2001, **46**, 15.
129. Kim, K. M., Inakura, T., Chujo, Y., *Polym. Bull.*, 2001, **46**, 351.
130. Kim, K. M., Adachi, K., Chujo, Y., *Polym. Prepr. Jpn.*, 2000, **49**, 1757.
131. Kim, K. M., Adachi, K., Chujo, Y., *Polymer*, 2002, **43**, 1171.
132. Kim, K. M., Chujo, Y., *J. Polym. Sci., Part A: Polym. Chem.*, 2003, **41**, 1306.
133. Keum, D. K., Kim, K. M., Naka, K., Chujo, Y., *J. Mater. Chem.*, 2002, **12**, 2449.

134. Itoh, H., Naka, K., Chujo, Y., *Nano Letters*, 2002, **2**, 1183.
135. Duchateau, R., Abbenhuis, H. C. L., van Santen, R. A., Meetsma, A., Thiele, S. K. H., van Tol, M. F. H., *Organometallics*, 1998, **17**, 5663.
136. Crocker, M., Herold, R. H. M., Orpen, A. G., *Chem. Commun.*, **1997**, 2411.
137. Duchateau, R., Cremer, U., Harmsen, R. J., Mohamud, S. I., Abbenhuis, H. C. L., van Santen, R. A., Meetsma, A., Thiele, S. K. H., van Tol, M. F. H., Kranenburg, M., *Organometallics*, 1999, **18**, 5447.

**< Section 3.5 >**

138. Joachim, C., Gimzewski, J. K., A. Aviram, J. K., *Nature,* 2000, **408**, 541.
139. Nishihara, H., *Handbook of Organic Conductive Molecules and Polymers* (Nalwa, H. S. ed.), vol.2, p.799, John Wiley & Sons, 1997.
140. Nishihara,H., *Bull. Chem. Soc. Jpn.*, 2001, **74**, 19.
141. Kurihara, M., Nishihara, H., *Coord. Chem. Rev.*, 2002, **226**, 125.
142. Nishihara, H. *Adv. Inorg. Chem.*, 2002, **53**, 41.
143. Barlow, S., Bunting, H. E., Ringham, C., Green, J. C., Bublitz, G. U., Boxer, S. G., Perry J. W., Marder, S. R., *J. Am. Chem. Soc.*, 1999, **121**, 3715.
144. Barlow, S., Marder, S. R., *Chem. Commun.*, **2000**, 1555.
145. Patai, S., ed., *The Chemistry of the Quinonoid Compounds Part 1 and 2*, John Wiley & Sons, Bristol, 1974.
146. Fukuzumi, S., Ishikawa, K., Hironaka, K., Tanaka, T., *J. Chem. Soc., Perkin Trans. 2*, **1987**, 751.
147. Graige, M. S., Paddock, M. L., Bruce, J. M., Feher, G., Okamura, M. Y. *J. Am. Chem. Soc.*, 1996, **118**, 9005.
148. Hu,Y.-Z., Tsukiji, S., Shinkai, S., Oishi, S., Hamachi, I., *J. Am. Chem. Soc.*, 2000, **122**, 241.
149. Wasielewski, M. R., *Chem. Rev.*, 1992, **92**, 435.
150. Gust, D., Moore T. A., Moore, A. L., *Acc. Chem. Res.*, 1993, **26**, 198.
151. Harriman, A., Sauvage, J.-P., *Chem. Soc. Rev.*, 1996, **25**, 41.
152. Gust, D., Moore, T. A., Moore, A. L., Lee, S.-J., Bittersmann, E., Luttrull, D. K., Rehms, A. A., DeGraziano, J. M., Ma, X. C., Gao, F., Belford, R. E., Trier, T. T., *Science*, 1990, **248**, 199.
153. Raymond, K. S., Grafton A. K., Wheeler, R. A., *J. Phys. Chem. B*, 1997, **101**, 623.
154. Gupta, N., Linschitz, H., *J. Am. Chem. Soc.*, 1997, **119**, 6384.
155. Murata, M., Fujita, T., Yamada, M., Kurihara, M., Nishihara, H., *Chem. Lett.* **2000**, 1328.
156. Murata, M., Ph.D. thesis, The University of Tokyo, 2002.
157. Adams, D. M., Hendrickson, D. N., *J. Am. Chem. Soc.*, 1996, **118**, 11515, and references cited therein.
158. Pierpont, C. G., Lange, C. W., *Prog. Inorg. Chem.*, 1994, **41**, 331.
159. Turbitt, T. D., Watts, W. E., *J. Chem. Soc., Perkin Trans. 2*, **1974**, 177.
160. Olah, G. A., Liang, G., *J. Org. Chem.* 1975, **40**, 1849.
161. Cully, N., Watts, W. E., *J. Organomet. Chem.*, 1979, **182**, 99.
162. Ceccon, A., Giacometti, G., Venzo, A., Paolucci, D., Benozzi, D., *J. Organomet. Chem.*, 1980, **185**, 231.
163. Behrens, U., *J. Organomet. Chem.*, 1979, **182**, 89.
164. Hulme, B. E., Land, E. J., Phillips, G. O., *J. Chem. Soc., Faraday Trans. 1*, 1972, **68**, 1992.
165. Murata, M., Yamada, M., Fujita, T., Kojima, K., Kurihara, M., Kubo, K., Kobayashi, Y., Nishihara, H., *J. Am. Chem. Soc.*, 2001, **123**, 12903.
166. Morrison, W. H., Jr. and Hendrickson, D. N., *Inorg. Chem.* 1975, **14**, 2331.
167. Shon, Y. S., Hendrickson, D. N., Gray, B., *J. Am. Chem. Soc.*,1971, **93**, 3603.
168. Murata, M., Kojima, K., Fujita, T., Kurihara M., Nishihara, H., in preparation.
169. Nishihara, H., Murata, M., Kojima, K., Fujita, T., Kubo K., Kurihara, M., *Polym. Mater. Sci. Eng.*, 2002, **86**, 87.
170. Kurihara, M., Sano, M., Murata, M., Nishihara, H., *Inorg. Chem.*, 2001, **40**, 4.

**< Section 3.6 >**

171. Pittman, C. U., Jr., C. E. Carraher, J., Zeldin, M., Sheats, J. E., Culbertson, B. M. eds. *Metal-Containing Polymeric Materials*, Plenum Press, New York, 1996 and references cited therein.
172. Nguyen, P., Gómez-Elipe, P., Manners, I, *Chem. Rev.*, 1999, **99**, 1515.
173. Rausch, M. D., Chiappenelli, D. J., *J. Organomet. Chem.* 1967, **10**, 127.
174. Rausch, M. D., Moser, G. A., Meade, C. F., *J. Organomet. Chem.*, 1973, **51**, 1.
175. Arimoto, F. S., Haven, A. C., *J. Am. Chem. Soc.*, 1955, **85**, 6259.
176. Aso, C., Kunitake, T., Nakashima, T., *Makromol. Chem.*, 1969, **124**, 232.

177. Chen, Y.-H., Fernandez-Refojo, M., Gassidy, H. G., *J. Polym. Sci.*, 1959, **15**, 433.
178. George, M. H., Hayes, G. F., *J. Polym. Sci.*: *Polym. Chem. Ed.*, 1975, **13**, 1049.
179. George, M. H., Hayes, G. F., *J. Polym. Sci.*: *Polym. Chem. Ed.*, 1976, **14**, 475.
180. Hayes, G. F., Young, R. N., *Polymer*, 1977, **18**, 1286.
181. Sasaki, Y., Walters, L. L., Hurst, E. L., C. U. Pittman, J., *J. Polym. Sci.*: *Polym. Chem. Ed.*, 1973, **11**, 1213.
182. Tinker, A. J., George, M. H., Barrie, J. A., *J. Polym. Sci.*: *Polym. Chem. Ed.*, 1975, **13**, 2133.
183. Simionescu, C., Lixandru, T., Neglescu, I., Mazilu, I., Tataru, L., *Makromol. Chem.*, 1973, **163**, 59.
184. Rasburn, J., Foucher, D. A., Reynolds, W. F., Manners, I., Vancso, G. J., *Chem. Commun.*, **1998**, 843.
185. Rulkens, R., Resendes, R., Verma, A., Manners, I., Murti, K., Fosssum, E., Miller, P., Matyjaszewski, K., *Macromolecules*, 1997, **30**, 8165.
186. Rulkens, R., J.Lough, A., Manners, I., Lovelace, S. R., Grant, C., Geiger, W. E., *J. Am. Chem. Soc.*, 1996, **118**, 12683.
187. Foucher, D. A., Tang, B.-Z., Manners, I., *J. Am. Chem. Soc.*, 1992, **114**, 6246.
188. Foucher, D. A., Edwards, M., Burrow, R. A., Lough, A. J., Manners, I., *Organometallics*, 1994, **13**, 4959.
189. Dement'ev, V. V., Cervantes-Lee, F., Parkanyi, L., Sharma, H., Pannell, K. H., *Organometallics*, 1993, **12**, 1983.
190. Pannell, K. H., Sharma, H. K., *Organometallics*, 1997, **16**, 3077.
191. Kapoor, R. N., Crawford, G. M., Mahmoud, J., Dementiev, V. V., Nguyen, M. T., Diaz, A. F., Pannell, K. H., *Organometallics*, 1995, **14**, 4944.
192. Southard, G. E., Curtis, D., *Organometallics*, 1997, **16**, 5618.
193. Heo, R. W., Somoza, F. B., Lee, T. R., *J. Am. Chem. Soc.*, 1998, **120**, 1621.
194. Buretea, M. A., Tilley, T. D., *Organometallics*, 1997, **16**, 1507.
195. Stanton, C. E., Lee, T. R., Grubbs, R. H., Lewis, N. S., *Macromolecules*, 1995, **28**, 8713.
196. Pudelski, J. K., Gates, D. P., Rulkens, R., Lough, A. J., Manners, I., *Angew. Chem. Int. Ed. Engl.*, 1995, **34**, 1506.
197. Compton, D. L., Rauchfuss, T. B., *Organometallics*, 1994, **13**, 4367.
198. Manners, I., *Adv. Organomet. Chem.*, 1995, **37**, 131.
199. Honeymam, C. H., Fourcher, D. A., Dahmen, F. Y., Rulkens, R., Lough, A. J., Manners, I., *Organometallics*, 1995, **14**, 5503.
200. Yamamoto, T., Maruyama, T., Zhou, Z.-h., Ito, T., Fukuda, T., Yoneda, Y., Begum, F., Ikeda, T., Sasaki, S., Takezoe, H., Fukuda, A., Kubota, K., *J. Am. Chem. Soc.*, 1994, **116**, 4832.
201. Wang, Q., Wang, L., Yu, L., *J. Am. Chem. Soc.*, 1998, **120**, 12860.
202. Walters, K. A., Trouillet, L., Guillerez, S., Schanze, K. S., *Inorg. Chem.*, 2000, **39**, 5496.
203. Peng, Z., Yu, L., *J. Am. Chem. Soc.*, 1996, **118**, 3777.
204. Schulz, G. E., Schirmer, R. H., *Principle of Protein Structure*, Springer-Verlag GmbH & Co. KG, Berlin and Heidelberg, 1979.
205. Krauthäuser, S., Christianson, L. A., Powell, D. R., Gellman, S. H., *J. Am. Chem. Soc.*, 1997, **119**, 11719.
206. Shankaramma, S. C., Singh, S. K., Sathyamurthy, A., Balaram, P., *J. Am. Chem. Soc.*, 1999, **121**, 5360.
207. Lumme, P. O., Knuuttila, H., *Polyhedron*, 1995, **14**, 1553.
208. Biradha, K., Desiraju, G. R., Braga, D., Grepioni, F., *Organometallics*, 1996, **15**, 1284.
209. Antsyshkina, A. S., Porai-Koshitz, M. A., Khandlovich, M., Ostrikova, V. N., *Koord. Khim.*, 1981, **7**, 461.
210. Hergold-Brundic, A., Kaitner, B., Kamenar, B., Leovac, V. M., Iveges, E. Z., Juranic, N., *Inorg. Chim. Acta*, 1991, **188**, 151.
211. Romero, M. A., Sánchez, M. P., Quirós, M., Sánchez, F., Salas, J. M., Moreno, M. N., Faure, R., *Can. J. Chem.*, 1993, **71**, 29.
212. Ulvenlund, S., Georgopoulou, A. S., Mingos, D. M. P., Baxter, I., Lawrence, S. E., White, A. J. P., Williams, D. J., *J. Chem. Soc., Dalton Trans.*, **1998**, 1869.
213. Price, C., Rees, N. H., Elsegood, M. R. J., Clegg, W., Houlton, A., *J. Chem. Soc., Dalton Trans.*, **1998**, 2001.
214. Hay, R. W., Pujari, M. P., Korybut-Daszkiewicz, B., Ferguson, G., Ruhl, B. L., *J. Chem. Soc., Dalton Trans.*, **1989**, 85.

215. Kumar, K., Chang, C. A., Francesconi, L. C., Dischino, D. D., Malley, M. F., Gougoutas, J. Z., Tweedle, M. F., *Inorg. Chem.*, 1994, **33**, 3567.
216. Maumela, H., Hancock, R. D., Carlton, L., Reibenspies, J. H., Wainwright, K. P., *J. Am. Chem. Soc.*, 1995, **117**, 6698.
217. Dupray, L. M., Meyer, T. J., *Inorg. Chem.*, 1996, **35**, 6299.
218. Taniguchi, M., Ueno, N., Okamoto, K., Karthaus, O., Shimomura, M., Yamagishi, A., *Langmuir*, 1999, **15**, 7700.
219. Storrier, G. D., Takada, K., Abruna, H. D., *Langmuir*, 1999, **15**, 872.
220. Hou, Y.-J., Xie, P.-H., Zhang, B.-W., Cao, Y., Xiao, X.-R., Wang, W.-B., *Inorg. Chem.*, 1999, **38**, 6320.
221. Hesek, D., Inoue, Y., Everitt, S. R. L., Ishida, H., Kunieda, M., Drew, M. G. B., *Inorg. Chem.*, 2000, **39**, 308.
222. Ohkubo, K., Hamada, T., Inaoka, T., Ishida, H., *Inorg. Chem.*, 1989, **28**, 2021.
223. Ohkubo, K., Hamada, T., Ishida, H., *J. Chem. Soc., Chem. Commun.*, **1993**, 1423.
224. Beer, P. D., Szemes, F., Balzani, V., Sala, C. M., Drew, M. G. B., Dent, S. W., Maestri, M., *J. Am. Chem. Soc.*, 1997, **119**, 11864.
225. Storrier, G. D., Colbran, S. B., Craig, D. C., *J. Chem. Soc., Dalton Trans.*, **1997**, 3011.
226. Bachas, L. G., Cullen, L., Hutchins, R. S., Scott, D. L., *J. Chem. Soc., Dalton Trans.*, **1997**, 1571.
227. Pickup, P. G., Osteryoung, R. A., *Inorg. Chem.*, 1985, **24**, 2707.
228. Gallagher, J., Zelenko, O., Walts, A. D., Sigman, D. S., *Biochemistry*, 1998, **37**, 2096.
229. Imahori, H., Hagiwara, K., Akiyama, T., Aoki, M., Taniguchi, S., Okada, T., Shirakawa, M., Sakata, Y., *Chem. Phys. Lett.*, 1996, **263**, 545.
230. Tamaki, K., Imahori, H., Nishimura, Y., Yamazaki, I., Shimomura, A., Okada, T., Sakata, Y., *Chem. Lett.*, **1999**, 227.
231. Braga, D., Grepioni, F., Sabatino, P., Desiraju, G. R., *Organometallics*, 1994, **13**, 3532.
232. Meurs, F. v., Koningsveld, H. v., *J. Organomet. Chem.*, 1974, **78**, 229.
233. Patel, B. P., Crabtree, R. H., *J. Am. Chem. Soc.*, 1996, **118**, 13105.
234. Severin, K., Bergs, R., Beck, W., *Angew. Chem. Int. Ed.*, 1998, **37**, 1635 and the cited references.
235. Ederer, T., Sunkel, K., Beck, W., *Z. Anorg. Allg. Chem.*, 1999, **625**, 1202.
236. Butler, I. R., Quayle, S. C., *J. Organomet. Chem.*, 1998, **552**, 63.
237. Okamura, T., Sakauye, K., Ueyama, N., Nakamura, A., *Inorg. Chem.*, 1998, **37**, 6731.
238. Lu, S., Strelets, V. V., Ryan, M. F., Pietro, W. J., Lever, A. B. P., *Inorg. Chem.*, 1996, **35**, 1013.
239. Okamura, T., Iwamura, T., Nozaki, K., Ohno, T., Yamamoto, H., Ueyama, N., *Mol. Cryst. Liq. Cryst.*, 2002, **379**, 431.
240. Newkome, G. R., Gross, J., Patri, K., *J. Org. Chem.*, 1997, **62**, 3013.
241. Lay, P. A., Sargeson, A. M., Taube, H., *Inorg. Synth.*, 1986, **24**, 291.
242. Yu, S. C., Gong, X., Chan, W. K., *Macromolecules*, 1998, **31**, 5639.
243. Drew, H. D. K., Pinkard, F. W., Preston, G. H., Wardlaw, W., *J. Chem.Soc.*, **1932**, 1895.
244. Gidney, P. M., Gillard, R. D., Heaton, B. T., *J. Chem.Soc., Dalton Trans.*, **1973**, 132.
245. Shaw, J. R., Webb, R. T., Schmehl, R. H., *J. Am. Chem. Soc.*, 1990, **112**, 117.

# 4
# Biological Macromolecular Complexes

# 4.1 Macromolecular Metal Complexes in Biological Systems

**Jan Reedijk**

## 4.1.1 Introduction

Metal ions and metal complexes are crucial in living systems, both for growth and in metabolism. The most well known and perhaps most prominent role is that of iron. As an essential metal iron has been known since the 18th century, whereas the role of other elements, such as molybdenum, manganese, cobalt, copper and zinc has been reported since the beginning of the 20th century.[1)]

A shortage of these metals can lead to serious health problems. It has also been known for a long time that an excess of these elements can be very dangerous. As a matter of fact, only a narrow safe concentration window exists for most of the so-called trace elements. Despite the fact that many elements of the periodic table are nowadays known and accepted to be essential, or beneficial for life on earth, the molecular role of many of these elements is only beginning to be understood.[1, 2)]

Very many other nonessential elements of the periodic table are known to have an influence on the quality of life, either as a toxic pollutant or as a drug to cure certain diseases. Several examples of these elements are well documented nowadays, although still little is understood about the details of their modes of action on the molecular level. An increasingly important challenge of modern bioinorganic chemistry is to better understand the molecular basis of such interactions.[3)] An even greater challenge is to apply this knowledge in, e.g., biology, catalysis, environmental sciences, medicine, polymer sciences and technology.

The present section, which also comprises an update of earlier overviews,[4, 5)] focuses on the coordination aspects of these metals in biopolymers, both in their natural environments (e.g., bound to proteins and nucleic acids) and on the effect of nonessential metal compounds on these biological macromolecular systems.

To introduce the reader to the field, some general aspects of bioinorganic chemistry will first be presented. In Table 4.1 a selection of important elements is listed, together with some statistical information and a few comments about their biological role.[5)] The list is restricted to some of the most important transition elements, but Ca and Mg have also been included.

Starting in the late 1960s, the field of bioinorganic chemistry has grown rapidly to a large field of science with a variety of specialist conferences on the

**Table 4.1** Some of the most important elements in biology. Occurrence, concentration of element and biological roles.

| Element name | Concentrations in human species(mg) | Daily required(mg)[a] | Some comments about role in biology |
|---|---|---|---|
| Calcium | $10^6$ | 800 | Bones, teeth; muscle activation |
| Magnesium | $4\times10^4$ | 350 | Photosynthesis; nucleic acid processes |
| Iron | 4500 | 10 | Many enzymes; respiratory proteins |
| Zinc | 2000 | 12 | Hydrolytic enzymes; nucleic acid synth. |
| Copper | 100 | 2 | Many enzymes; dioxygen transport |
| Manganese | 20 | 3 | Enzyme activation, photosynthesis |
| Molybdenum | 5 | 0.2 | Redox enzymes; nitrogenase (plants) |
| Cobalt | 1 | 0.3 | Vitamin $B_{12}$ as most known example |

[a] Generally accepted daily amounts required for humans.

international level. Scientists from several disciplines now contribute to this interdisciplinary area. The role of metal ions, metal-containing proteins and other metal compounds in relation to living systems is a subject of common interest in this field. Bioinorganic chemistry has been subdivided in various ways as follows.

- Study of the structural details at atomic resolution of the metal coordination environment in metalloproteins, nucleic acids, carbohydrates, phospholipids, membranes;
- Study of the mechanism of reactions that occur at a metal center in an enzyme;
- Studies towards changes (structure, reactivity) in genetically modified active-site structures of metalloproteins;
- Design and synthesis of synthetic analogues for the active sites in metalloproteins (design, synthesis, structure, spectroscopy and catalytic reactions);
- Synthesis and application of metal-containing drugs to prevent or to cure diseases (including design of new compounds and study of their mechanism of action);
- Removal and controlled transport of metal ions and metal compounds to and from living systems (detoxification studies);
- Study of the process of biomineralization, i.e., the build in of metal ions in solid materials and their control by macromolecules.

The molecular interaction between metal ions and biomacromolecules is an essential element in this area. The metal and its surrounding neighbors (called ligands) are of importance for the structure, the stability and the processes regulated, or distorted, by the metal species. The study of metal ions and their interactions with ligands is usually called coordination chemistry, or more specifically "biocoordination chemistry," when either the metal or the ligand has a biological role.

Metal-ligand interactions in biological systems play a key role in almost every important event that takes place during biological processes, both natural processes and human-induced processes (drugs, pollution, cleaning). Metals are kept fixed in their position through coordination bonds with the ligands. Ligands are usually oriented, polarized, activated, etc. through specific interaction with certain metal ions. Excesses of (natural or non-natural) metals can be removed by chelation with natural or specifically designed chelating ligands.

Certain (so-called xenobiotic) ligands may interact with metals in living tissue, thereby blocking coordination sites and hampering normal reactions, in some cases with dramatic consequences. Certain toxic metals may substitute nontoxic metal ions in metalloproteins, or may bind to ligands *in vivo* (protein side groups, nucleic acids), changing their reactivities, or also by fixation of certain conformations. Metal compounds may be used as drugs with the specific purpose of introducing either the metal, or the ligand, or both, to certain positions in the living system (metal-containing drugs; metal-containing antibodies; metallo-radiopharmaceuticals). To understand some of these processes in more detail, a classification of metals, ligands and their mutual interactions is desired, and will be presented below.

It should be mentioned here that in certain areas of the life sciences, the use of the word "ligand" has also been introduced for a (small) substrate that binds to a larger molecule, like DNA or a protein. In such cases, $Ca^{2+}$, for example, can be a ligand for the protein thermolysin to stabilize its structure; however, at the same time the carboxylate side chains of the thermolysin backbone are ligands for the calcium ions! In this chapter, the use of ligand will only be in combination with a metal.

## 4.1.2 Small Ligands and Polymeric Ligands Present in Biology

### A. Coordination Behavior of Ligands towards Metal Ions

A quite important, if not dominant, property of the metal coordination compounds is the fact that the M-ligand bond, which has the thermodynamic strength of a typical coordination bond (say 100 kJ/mole or below), and these are much weaker than (covalent) C-C and C-N or C-O single and double bonds (which are between 250 and 500 kJ/mole).[6)]

Underlying principles to describe different ligand preferences for different metals have been known for years, both theoretically and empirically. For a clear understanding the so-called HSAB theory[7)] is a quite useful approximation. It is sufficient to know that ions such as Ca, Mg, Na, K, Mn belong to the ionic (class A) group, whereas metal ions such as Pt, Hg, Cd, Pb belong to the more polarizable group (class B). Transition-metal ions such as Ni, Cu and Zn are considered to be intermediate between A and B. Ligands usually are classified according to their donor sites (i.e., the metal-binding atoms): ionic or polar ligands are those with an oxygen-donor group (carboxylate, alcohol), the polarizable ones are those with sulfur or phosphorus donor atoms (thioethers, thiolates, phosphanes). Nitrogen-donor ligands are generally considered to be ligands with intermediate properties.

As a basic rule of thumb it is practical to consider the following guidelines:

· The more ionic metals preferentially bind with ionic ligands; and

· The more polarizable metal ions preferentially bind with polarizable ligands.

In all cases of metal binding to ligands one should distinguish (reversible) kinetically fast and kinetically slow binding metals. Here the pioneering work of Taube[8] has been crucial for the development of the understanding. Taube showed and nicely explained that ligand-exchange processes for the same ligands, but for a variation of metal ions, can be at least 14 orders of magnitude different. In the case of small ligands, like water, the ligand exchange for metals ions like Pt(II), Ru(II), Ru(III) and low-spin Co(III) is on the order of magnitude of a few hours, whereas for ions like Ni(II), Cu(II) this process is on the order of magnitude of microseconds. Of course when chelating, large ligands, like polymers, totally surround the metal ions, the ligand-exchange kinetics can be significantly lowered.

In biological systems metals can be coordinated to a variety of biomacromolecules, including

1) proteins, and in particular to their polar side chains (N, O, S);
2) nucleic acids, both at phosphates and at their nucleobase N-donor atoms;
3) carbohydrates, lipids (C=O ; P=O).

Although not considered as a part of coordination chemistry, in a sense also in inorganic oxidic solids (bone, teeth, deposits like kidney stones, magnetite), ligands do bind to metal ions.

## B. Brief Overview of Ligand Groups that Bind to Metals in Biological Systems

Frequently found metal-binding ligands in biological systems are:

a) polar protein side groups, such as imidazole, alcohol, thiolate, carboxylates from protein side chains; some of these are depicted in Fig. 4.1 with their donor sites; in several cases dehydronation (i.e., loss of a hydron, $H^+$) takes place upon binding to a metal, even at pH=7;
b) prosthetic groups, such as tetrapyrroles and cofactors, like PQQ and FAD (in many cases these cofactors are covalently attached to the protein, and therefore form a part of the biopolymer); a few structures are redrawn in Fig. 4.2.
c) small ions, such as $OH^-$, $O_2^{2-}$, $O_2^-$ (a radical; so a better formula would be $O_2^{\bullet-}$), $S_2^{2-}$, $OOH^-$, $CO_3^{2-}$ and small molecules (like $O_2$, $H_2O$);
d) nucleic acid bases, such as guanine and adenine (see Fig. 4.3, for some structures and the preferred metal binding sites);
e) phosphates and alcohol groups such as those present in nucleic acids and in ADP, ATP and GTP.

In Table 4.2 a brief overview of the ligands is given with examples of their biological presence in enzymes.

A somewhat different classification of ligands and metal ions in biological systems, is given in Table 4.3, where the known cases are tabulated in a matrix form, illustrating the relative high frequency and general occurrence of certain ligands like the imidazole group.

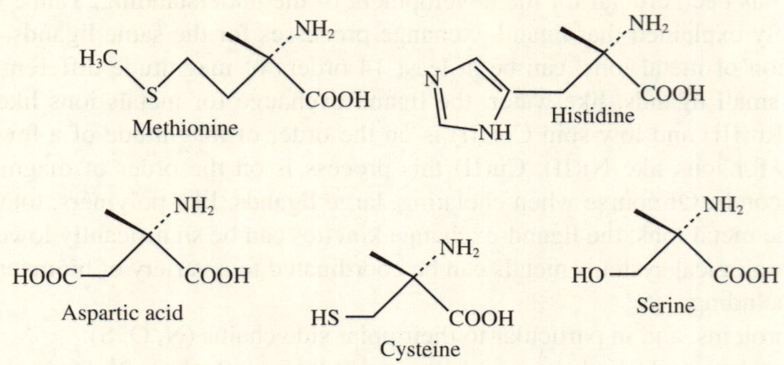


**Fig. 4.1** Selection of ligands and ligand groups (amino acids with coordination side chains) that are frequently found in metalloproteins.

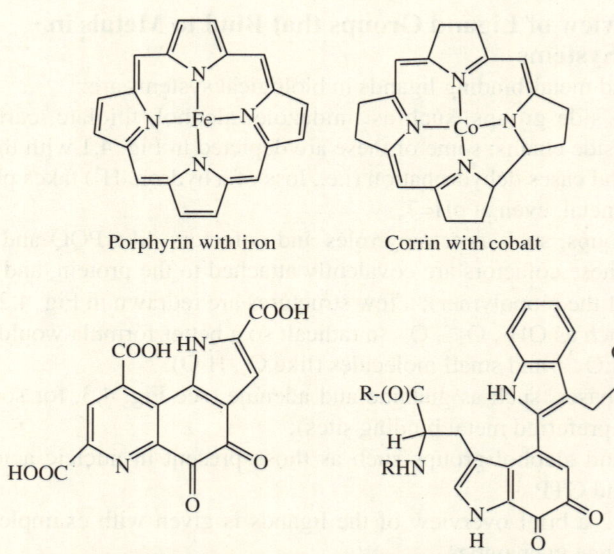


**Fig. 4.2** Selection of prosthetic groups that are frequently found in metalloproteins. Only the basic skeleton is shown, and substituents have been deleted.

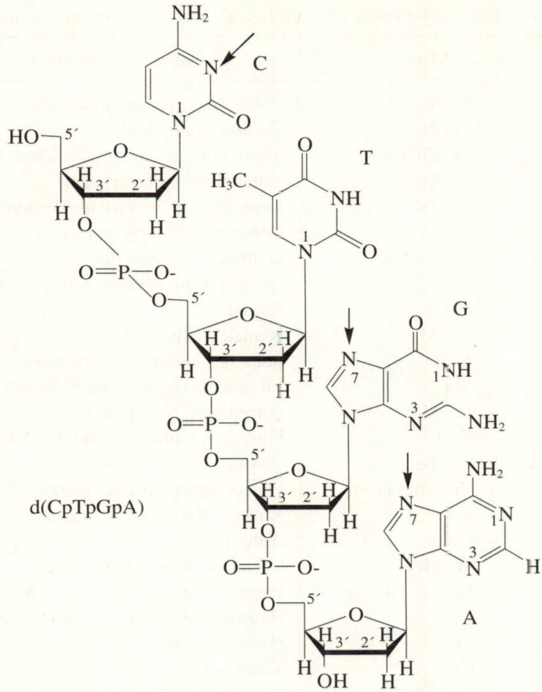


**Fig. 4.3** The four major bases in DNA connected in a single-strand chain; the most important metal-binding sites at neutral pH are indicated with arrows.

### 4.1.3 Reactivity of Metals Coordinated to Biological Ligands General Aspects

Metal ions and ligands that undergo reactions *in vivo* are generally part of a complex system of equilibria, transport and storage. Kinetics and thermodynamics both play key roles in these reactions. The simple presence of a metal and a ligand in a biological system does not necessarily result in a reaction. Either the coordination or the catalytic reaction is far too slow (such as the introduction of a metal into a tetrapyrrole system), or the number of competing reactions with other metal ions or ligands leads to a variety of products. The kinetics of biological reactions dealing with Fe, Cu and Co are beginning to be somewhat understood. Reactions with other metal ions, are far less well known. An example of a biomedical reaction which is beginning to be understood is the mechanism of action of platinum antitumor compounds (*vide infra*). Starting with drug administration and transport through the body, the drug enters the cell, is hydrolyzed and binds – mainly, but not exclusively – to DNA. Details of such a case will be presented below.

**Table 4.2** Important biologically relevant ligands and their presence in natural systems

| Ligand Group | Metal | Detected or proven in: |
|---|---|---|
| =O, $O^{2-}$ | Fe, V | P-450 enzymes, haloperoxidases |
| $-OH^-$ | Fe, Zn | Carbonic anhydrase |
| $H_2O$ | Fe, Zn, Ca | Many proteins; as an additional, labile ligand |
| $O_2^{2-}$ | Fe, Cu | Hemocyanin, hemerythrin |
| $O_2^{\bullet -}$ | Cu, Fe | Hemoglobin, Superoxide dismutase |
| $-OOH^-$ | Fe | Hemerythrin, Lipoxygenases |
| Tyrosine | Fe | Oxidases, Transferrin |
| Glu (and Asp) | Fe | Hemerythrin, ribonucleotide reductase |
| Carbamate ($-NHCO_2^-$) | Ni | Urease |
| $OPO_2R^-$ | Ca, Mg | Nucleic acids; ATP |
| $NO_3^-$, $SO_3^{2-}$ | Mo | Substrate in several reductases |
| $-Cl^-$ | Mn | Mn cluster in photosynthesis (PS-II) |
| $-S^{2-}$ | Fe, Mo | Ferredoxin; Nitrogenase |
| $-SR^-$ (cysteine) | Fe, Cu | Ferredoxin, plastocyanin, P-450, azurin |
| Me-S-R (methionine) | Cu, Fe | Plastocyanin, cytochromes |
| Imidazole | Cu, Zn, Fe, Mn | Plastocyanin, insulin, hemoglobin, hemocyanin |
| Benzimidazole | Co | Vitamin $B_{12}$ |
| $(N<)^-$ (peptide) | Cu | Albumin |
| Pterin | W, Mo | Co-factor in several oxidases |
| Tetrapyrroles | Fe, Co, Ni, Mg | Prosthetic groups, such as in hemoglobin |
| CO | Fe | Hydrogenases; Toxic ligand for cytochrome oxidase |
| $CN^-$ | Fe | Hydrogenases |
| $(CH_2\text{-}R)^-$ | Co | Vitamin $B_{12}$ |

## A. Metal Ions May Have Only a Structural or Structure-organizing Role

The role of a metal ion in biomacromolecules, such as those present in a biological system, can be limited to a structural function, i.e., holding together certain biomolecules (or fragments of them) in a more or less rigid structure. Well-known examples are $Ca^{2+}$ in thermolysin (a protein which has Zn at the active site), Zn in bovine superoxide dismutase and in liver alcohol dehydrogenases (both at noncatalytic sites) and in several solid structures, such as bones and teeth. The structural role of magnetite ($Fe_3O_4$) found in bacteria, honey bees and more recently also in pigeons and higher organisms, like humans, is worthwhile mentioning here. It appears that single crystals of $Fe_3O_4$ are used—probably in combination with other objects—for orientation, using the earth's magnetic field.[9)] The role of zinc fingers in gene expression is the result of Zn-induced protein folding. This topic has been receiving increasing attention during the last decade, when high-resolution NMR structures and even a number of X-ray structures became available[10)] of the so-called zinc fingers and their adducts with pieces of DNA.[11)]

## B. Transport and Storage of Metal Ions by Biomacromolecules

Living systems using metal ions in several locations require efficient transport of metal ions. Well-known examples are transferrin for iron transport in humans and

**Table 4.3** Overview of ligands in biological systems: yes = proven binding site; prob = probable binding site; ? a suspected binding site

| Metal :<br>Ligand : | Mg | Ca | Mn | Fe | Co | Ni | Cu | Zn | Mo | W | V |
|---|---|---|---|---|---|---|---|---|---|---|---|
| $O^{2-}$ | | | ? | yes | | | | | yes | prob | yes |
| $OH^-$ | | | yes | yes | | prob | ? | yes | yes | | |
| $OOH^-$ | | | prob | ? | | | ? | | | | |
| $O_2^{2-}$ | | | prob | yes | | | yes | | | | |
| $O_2^-$ | | | ? | yes | | | yes | | | | |
| $O_2$ | | | ? | | | | | | | | |
| $H_2O$ | yes | yes | yes | yes | | yes | yes | yes | | | |
| $H_2$ | | | | prob | | prob | | | | | |
| $N_2$ | | | | prob | | | | | ? | | |
| $RNH_2$ | | | | yes | | | | | | | |
| imidazole | ? | yes | yes | yes | yes | yes | yes | yes | yes | | yes |
| porphyrin | yes | | | yes | | | ? | | | | |
| corrin | | | | | yes | | | | | | |
| corphin | | | | | | yes | | | | | |
| $RCOO^-$ | yes | yes | yes | yes | | yes | | yes | yes | | |
| $SO_3^{2-}$ | | | | | | | | | ? | | |
| $NO_3^-$ | | | | | | | | | ? | | |
| $RPO_3^{2-}$ | yes | | | ? | | | | ? | | | |
| $S^{2-}$ | | | | yes | | ? | yes | | yes | yes | |
| $RS^-$ | | | | yes | | yes | yes | yes | yes | yes | |
| $RSCH^3$ | | | | yes | | | yes | | ? | | |
| $O(Tyr)^-$ | | | ? | yes | | | yes | | | | |
| OH(ser) | yes | yes | | | | | | | yes | | |
| OH(thre) | yes | yes | | | | | | | | | |
| C=O | yes | yes | yes | | | ? | yes | | | | |

albumin for copper transport, as well as ferritin for iron storage.[12,13] Apart from these natural transporting proteins, nature also uses efficient systems to remove excesses of toxic metal ions. Examples are the metallothioneins that take care of metal ions such as cadmium,[14] and excesses of zinc and copper.[2]

## C. Catalysis by Metal Ions Present in Biomacromolecules

In many cases the metal ion and the binding of the ligands play an important catalytic role, and the study of this role is the main theme of the present monograph. This catalysis role can be in electron transfer only; it can also be a complicated redox reaction, or a relatively simple acid-base reaction. The presence of the metal at the particular site usually results in activation reactions of substrates by the metal and the surrounding ligands.

All enzymes were classified many years ago into six major groups,[15] an important classification which also holds for the many metal-containing enzymes and which is used worldwide.

**Table 4.4** Overview of important reactions of metalloenzymes with small molecules

| Enzyme | Substrate | Reaction | Relevant metals |
|---|---|---|---|
| Cytochrome oxidase | $O_2$ | $O_2 + 4\,H^+ + 4e^- \rightarrow 2\,H_2O$ | Cu, Fe, Zn, Mg |
| Superoxide dismutase | $O_2^-$ | $2\,O_2^- + 2\,H^+ \rightarrow H_2O_2 + O_2$ | Cu, Zn |
| Peroxidases | $H_2O_2$ | $2\,H_2O_2 \rightarrow 2\,H_2O + O_2$ | Fe |
| Nitrogenase | $N_2$ | $N_2 + 8H^+ + 8e^- \rightarrow H_2 + 2\,NH_3$ | Fe, Mo |
| Methane monooxygenase | $CH_4$ | $2\,CH_4 + O_2 \rightarrow 2\,CH_3OH + H_2O$ | Fe |
| Photosystem-II | $H_2O$ | $2\,H_2O \rightarrow 4\,H^+ + 4e^- + O_2$ | Mn, Ca |

1. Oxidoreductases (catalyzing oxidation-reduction reactions); in this group one usually distinguishes between dehydrogenases (all used dioxygen or hydrogen peroxide is converted into water, oxygenase (one of the oxygen atoms is introduced in the substrate) and dioxygenases (two oxygen atoms are used to be built into the substrate);
2. Transferases (transferring an organic group from a donor molecule to an acceptor molecule);
3. Hydrolases (hydrolytically cleaving of C-O or C-N bonds);
4. Lyases (cleaving C-N, C-O or C-C bonds by elimination);
5. Isomerases (geometric or structural change within a molecule);
6. Ligases (these join together two molecules or fragments; often coupled with pyrophosphate hydrolysis).

Usually enzymes are classified by a so-called EC number, consisting of 3 or 4 digits separated by dots, to indicate subgroups and sub-subgroups. Well-known metal-containing examples are nitrogenase (EC 1.18.6.1), bovine superoxide dismutase (SOD, EC 1.15.1.1) and urease (EC 3.5.1.5).

It is evident from the lists of classified enzymes[15)] that very often a metal is involved, either directly at the active center, or indirectly at another place, for instance in the electron transfer process. In fact, when a difficult chemical job has to be done in a natural system, it will in almost all cases involve the use of metals in a biopolymer. Table 4.4 is a list of important reactions, often requiring more than one metal ion. A recent monograph on bioinorganic catalysis is available[16)] for more details and references.

## D. Toxic and Medicinal Functions of Metal Ions Related to Biomacromolecules

Metal ions not only play a role in structure, transport or catalysis, but they may also have deleterious effects as a toxic metal (see above), or play an important role as a metal-containing drug or diagnostic agent. Examples of the last category are:

- Gold coordination compounds with S-donor ligands in the treatment of arthritis;
- Bismuth carboxylate salts in the treatment of ulcers;
- Platinum-amine compounds as antitumor drugs;

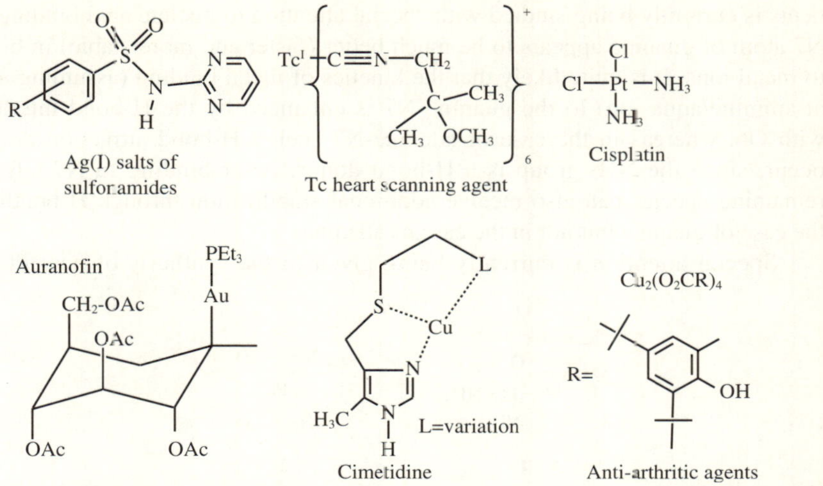


**Fig. 4.4** Examples of metal coordination compounds which are used in medicine: copper, silver, gold, platinum drugs and a Tc scanning agent.

· Radiopharmaceutical imaging with technetium coordination compounds;
· Magnetic resonance imaging with gadolinium compounds.

A selection of such compounds has been redrawn in Fig. 4.4. Reviews on this matter are available.[17-19] An example of our own work deals with the binding of platinum-antitumor compounds to DNA, and the resulting distortion of the DNA and its biological consequences will be discussed in some detail in the next section.

## E. Binding of Pt-antitumor Drugs to DNA

The simple coordination compound *cis*-$PtCl_2(NH_3)_2$, now also known as "cisplatin," is a quite successful antitumor drug in the treatment of urogenital and other tumors. In fact it is the agent of choice for the treatment of testicular and ovarian cancer, and is increasingly used against cervical, bladder and head/neck tumors. Derivative compounds of this type and in particular complexes belonging to the *cis*-$PtCl_2$(amine)$_2$ structural class, show equal biological activity when tested on tumor cell lines.[20,21]

New structural classes of Pt antitumor drugs that possess chemical and biological properties quite different from those of cisplatin are also now known, and such compounds are expected to be of great use in the clinic to overcome the problems of resistance that certain patients develop against cisplatin treatment. Some recently discovered examples of such a new class and which may provide the basis for expanding the spectrum of activity of Pt-based drugs are asymmetric Pt compounds, compounds with larger leaving groups and even dinuclear Pt compounds.[22-24] A selection of these new compounds is depicted in Fig. 4.5.

The specific interaction of these platinum compounds with cellular components is currently being studied with special attention to nucleic acid binding. The N7 atom of guanine appears to be much better (faster and more stable) in binding to metal ions. It is quite likely that the kinetics of metal binding (assuming a metal-ammine/aqua ion) to the guanine-N7 is enhanced by the H-bond interaction with O6, whereas in the case of adenine-N7 such a H-bond attraction does not occur, since the $NH_2$ group is a H-bond donor. After binding to N7 sites, the remaining species can also receive additional stabilization through H-bonding in the case of guanine but not in the case of adenine.

Special attention is currently being given to the synthesis of new Pt com-

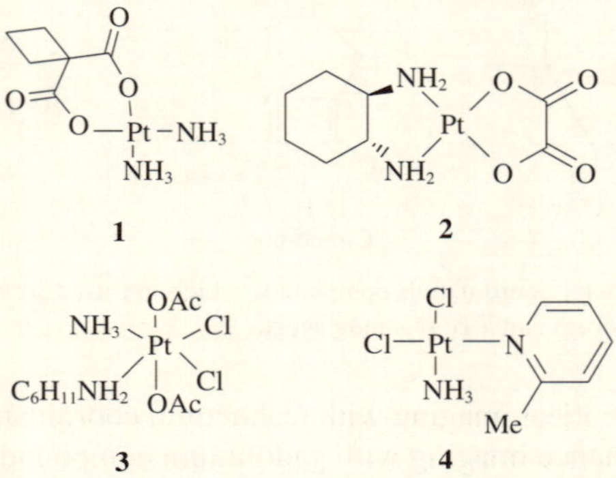

**Fig. 4.5** Examples of some other platinum coordination compounds that are antitumor active. Compound 1: Carboplatin; 2: oxaliplatin; 3: JM-216; 4: AMD-473.

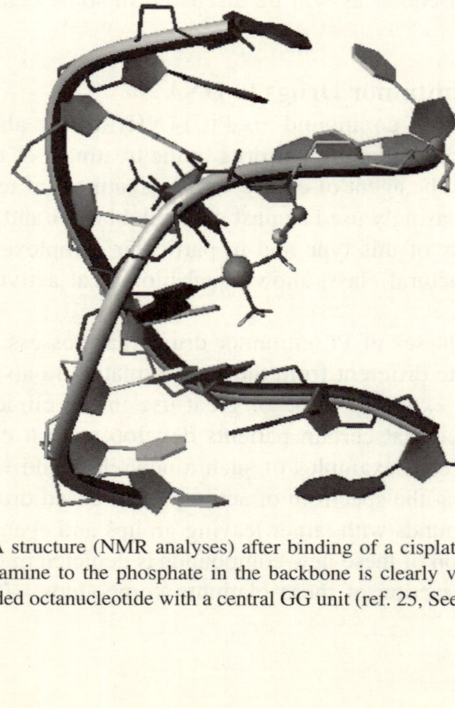

**Fig. 4.6** Kinked DNA structure (NMR analyses) after binding of a cisplatinum unit. The Hydrogen bond of the amine to the phosphate in the backbone is clearly visible. Data taken from a double stranded octanucleotide with a central GG unit (ref. 25, See text).

pounds and to their specific interaction with nucleic acids and their synthetic fragments. Structural changes in DNA (fragments) after the binding of a metal currently enjoy special attention, since Pt is known to bind to DNA, and this is generally accepted to be the biological target leading to cell killing. A summary of our earlier work dealing with the distortion of the DNA is as follows.[8,20,21]

1. In the DNA adducts of cisplatin the sites with AG and GG (A=adenosyl, G=guanosyl). are most frequently chelated, and the two bases are coordinated through N7 to *cis*-Pt in a "head-to-head" orientation. The conformational characteristics hardly change upon platination; in fact only the deoxyribose moiety of the 5'-guanine has adopted an almost pure N-type conformation, compared to the S-type conformation in free d(GpG).
2. The distortion of double-stranded DNA oligonucleotides after cisplatin binding is rather small, as deduced from analysis of proton and phosphorus nuclear magnetic resonance (NMR) spectra and consideration of the circular dichroism (CD), and later X-ray diffraction. The conclusion was reached that the double helix is somewhat destabilized after the platination, as reflected by the decrease in the melting temperature of the duplex by 10−20 °C with concentrations as used in NMR (3 mM). Basepair formation by hydrogen bonding appears still possible after the platination, as reflected by the appearance of imino-proton resonances. For the central GG sequence, these signals are only observed at low temperature, although shifted to lower field and broadened.
3. Chemical shift and coupling constant analysis from the NMR spectra followed by conformational analysis suggests a rather small but significant distortion of the double helix of the larger oligonucleotides upon platination, which has been described as a "kink" of about 40−70 ° in the helical axis at or around the GG lesion. A structure of such a distortion[25] is presented in Fig. 4.6.
4. Comparison of CD spectra and $^{31}P$ NMR spectra of both platinated DNA from several sources and the platinated double-stranded (ds) decanucleotide strongly suggests similar distortions for both DNA, plasmids and oligonucleotides.
5. The interactions with the 5'-phosphate group in the DNA seem to be important, i.e., it is involved in a hydrogen bond with an $NH_3$ ligand of platinum. This could be the origin for the observation that active platinum antitumor drugs need an acid N-H group. The phosphate-ammonia interaction might quite well induce and/or stabilize DNA distortions, thereby interfering with the replication process. Later work of Lippard has shown that certain proteins specifically recognize this kinked-platinated DNA structure and bind to them.[26]

### 4.1.4 Concluding Remarks

It has been a long-standing challenge for scientists to make use of the principles of metal-ion binding by biological macromolecules. A few more examples based on our work will be briefly mentioned below. For references to other work, the reader is referred to the indicated general references.

The first example deals with selective ion-exchangers in which chelating groups specific for, for example, Cu(II), are covalently attached to polymers, or other solid carriers like silica. Our early work was directed to hydrophobic polystyrene-type polymers,[27] whereas more recently we have been focusing on hydrophilic polymers and silica.[28] By making use of imidazole groups—excellent ligands for copper ions—large selectivities could indeed be observed for Cu(II) over zinc, cobalt, nickel and cadmium.[28] In a subsequent study we gave attention to Cd selectivity.[29]

A second example deals with proteins covalently attached to an electrode. The covalent link is made by attaching an imidazole ligand to a surface of glassy carbon. This ligand then coordinates to a genetically engineered protein (in this case a blue copper protein), in which a ligand histidine close to the surface of the protein has been mutated to a glycine. This allows the detection of a redox reaction in solution by following the reaction amperometrically.[30]

The last example to be mentioned deals with the application of coordination compounds attached to polymers and their use as "immobilized catalysts." This technique has been used for a long time in organometallic catalysis. Similar reactions with biomimetic catalysts, as with Cu(II) oxidases, are less well known, and a review for polymeric copper imidazole complexes used in oxidative phenol coupling is available.[31]

The great variety of exciting results about metal-ion binding to biopolymers studied by us and others over the last few decades have generated many new ideas and questions, directed towards the principle of "mimicking nature," as well as "applying principles from nature." Although some successes have been booked, a chemical reaction which is simple on paper, like the nitrogenase reaction in Table 4.4, cannot yet be done by any of the artificial systems, either alone, or bound to a polymer.

Of course, before any application comes into reach, an understanding of the principles must come first. For better understanding, mimicking of biological systems is important, and it is only after that recreation of the building blocks and reaction steps might come within reach.

However, many other disciplines will also stimulate this type of research: medical research, where better chelating agents for selective removal of excesses of toxic metal ions will remain of great importance; one could also think of better understanding of biomineralization processes (surface coordination on macromolecules), of improved diagnostic reagents (monoclonals with an attached metal or other surface recognition agents), environmental cleaning processes (metallothionein-like chelating ligands) and mutation research (role of certain metal ions on mutation and repair mechanisms). Macromolecular coordination chemistry in living systems is more than just a matter of M-L bond formation and M-L stability. The polymer part is crucial, and in fact terminology using the adjective supramolecular, although originally meant for other classes of chemistry, is now commonly used in this field.

For further references in this field numerous excellent review papers on bioinorganic chemistry with a focus on the coordination chemistry around the metal ion appear regularly, e.g., from the series by Sigel & Sigel.[32]

## Acknowledgments

The work in Leiden mentioned above was successfully conducted thanks to the dedication, inspiration, quality and hard work of undergraduates, graduate students, postdocs, visiting scientist and collaborators over the last two decades. Their names appear as co-authors in the references from our laboratory. Local support from the Graduate Research School BIOMAC is acknowledged.

Continuous financial support from Leiden University and the Councils of Chemical Research and Applied Sciences of the Netherlands Organisation for Research (NWO) is gratefully acknowledged, as are the highly appreciated loan schemes $K_2PtCl_4$ and $RuCl_3$ from Johnson and Matthey (Reading UK).

European help (for postdocs and research trainees) from Framework Programmes 3, 4 and 5 since 1992 is thankfully mentioned, as is COST support since 1992, allowing regular research exchanges, including short visits of students, to partner laboratories inside EU countries.

# 4.2 Direct Observation of Supramolecular Structures of Biorelated Materials by Atomic Force Microscopy

**Hiroyasu Yamaguchi and Akira Harada**

## 4.2.1 Introduction

In biological systems, the life process is led by the unique behavior of macromolecules such as DNA and proteins. Molecular recognition by the macromolecules plays an especially important role, for example, in translation and transcription of the genetic code, substrate specificities of enzyme and antigen-antibody reactions in human life. We have focused our attention on the special behavior of monoclonal antibodies because of their unique structures and functions. They can recognize a larger and more complex compound with high specificity.[33-36] Monoclonal antibodies have become more and more important with high possibility as new chemical materials,[37,38] e.g., catalysts[39] and sensors.[40] In recent years, much attention has been focused on supramolecular science, the science of noncovalent assembly, because of the recognition of the importance of specific noncovalent interactions in biological systems and in chemical processes.[41] In particular, the artificial supramolecular complexes constructed by the biofunctional molecules as building blocks can be expected to have novel properties not seen in natural systems or to improve the properties of various components. We describe here the preparation of supramolecular complexes composed of biomacromolecules such as circular DNAs or immunoglobulins and observation of their topological structures by atomic force microscopy (AFM).

The first topic discussed in this section is the imaging of the individual antibody molecules[42] and antibody supramolecules.[43,44] Immunoglobulin G (IgG)[45,46] is a basic type of antibody which is generated in the final stage of immunization with high selectivity for antigens. IgM has a pentameric structure of IgG and ten antigen binding sites in a single molecule.[47] IgM is generated in the initial stage of immunization with a somewhat lower specificity for antigens than IgG. In order to design an antibody system with high specificity and high affinity, a combination of the functions of both IgG and IgM appears to be important. We have designed and prepared dendritic antibody supramolecules in which IgM is placed in the core and many IgGs are bound around the IgM as branches.

The second topic is the direct observation of topological structures of DNA supramolecules.[48-50] Catenated DNA molecules,[51-55] i.e., multiply linked DNA rings, are the product of DNA replication and recombination involving circular genomes and the substrate for topoisomerases.[56] Despite the importance of catenated DNA, little is known about its structure in solution. It is crucial to determine the complex structure of catenanes to obtain important information about the processes that generate them and the structure of DNA. [*n*]Catenanes can be expected to have novel physicochemical properties.

AFM[57] is a useful technique for imaging biorelated materials on a nanometer scale under conditions close to their natural environments without complicated sample preparation techniques, such as staining, used in transmission electron microscopy.[58-61] Since specimens for AFM do not have to be electrically conductive, it is the preferred choice for biological applications.[62-67] Although AFM images of proteins and DNA are larger than the real structure because the images are created by the convolution of the protein or DNA itself and the radius of the curvature of the tip, the effect is negligible for length measurements in the case of DNAs or large protein assemblies whose length or volume is much greater than the curvature of the tip.[68-79] Thus AFM can be potentially applied for investigating microscopic structures of the complicated supramolecular complexes of biomacromolecules by taking advantage of the height information.

## 4.2.2 Antibody Supramolecules

### A. Single Molecular Images of Antibodies

A typical antibody molecule (IgG) was observed deposited from a solution onto a freshly cleaved HOPG surface by AFM at room temperature. IgG molecules consist of two heavy peptide chains with a molecular weight of about 50,000 and two light chains with a molecular weight of about 25,000. There are two identical binding sites at the top of the Fab fragments of IgG which are bound by flexible hinges with a single constant stem (Fc). Fig. 4.7 shows AFM images of a IgG molecule of molecular weight 150,000. Characteristic T or Y shape molecules[45,46,80] can be seen. The overall lateral dimension is 40–50 nm as shown A, B and C in Fig. 4.7, which is in good agreement with the expected values.[81] Although these molecules are monoclonal antibodies, the antibody molecular images are not identical. There are two fragments whose height are the same and one fragment whose height is different with the other two fragments in each molecular image. Fab fragments and an Fc fragment can be differentiated by comparing the height of the three fragments. The angles between the Fab fractions of each immunoglobulin are different. Each antibody molecule takes on a somewhat different shape. This is probably due to the flexibility of its hinge region. Fig. 4.8 shows IgM images. Two kinds of images are evident even though the object is a single species. The larger one is a flat pentamer and the smaller one has a higher center. These results are consistent with those obtained by cryo-AFM measurement.[82,83]

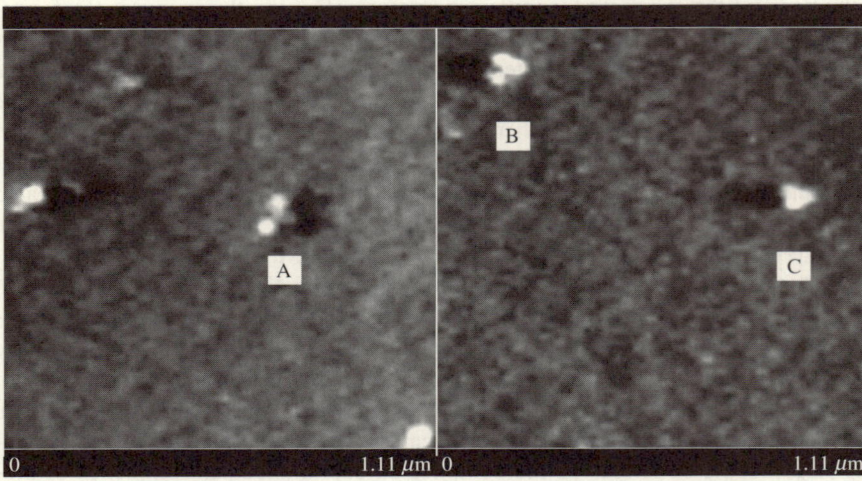


**Fig. 4.7** AFM images of the individual antibody (IgG) molecules.

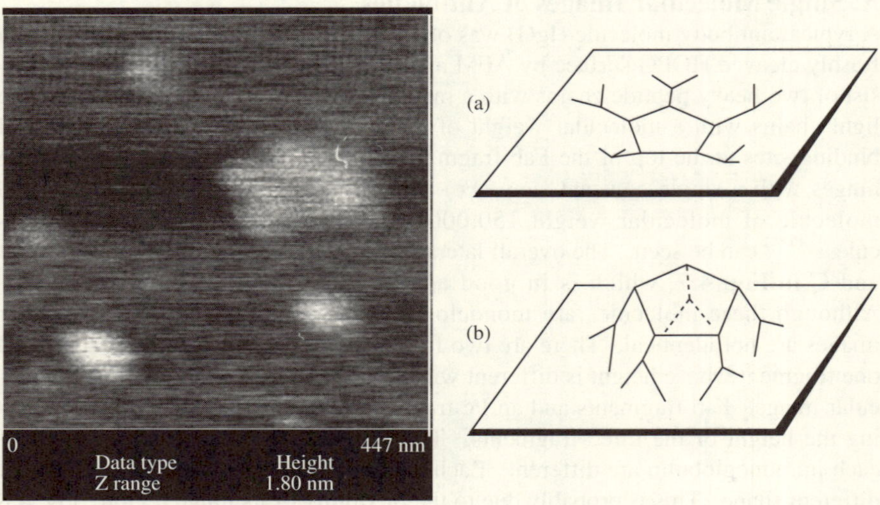


**Fig. 4.8** Molecular images of the monoclonal IgM on a cleaved highly orientated pyrolytic graphite (HOPG) surface and schematic representation of the images for IgM. A flat pentagram (a) and a smaller object with higher center (b).

## B. Direct Observation of Supramolecules between Antibodies and Porphyrin

Specific binding of an antibody (divalent) and a divalent antigen produces supramolecular complexes with linear or cyclic structures as shown in Scheme 4.1. The structure of the complex between one of the monoclonal antibodies for water-soluble porphyrins and porphyrin dimer was observed by AFM. Fig. 4.9 shows one of the images of the complex obtained by mixing the IgG with the divalent antigen. This image suggests the presence of the supramolecular structures of the antibodies connected by the divalent antigens, e.g., dimer, trimer, pentamers, and so on. The molecular images of trimers and pentamers are clearly observed by selecting each image and expanding them as shown in Fig. 4.10(a) and (b), respectively. From the 3D image, it is suggested that this pentamer is the precursor of a cyclic pentamer.

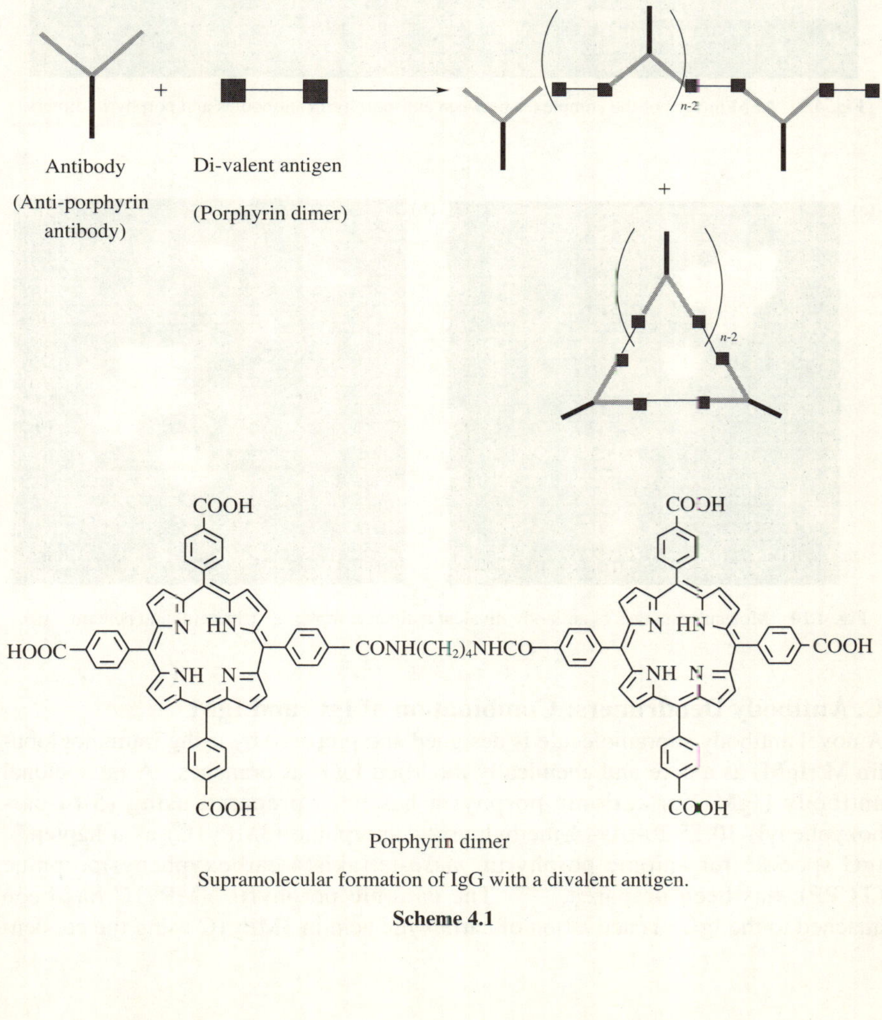


Supramolecular formation of IgG with a divalent antigen.

**Scheme 4.1**

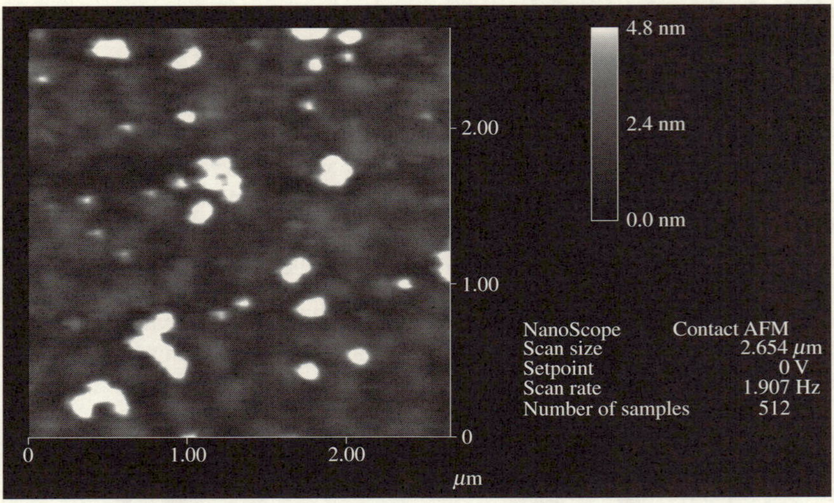


**Fig. 4.9** AFM images of the complexes between anti-porphyrin antibodies and porphyrin dimers.

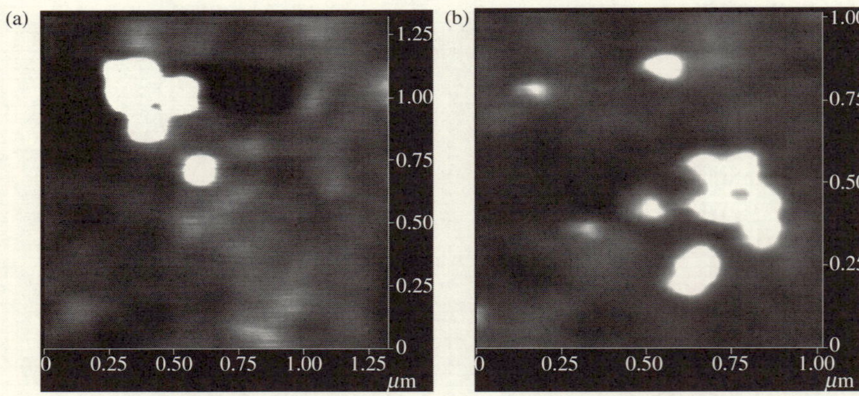


**Fig. 4.10** Molecular images of antibody-divalent antigen complexes: trimer (a) and pentamer (b).

## C. Antibody Dendrimers: Combination of IgG and IgM

A novel antibody supramolecule is designed and prepared by using immunoglobulin M (IgM) as a core and chemically modified IgGs as branches. A monoclonal antibody (IgM) for cationic porphyrin has been prepared using [5-(4-carboxyphenyl)-10,15,20-tris-(4-methylpyridyl)]porphine (3MPy1C) as a hapten.[44] IgG specific for anionic porphyrin, *meso*-tetrakis(4-carboxyphenyl)porphine (TCPP), has been prepared.[33-36] The cationic porphyrin 3MPy1C has been attached to the IgG via activation of carboxylic acid in 3MPy1C using the condensation agent carbonyldiimidazole. The IgG-cationic porphyrin conjugate was purified by a column chromatography using Sephadex G-150 to remove the porphyrins that did not react to the antibody. The characteristic binding ability and specificity of IgG were found to remain during the chemical modification of IgG[84] with 3MPy1C. Scheme 4.2 shows the route for the construction of the dendritic supramolecules. When IgM for 3MPy1C is treated with IgG covalently bound cationic porphyrin, IgM binds the cationic porphyrin attached on the IgG to give a dendritic antibody supramolecule "antibody dendrimer"(**G1** in Scheme 4.2).

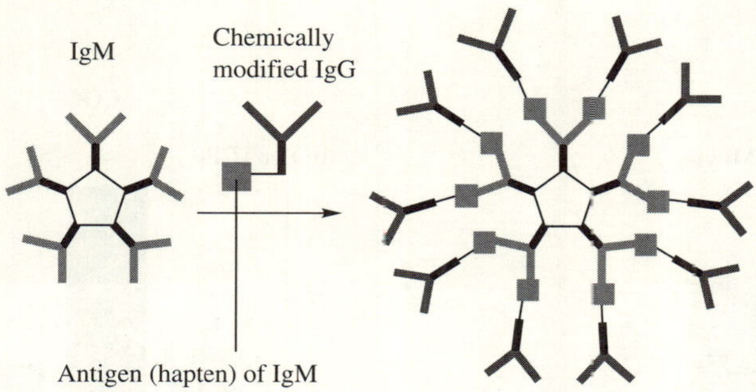


**G1**

The synthetic route of the complete antibody dendrimer. An ideal structure of the dendritic supramolecule is shown as **G1**.

**Scheme 4.2**

The binding property of the antibody dendrimer (**G1**) with a cationic- or anionic-porphyrin was measured by the enzyme-linked immunosorbent assay (ELISA). Fig. 4.11(a) shows the binding properties of the IgG, IgM and **G1** with the cationic porphyrin *meso*-tetrakis(4-methylpyridyl)porphine (TMPyP). Although IgG did not bind the cationic porphyrin and IgM bound the cationic porphyrin, the dendrimer did not bind TMPyP. These results show that the cationic porphyrin attached to IgG occupies the binding sites of IgM in the dendrimer; thus there are no free binding sites against TMPyP on IgM. Fig. 4.11(b) shows the binding of IgG, IgM and **G1** to TCPP. The IgM used in this study can bind both anionic- and cationic-porphyrins due to the low specificity of IgM against porphyrins. Both IgM and IgG bound TCPP, while **G1** bound TCPP more efficiently than IgM or IgG. The increase in affinity of **G1** for the anionic porphyrin indicates that many IgG molecules attach to the surface of the IgM molecule. The biosensor technique based on surface plasmon resonance (SPR)[35] shows that the antibody dendrimer has the advantage of the amplification of detection signals for antigens. The signal intensity increased by the injection of the antibody dendrimer was sufficiently larger than that of simple addition of IgG. Taking into account the change in the binding property of the antibody dendrimer for porphyrins with the increase in the amount of bound antibody to the anionic porphyrin on the SPR

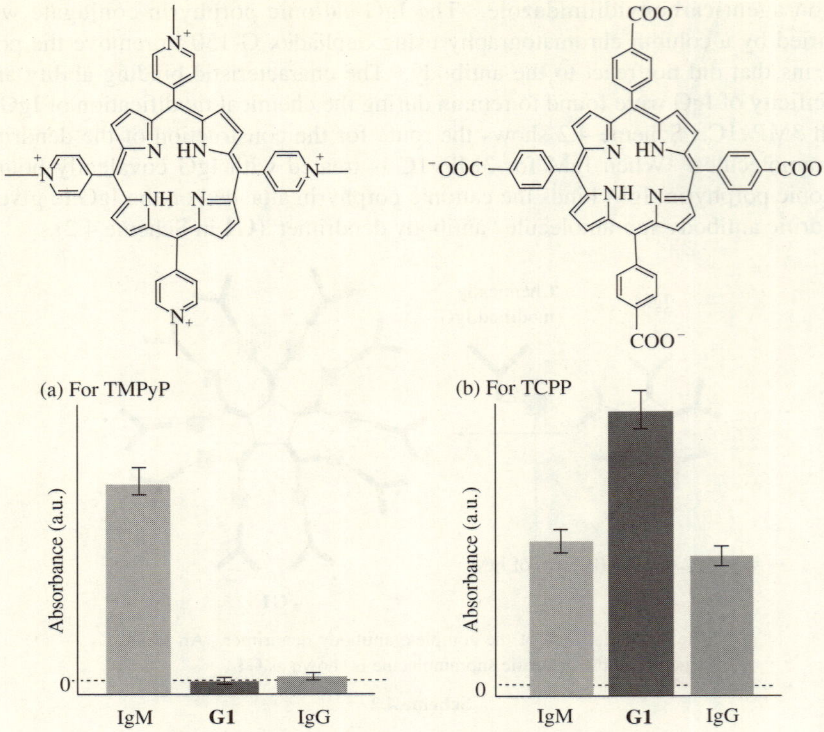


**Fig. 4.11** Binding properties of IgG, IgM and the antibody dendrimer (**G1**) with the cationic porphyrin (TMPyP, (a)) and those with the anionic porphyrin (TCPP, (b)) estimated by ELISA.

biosensor, the antibody dendrimer has many IgG molecules successively bound to the IgM molecule.

The structural observation of the antibody dendrimer was carried out by AFM. The sample surface was observed under conditions such that any damage caused by scanning the cantilever was minimized and that no nonspecific assembly among antibodies occurred.[42] Fig. 4.12 shows AFM images of the dendrimer and the starting IgM. The image of the dendrimer was twice as large as that of the starting IgM. Some branches (IgGs) can be seen outside the IgM core. Such an assembled structure was not observed in a chemically-modified IgG solution or an IgM solution alone.

## 4.2.3 DNA Supramolecules

### A. Imaging of Individual Plasmid DNA Molecules

The two main methods used for determining the topology of DNA catenanes are agarose gel electrophoresis and AFM. Although only AFM gives the complete stereostructure of individual molecules, electrophoresis provides a more rapid and

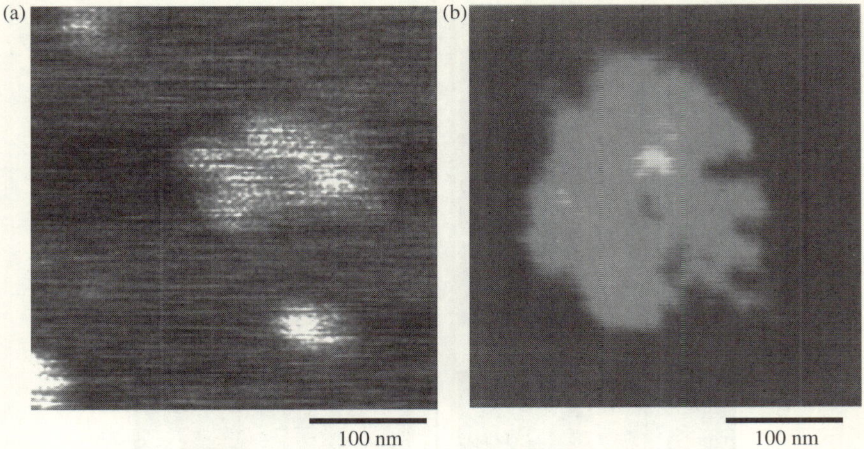


**Fig. 4.12** AFM images of IgM (a) and the antibody dendrimer (b). A total 2 μL of solutions of the antibodies ($3.0 \times 10^{-9}$ M) in 0.1 M phosphate borate buffer (pH 9.0) was deposited onto the surface of HOPG and air-dried.

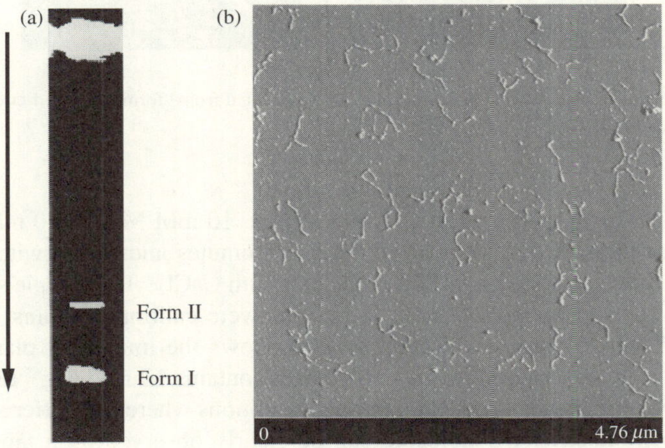


**Fig. 4.13** Agarose gel electrophoresis (a) and the tapping mode AFM image (b) of plasmid DNA pBR322. Downward-pointing arrow shows the direction of electrophoresis.

quantitative overview of the population. Therefore, it is best to use both methods in combination. Fig. 4.13(a) shows electrophoresis of plasmid DNA pBR 322 using 1.0% agarose gel. Two bands are clearly separated. The upper and lower bands correspond to form I (closed-circular DNA) and form II (open-circular DNA), respectively. The ratio of illumination intensities of form I band and form II band is *ca*. 4. To prepare samples for visualization, 4 μL total of DNA solution

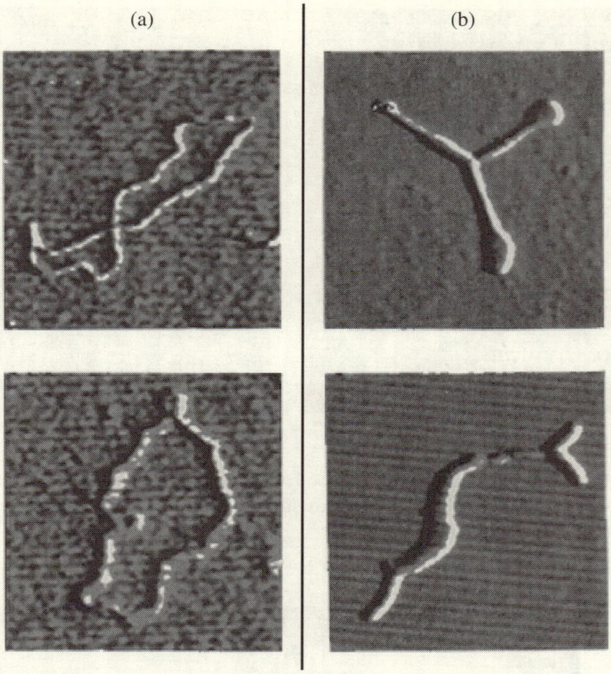


**Fig. 4.14** Single molecular images of plasmid DNA extracted from "form I" (closed-circular DNA, (a)) and "form II" (open-circular DNA, (b)).

(0.5 ng / μL in 10 mM Tris-HCl, 1 mM EDTA, 10 mM $MgCl_2$, 10 mM HEPES) was placed onto freshly cleaved mica for 15 minutes and rinsed with deionized water, then dried for six hours in a desiccator with $CaCl_2$. The sample surface was observed by tapping mode AFM. All scales were calibrated against a standard sample (5 μm × 5 μm grid). Fig. 4.13(b) shows the images of plasmid DNA pBR322 on freshly cleaved mica. The buffer contains 10 mM $Mg^{2+}$ to bind individual DNA molecules on substrate under conditions where the buffer salts can be washed off. Each DNA molecule can be clearly observed by a tapping mode AFM under these conditions. The single molecular images of DNA molecules extracted from form I band and those of form II band are shown in Fig. 4.14(a) and Fig. 4.14(b), respectively. There are two types of topological structures of plasmid DNA. The ratio of the number of molecules in Fig. 4.14(a) and (b) equals that of the intensities of form I and form II bands in agarose gel electrophoresis. The structural information obtained by AFM measurement corresponds to that of electrophoresis. The AFM images of DNA molecules in this report are indicated to show the DNA structures in solution. The topological structures of plasmid DNA are considered to remain during the extraction procedure and the sample preparation for AFM.

### B. Preparation of Nicked DNA by the Addition of DNase I to Plasmid DNA pBR322

Nicked DNA was prepared by the addition of DNase I (in 10 mM Tris-HCl (pH 7.5), 10 mM $MgCl_2$, 10 mM $CaCl_2$ and 50 % glycerol) to the aqueous solution of pBR322. DNase I can cut one of the two chains on double stranded DNA. Plasmid DNA becomes an expanded circular molecule by the enzymatic reaction. The increase in the concentration of DNase I leads to the production of linear DNA or partially separated smaller DNA molecules. The reaction products were analyzed by 1.0 % agarose gel electrophoresis using Tris / Acetate / EDTA buffer at 5 V / cm for 90 minutes. Gels were stained with ethidium bromide (20 mg / mL) and photographed using ultraviolet illumination (312 nm). Fig. 4.15 shows the electrophoresis of pBR322 in the presence of various concentrations of DNase I. The concentration of form I DNA decreased by the increase of the concentration of DNase I. The faster migrated bands are obtained as shown in lane 9 and lane 10 at [DNase I] > $2.0 \times 10^{-6}$ units / μL. The DNase I concentration dependency on the topological structure of nicked DNA is also monitored by AFM using a tapping mode. Fig. 4.16 (a) shows the AFM image of the DNA after the enzymatic reaction of $1.0 \times 10^{-6}$ units / μL DNase I. The topological structure of pBR322 is mainly form II (open circular structure) compared with the original structure of plasmid DNA (mixture of open and closed circular DNA). Linear and shorter DNAs are observed by AFM in the presence of higher concentration of DNase I (Fig. 4.16(b) and (c)). The structural information obtained by electrophoresis is confirmed by AFM measurement.

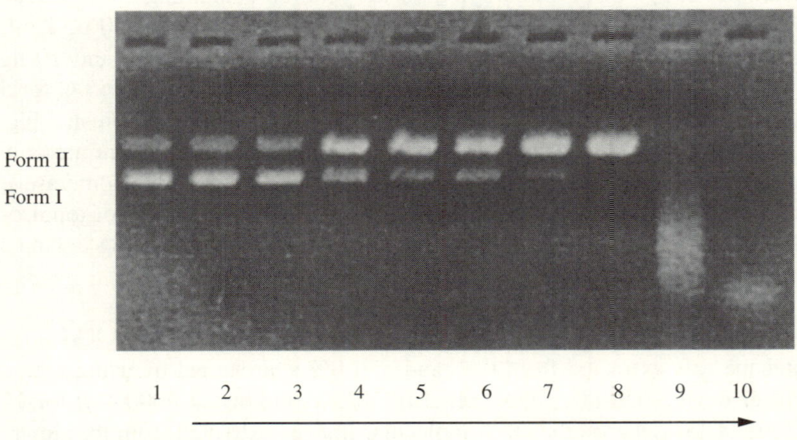


**Fig. 4.15** Electrophoresis of the reaction products of plasmid DNA pBR 322 with DNase I. The concentration dependence of DNase I on the structures of nicked DNAs. The concentration of DNA was fixed at 20 ng / μL. [DNase I] (units / μL): 0 (lane 1), $10^{-8}$ (lane 2), $10^{-7}$ (lane 3), $1.0 \times 10^{-6}$ (lane 4), $2.0 \times 10^{-6}$ (lane 5), $4.0 \times 10^{-6}$ (lane 6), $8.0 \times 10^{-6}$ (lane 7), $10^{-5}$(lane 8), $10^{-4}$ (lane 9), $10^{-3}$ (lane 10).

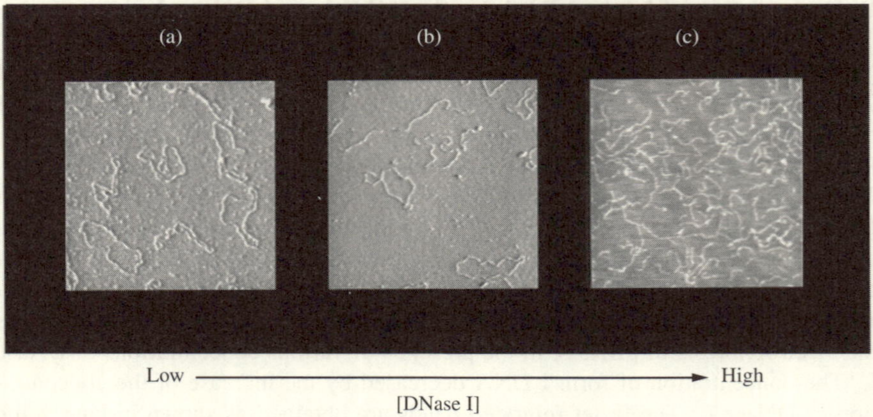


**Fig. 4.16** AFM images of the reaction products of pBR322 with DNase I used in Fig. 4.15, lane 4 (a), lane 7 (b) and lane 9 (c).

## C. Catenation Reaction Using Topoisomerase I

Catenation reaction was performed as described in Scheme 4.3. An open circular DNA was obtained by mixing pBR322 DNA (12.5 ng / μL) and DNase I ($7.4 \times 10^{-6}$ ng) in a 16-μL solution containing 40 mM Tris-HCl (pH 7.5), 6 mM $MgCl_2$ and 6 mM $CaCl_2$ at 37°C for 30 minutes. Reactions were terminated by heating at 65 °C for 5 minutes. Catenation reactions between nicked DNA and circular DNA were carried out at 37 °C for 4 hours in a 10-μL solution containing 20 mM Tris-HCl (pH 7.5), 20 mM KCl, 5 mM spermidine, 2 mM dithiothreitol, 50 μg / mL bovine serum albumin, 10 % glycerol, 12.5 ng / μL pBR322 DNA and 10 ng topoisomerase I. The enzymatic reaction was terminated by heating at 65°C for 5 minutes. The reaction of topoisomerase I was monitored by electrophoresis. Fig. 4.17 shows the results of electrophoresis of the reaction products. There are only two bands (lanes 1 and 2) in the presence of 0.05 unit / mL of topoisomerase I. However, a new band is observed in the presence of two-fold amounts of topoisomerase I to the previous conditions. The slower migrated band appearing in lane 3 is considered to be a [2]catenane.

## D. Single Molecular Images of DNA Catenanes Observed by AFM

DNA catenane was extracted from the band of 1.0% agarose gel by using a centrifugal filter device, Ultrafree-DA (MILLIPORE), spinning at 6000 $\times g$ for 10 minutes. Fig. 4.18(a) shows the DNA molecular images extracted from the slower migrated band appearing in lane 3 in Fig. 4.17. Parts of two circles fused over a distance of ~ 90 nm can be seen. Each circle is assigned to be a pBR 322 plasmid whose length is about 1.7 μm comparable in length to the 4361 bps. This molecular image is ascribed as a dimeric DNA catenane for the following reasons: (1) this fused DNA comprises two plasmid DNA and (2) the molecules seen in Fig. 4.18 (a) were extracted by the higher molecular weight band in electrophore-

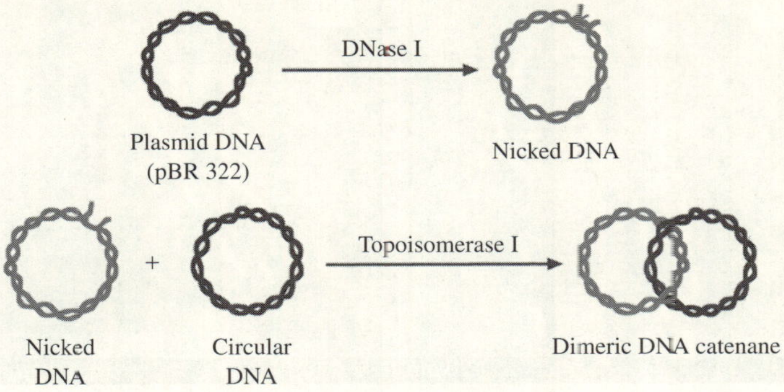


Preparation of dimeric DNA catenane by the reaction of topoisomerase I to the nicked DNA in the presence of circular DNA.

**Scheme 4.3**

**Fig. 4.17** Agarose gel electrophoresis of the reaction mixture between nicked DNA and circular DNA in the presence of topoisomerase I. [Topoisomerase I]: 0.05 unit/μL (lanes 1 and 2), 0.1 unit/μL (lanes 3 and 4). Lane 1 and lane 3 include nicked DNA. The concentration of circular DNA was fixed at 14 ng/μL.

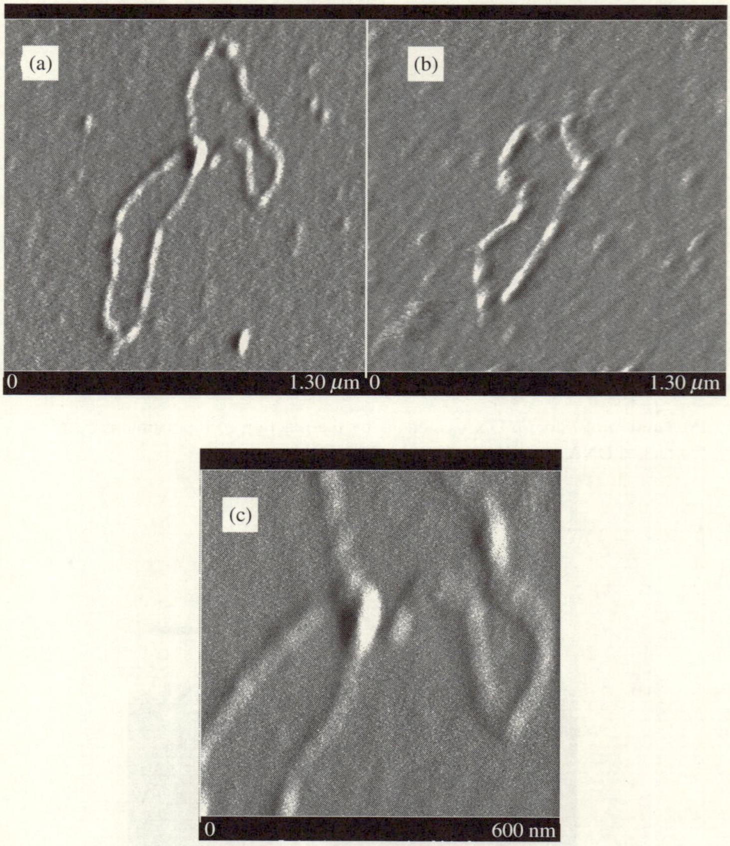


**Fig. 4.18** Tapping mode AFM images of dimeric DNA catenane (a) and plasmid DNA (DNA monomer) (b) and an enlarged image of the fused part of a catenated DNA molecule (c).

sis, which included no DNA monomer (Fig. 4.18(b)). The interlocked structure is clearly observed by expanding the fused part as shown in Fig. 4.18(c). Overpassing and underpassing strands can be seen at crossings.

## E. DNA [*n*]Catenanes Prepared by Irreversible Reaction with DNA Ligase

DNA catenanes were prepared by the addition of T4 DNA ligase to the mixture of linear DNA and nicked DNA as shown in Scheme 4.4. The linear and nicked DNAs were prepared by the addition of *Eco*RI or DNase I to a solution of pBR322 plasmid DNA, respectively. The advantage of using T4 DNA ligase is an irreversible recombination between the complementary ends of DNA, in contrast with the reversible one for topoisomerase I. Therefore we can expect an increase in the yield of DNA catenanes. Many slower migrated bands were observed in elec-

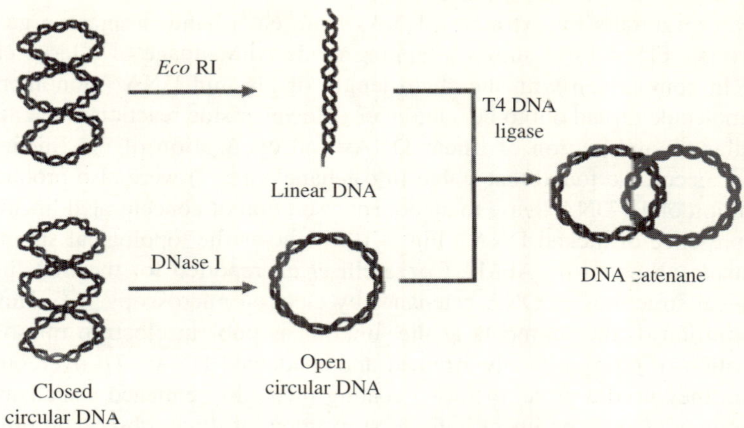


DNA catenanes prepared by the addition of T4 DNA ligase to a mixture of linear and nicked (open circular) DNAs.

**Scheme 4.4**

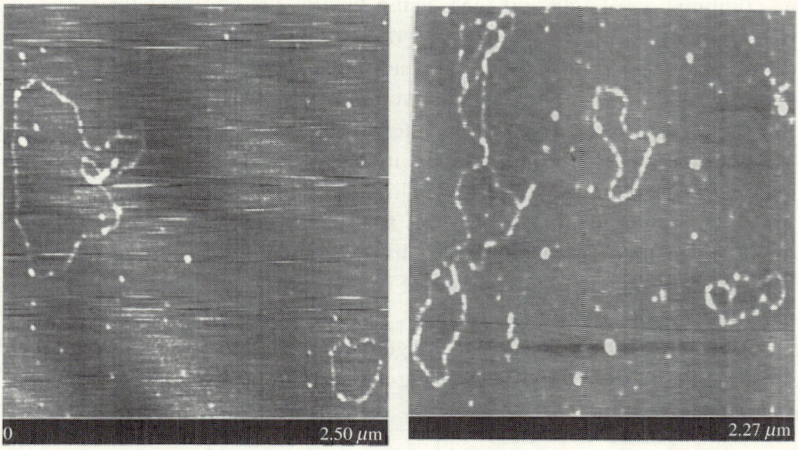


**Fig. 4.19** Tapping mode AFM images of the large circular DNA (a) and [4]catenane (b) prepared by the reaction of ligase. Specimen: a total 4 μl of DNA solution including reaction products (0.5 ng/ml in 10 mM Tris-HCl, 1 mM EDTA, 10 mM $MgCl_2$, 10 mM HEPES) was placed onto freshly cleaved mica for 15 min and rinsed with deionized water, then dried for 6 h in a desiccator with $CaCl_2$.

trophoresis by the addition of ligase to the mixture of linear and nicked DNA. These bands were found to include DNA catenanes, large circular DNAs and linear DNA oligomers by extracting DNAs from each band in agarose gel electrophoresis. Fig. 4.19(a) shows a tapping mode AFM image of a large circular DNA. In comparison with the chain length of plasmid DNA (monomer), this cyclic molecule turned out to be a tetramer. However side reactions such as intermolecular recombination of linear DNAs and cyclization of the linear-DNA oligomer occur, we found that stable [*n*]catenanes ($n > 2$) were also produced by the addition of T4 DNA ligase to an aqueous solution of concentrated linear DNA in the presence of nicked DNA. Fig. 4.19(b) shows the topological structure of [4]catenane observed by AFM. Cozzarelli *et al.* reported for the first time the topological structures of DNA catenanes by electron microscopy.[51-55] However, the resolution of the segments at the junction is poor in electron microscopic examination of conventionally strained and shadowed DNA. To overcome this problem, they used a protein (RecA) coating method to enhance visualization of the crossings. Our experiment is the first example of direct observation of DNA catenanes.

### 4.2.4 Conclusions and Summary

We have succeeded in observing the structures of the single antibody molecule and linear-, cyclic- and dendritic-antibody supramolecules at room temperature by AFM. New dendritic antibody supramolecules have been designed and prepared by the combination of IgG and IgM, using IgM as the core and IgG as branches. Many binding sites of IgG are arranged radially on the surface of one object, and the resulting artificial antibodies bind antigens more selectively than IgM and more strongly than IgG. The characteristic features of the antibody dendrimer are as follows. (i) they are composed of proteins; (ii) they have large size with molecular weight of about 2 million; (iii) they are constructed by noncovalent bonds and (iv) they bind strongly to antigens with high specificity. The antibody dendrimer will be used as functionalized materials for sensitive detection of many kinds of chemicals, for diagnosis and for drug delivery systems. DNA catenanes were prepared by the addition of topoisomerase I to the mixture of nicked DNA and plasmid DNA or by the addition of T4 DNA ligase to the mixture of linear DNA and nicked DNA. The single molecular images of *naked* DNA catenanes were clearly observed for the first time by AFM using a tapping mode at room temperature and in an ambient atmosphere.

# 4.3 Analyses of the Nanostructure of Calcium Carbonate Composite in Biominerals of *Pinctada fucata*

**Hitoshi Yamamoto and Norikazu Ueyama**

## 4.3.1 Introduction

The nacreous layer of pearl consists of a composite of $CaCO_3$ and biopolymers forming a mutilayered structure made up of the alternate stacking of an aragonite crystal layer and a biopolymer layer.[86-88 a)] The biopolymer in the nacreous layer is thought to control the morphology of calcium carbonate as aragonite. Exhaustive work on the biominerals proved that the organic molecules can control the crystal growth through *in vitro* experiments.[89-96)] However, information about the chemical state of the biopolymers in situ is quite limited due to the difficulty of analysis. We have found that the biopolymers could be observed by $^{13}C$ solid-state NMR and that they have anomalous short relaxation rates and highly efficient cross-polarization compared with pure calcium carbonate crystals. These results indicate that proton magnetization donors exist quite close to carbonate in the aragonite. By using high-resolution TEM and EELS, we found that the aragonite is composed of an assembly of nanocrystals of aragonite and biopolymers, not single crystals[88 b)]. Here we show the ultra-structure of the nacreous layer, revealing that the biomineralization process is controlled at the nano-level.

A pearl is composed of a thin nacreous layer (30 to 1000 μm thick) on a nucleus of a diameter of 3 to 8 mm. The nacreous layer is composed of translucent aragonite crystal layers (300 to 800 nm thick) and biopolymer membranes 5 to 20 nm thick stacked alternately about 60 to 1000 times.[87)] The aragonite layer consists of small aragonite bricks. It is well known that calcium carbonate exists in three different crystal morphologies: calcite, aragonite and vaterite.[97)] The aragonite phase is stable at high pressure but the calcite is in its most stable phase in the ambient condition.[97)] The most interesting point is that pearl oysters produce aragonite despite the instability of aragonite under ambient conditions. The biopolymer layer between the aragonite layers is thought to play a central role in controlling the crystal morphology.[95,96)] To understand the control mechanism, many in vitro experiments were carried out on extracts from biominerals and mimicked organic molecules[98-100)] to clarify the mechanism of controlling the crystal growth.[92-96)] Actually, the extracts from biominerals which contain aragonite induce the production of aragonite crystals, and the extracts from biominerals which contain calcite produce calcite.[95,96] The difficulty of *in situ* analysis of biopolymers in biominerals, however, limits progress in understanding the exact mechanism of biomineralization because of the low content of biopolymers ($<$ 5%) relative to the large quantity of inorganic components.

### 4.3.2 Solid State $^{13}$C-NMR Analysis of the Nacreous Layer

To obtain *in situ* chemical information about the organic components in the biominerals, we tried using $^{13}$C solid state NMR.[105,106] Fig. 4.20(a) shows the solid state $^{13}$C NMR spectrum of the nacreous layer of the pearl. Only the signal from carbonate carbon in the aragonite was observed. Remarkably, we found that the organic components in the nacreous layer could be satisfactorily observed by applying a short recycling delay time (5 s) to saturate the huge signal originating from the carbonate carbon in the aragonite and by enhancing the signals through cross-polarization (Fig. 4.20(b)). The signals observed at around 22, 44 and 51 ppm together with the broad signals of carboxylate carbons around the sharp carbonate signal at 171 ppm are assigned to the biopolymer in the nacreous layer. $T_1$ values for each peak were measured using Torchia's method.[107] The organic components have shorter $T_1$ values (5, 15 and 16 s for the signals at 22, 44 and 51 ppm, respectively), while the carbonate carbons at 171 ppm have a longer $T_1$ value (90 s). Cross-polarization (CP) is also effective for enhancing the signals of the organic components in the nacreous layer. In the CP/MAS experiment, the proton magnetization transfers to the carbon nucleus during the contact time period. The efficiency of the magnetization transfer depends on the distance between the proton and the carbon nuclei. Thus CP should be effective for the organic components which have many protons, and the carbonate carbons in the aragonite should not be effectively enhanced by CP. During the CP study to find the optimum contact duration, we found that carbonate carbons in the aragonite layer had an anomalous efficiency of magnetization transfer. As shown in Fig. 4.21(a), the signal of the carbonate carbons in aragonite is still observed even during very short contact times (0.1 ms). The carbonate carbon signals of pure calcium carbonate (Fig. 4.21(b)) and in-house prepared polycarboxylate-calcium carbonate composites (Fig. 4.21(c)), in which the polymer ligands attach to the surface of $CaCO_3$, completely disappear even after 0.2 ms of contact time. As shown in Fig. 4.21(d), (e) and (f), the size of each calcium carbonate crystal is quite similar. In addition to the effectiveness of CP, the carbonate carbon in the nacreous layer has a shorter $T_1$ value (90 s) than that in pure calcium carbonate (180 s). Considering the distance dependency of CP and the longitudinal relaxation rate, aragonite brick seems to have proton magnetization very close to the carbonate carbons inside the crystals. This strongly suggests that the aragonite crystals in the nacreous layer are not single crystals but have finer structure.

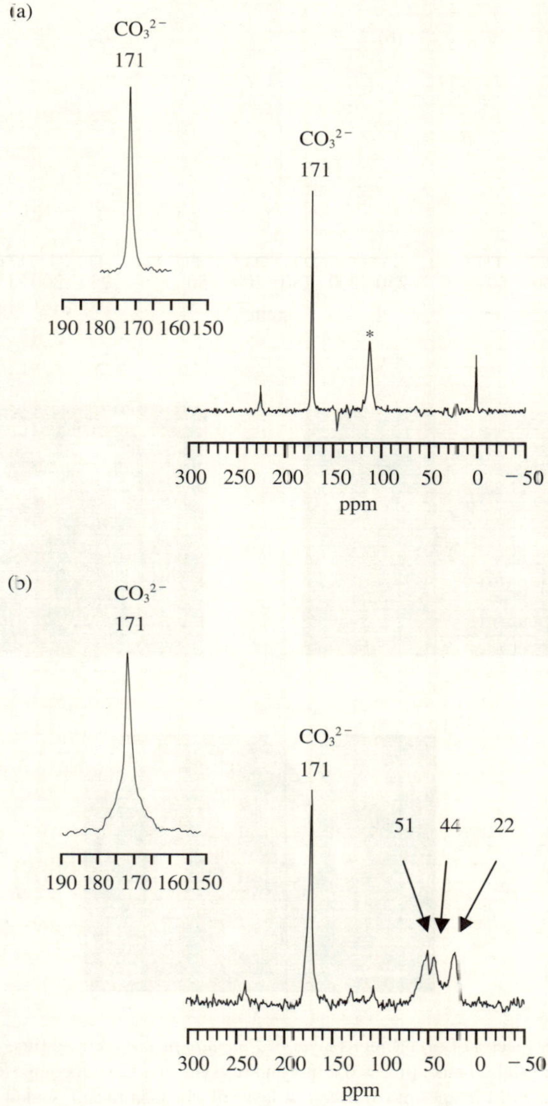


**Fig. 4.20** $^{13}$C MAS spectra of nacreous layer of *Pinctada fucata* (a) without CP and (b) with CP. The insets are expanded spectra of CO region. The broad signal around the sharp signal at 171 ppm (b) originates from the carboxylate in the biopolymer.

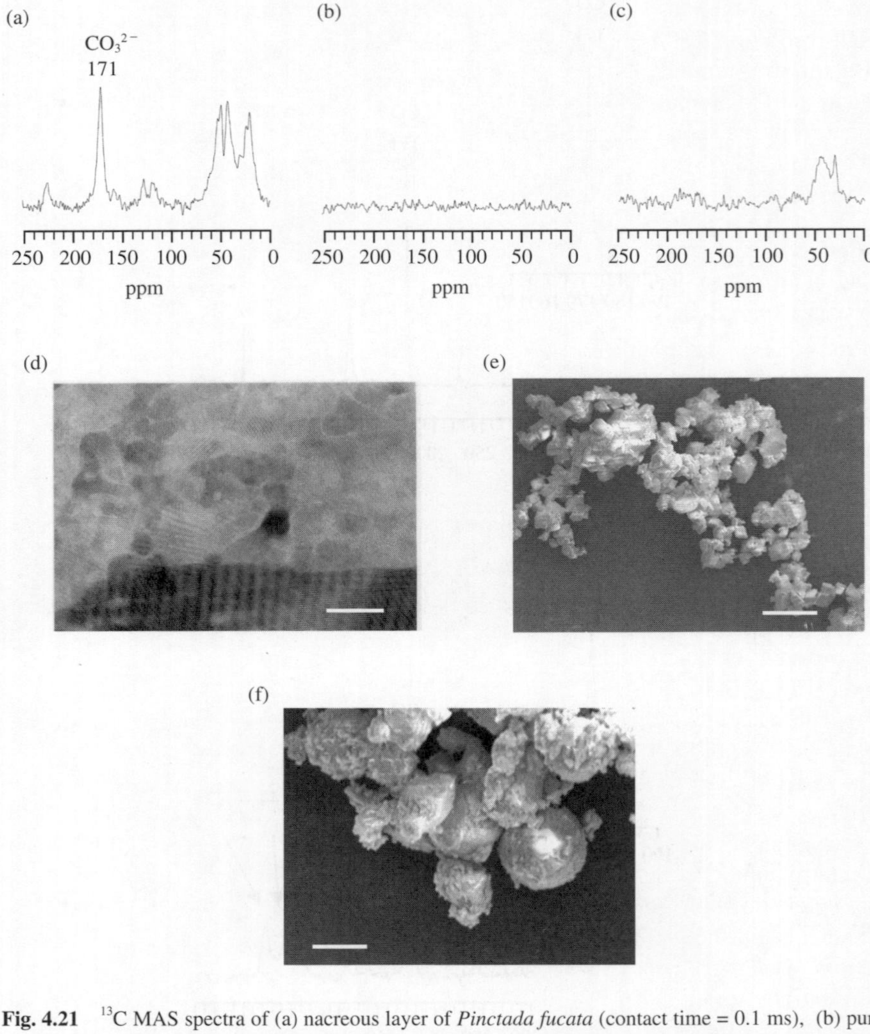


**Fig. 4.21** $^{13}$C MAS spectra of (a) nacreous layer of *Pinctada fucata* (contact time = 0.1 ms), (b) pure $CaCO_3$ crystal (contact time = 0.2 ms) and (c) synthetic $CaCO_3$ composite with carboxylate polymer. SEM images of (d) nacreous layer of Pinctada fucata (scale bar size: 15 nm), (e) pure $CaCO_3$ crystal (scalebar size: 10 μm) and (f) synthetic $CaCO_3$ composite with carboxylate polymer[108] (scale bar size: 2.5 μm).

### 4.3.3 High Resolution TEM Analysis of the Nacreous Layer

To prove the possibility described above, high-resolution transmission electron microscopy (TEM) analyses were performed. As the aragonite crystals do not have a cleavage plane, it is difficult to prepare a thinly sliced sample suitable for high-resolution TEM analysis because the force of the scraper disturbs the fine structure of the minerals. We utilized an FIB (focused ion beam) micro processing system to prepare a thinly sliced nacreous layer to avoid damage caused by the force from the scraper. By using a Ga ion FIB, the nacreous layer was processed to a small chip ($10 \times 6 \times 0.05$ μm) suitable for TEM analysis. The whole image of the thinly sliced nacreous layer is shown in Fig. 4.22(a). The nacreous layer of pearl is composed of layered aragonite bricks which have *ca.* 0.5 μm thickness and interlayer substrates of *ca.* 30 nm thickness. The intercrystal substrates described by Wada *et al.* for the nacreous layer were not observed for this sample.[87)] The existence of stripes on each aragonite brick due to the interference of the electron beam indicates the calcium carbonate in the aragonite brick is highly oriented. Electron microdiffraction experiments showed a clear diffraction pattern of aragonite crystals (Fig. 4.22(b)). These results are consistent with previous works on the nacreous layer except for the lack of intercrystal substrates. As shown in Fig. 4.22(c), however, the nanoscaled structure in the aragonite brick was observed at very high magnification (×2,000,000). The aragonite brick was not a single crystal, but was composed by nanocrystalline particles of calcium carbonate of 5 to 15 nm in diameter. Detailed analysis of the high magnification TEM image revealed that each nanocrystalline particle in a brick has its crystal axis pointed in almost the same direction and the direction of the axis is different for different bricks. Some of the nanocrystalline particles exist in the biopolymer layer (Fig. 4.22(c), arrows).

The aragonite bricks in mollusk shell have been believed to be single crystals so far because of their translucent appearance and the results of diffraction studies. The ordered assembly of nanocrystalline particles revealed by our TEM analysis behave as single crystals do when subjected to diffraction analysis. To obtain information about the ordering mechanism of the nanocrystalline particle axis, elemental mapping using EELS (electron energy loss spectroscopy) was carried out. Fig. 4.23 shows the elemental mapping of the nacreous layer for Ca, C, O and N. The maps are shown in the gray scale and the bright areas indicate a high concentration of each element. As shown in Fig. 4.23(a), there are some spots of highly concentrated calcium. The concentrations of oxygen and carbon were low at the "calcium spot" (Fig. 4.23(b) and (c)). The inhomogeneity of calcium, carbon and oxygen supports the supposition that the aragonite brick is not a single crystal. The most striking feature of this analysis is that a significant amount of nitrogen was observed in the aragonite brick (Fig. 4.23(d)). This indicates that organic molecules (likely proteins and/or glucosamines) may be incorporated into the aragonite brick, and if the biomineral is defined by the biologically produced organic-inorganic composites, the aragonite brick in the nacreous layer is already a biomineral itself. These organic components would control the generation of

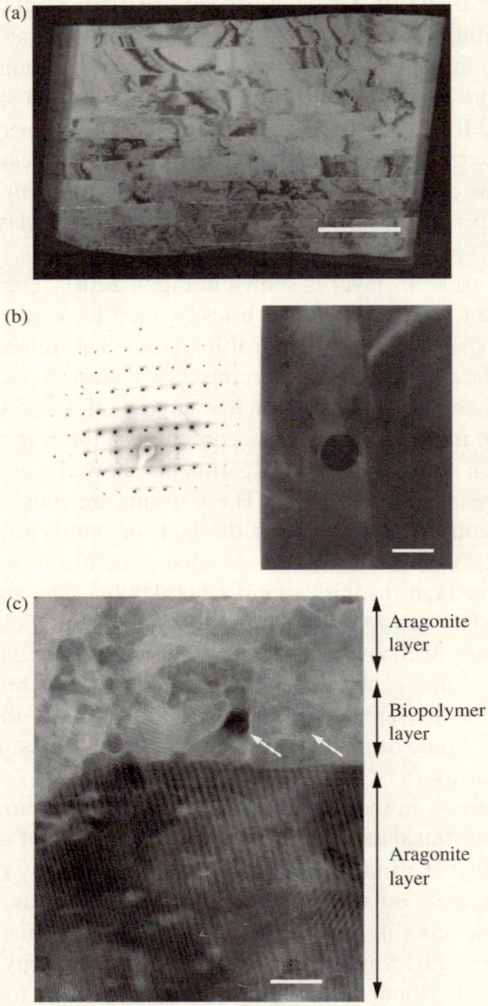


**Fig. 4.22** Transmission electron microscopy images and electron micro diffraction image of the nacreous layer. (a) Whole image of FIB processed nacreous layer. The stripes on the aragonite bricks are due to the interference of electrons. Scale bar size: 2μm. (b) Diffraction pattern of electron micro-diffraction analysis (left) and the area that the electron beam was irradiated (right, dark spot). A aragonite brick shows clear diffractions of aragonite. (c) High-resolution image of nacreous layer. Low density biopolymer layer is sandwiched between two aragonite bricks. The aragonite bricks are composed of assembly of nanocrystalline particles of $CaCO_3$. The particles were also found in the biopolymer layer (arrows). The difference in color density between two aragonite bricks indicates a difference in crystal axis direction. Scale bar size: 15 nm.

[Reproduced by permission of The Royal Society of Chemistry, K. Takahashi, H. Yamamoto, A. Onoda, M. Doi, T. Inaba, M. Chiba, A. Kobayashi, T. Taguchi, T. Okamura and N. Ueyama, *Chem. Commum.*, **2004**, 997.]

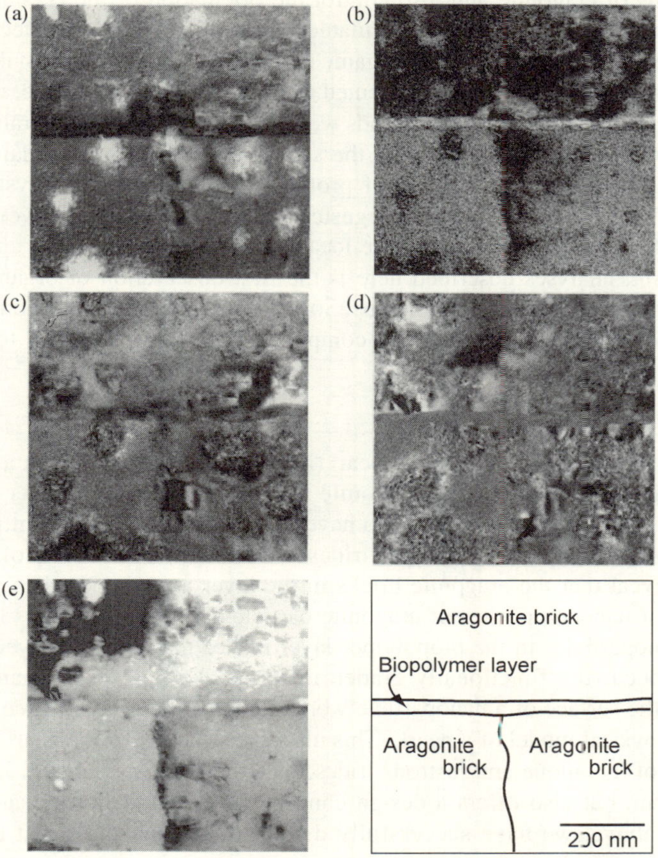


**Fig. 4.23** Elemental mapping by electron energy loss spectroscopy of the nacreous layer. The density of each element is indicated by the gray scale. The bright area means high concentration of the element. (a) The elemental mapping of calcium. The calcium atom is located mainly in the aragonite bricks and the concentration in the biopolymer layer is very low. There are some high density spots in the aragonite layer, indicating the nonhomogenity of aragonite brick. (b) The elemental mapping of oxygen. The oxygen atom is also found all over the analyzed area. However, the concentration around the "calcium spot" is low. These observations indicate that the calcium spot is not $CaCO_3$. (c) Carbon distribution image of nacreous layer. The content of carbon atom is high in the biopolymer layer. The carbon atom also exists in the aragonite as $CO_3^{2-}$. However the concentration of carbon around the "calcium spot" is low. (d) The elemental mapping of nitrogen. Significant amount of nitrogen atom was found in the aragonite bricks. This indicates that biomolecules are included in the aragonite brick. (e) Zero loss image (left) and schematic drawing (right) of the analyzed area.

[Reproduced by permission of The Royal Society of Chemistry, K. Takahashi, H. Yamamoto, A. Onoda, M. Doi, T. Inaba, M. Chiba, A. Kobayashi, T. Taguchi, T. Okamura and N. Ueyama, *Chem. Commun.*, **2004**, 997.]

nanocrystals of aragonite and also control the orientation of the crystal axis, i.e., the biomineralization process is regulated at the nanometer or molecular level. The possibility of the existence of organic components inside the crystals was suggested by several researchers, represented as intracrystalline molecule, water soluble matrix, etc.[101,102] These compounds were extracted from biominerals, and they induce $CaCO_3$ crystals *in vitro* with the same morphology of original biominerals.[101,102] The possible existence of proteins inside the calcite crystals of sea urchin spine, in particular, was suggested by Berman *et al.* However, it was obscure where these compounds were located in the biominerals.[89-96] The results of our *in situ* analyses described here is the first observation of organic components inside the inorganic components. Some of the reported water-soluble matrices may be located inside inorganic components that were assumed to be single crystals.

### 4.3.4 Conclusions

According to the optical model of a pearl (illuminant model) based on a computer-assisted visual simulation, the aragonite layer in the nacreous layer of a pearl should be highly transparent and also have point light source transmitting rays in all directions to explain its unique iridescence.[103,104] The results of our TEM analysis reveal that the aragonite bricks in the layer are composed of an ordered assembly of nanocomposites of aragonite particle and biomolecules. The dispersion of nanoparticles in the biopolymer layer makes the nacreous layer act like a naturally occurring functionally graded material like high performance optical fiber, and the surface of a nanoparticle works like the point light source as expected in the physical model of a pearl. This ultrafine structure of the pearl is not only the origin of its unique and lustrous iridescence that is appreciated in its role as a jewelry item, but also offers a design concept for new functional materials. As described above, we have successfully demonstrated that the finest unit of the nacreous layer is the ordered nanocomposite of aragonite particles and biomolecules. The traditional way of thinking regarding the mechanism for generating the nacreous layer (crystal grows on the biopolymer sheet that has oriented periodic acidic groups) should be reconsidered to incorporate the ordered nanocomposite of aragonite particles and biomolecules.

## Acknowledgment

The authors thank Mr. Takashi Inaba and Mr. Masahiro Chiba of Hitachi Kyowa Engineering Co., Ltd., for the FIB and TEM analyses.

# 4.4 A Novel Triple Helix Consisting of One Polynucleotide and Two Polysaccharide Chains

**Kazuo Sakurai, Kazuya Koumoto and Seiji Shinkai**

## 4.4.1 Introduction

From olden times, Chinese medicine men have known that some funguses are useful to cure gynecological diseases.[109] Modern chemistry has revealed that the active ingredient in the funguses is a polysaccharide belonging to the β-(1→3)-glucan family.[110] Subsequent studies showed that β-(1→3)-glucans can activate the immune system by promoting secretion of interleukins.[111] Now some of the glucans[112-114] have been commercialized as a medicine against uterine cancers. However, the detailed molecular mechanism of the biological process has not yet been clarified and remains a mystery.

The chemistry of β-(1→3)-glucans has a long history and demonstrates that the glucans have profound potentiality as a material for nanotechnology and supramolecular chemistry.[115,116] The main chain of β-(1→3)-glucans consists of repeating glucose units joined by a β-linkage between the first and third carbons of the glucose ring. Fig. 4.24 (a) shows, for example, the repeating unit of one of the β-(1→3)-glucans: schizophyllan that has the side chain at every three main chain glucose. Curdlan is a member of the β-(1→3)-glucans and has no side chain.[117] Lentinan has two β-(1→6)-glucosyl side chains linked to the five glucoses in the main chain.[115]

Natural curdlan is not water-soluble at all. Even when natural curdlan (Wako) was acid-hydrolyzed to be converted to a low molecular weight fraction of $M_W = 2000$, where $M_w$ is the weight-average molecular weight, its solubility is not improved.[118] On the other hand, lentinan and schizophyllan are water-soluble due to presence of the glucosyl side chain. When natural funguses and yeasts produce the β-(1→3)-glucans, the glucans take on the form of a right-handed triple helix as shown in Fig. 4.24(b). X-ray crystallography suggests that the three different chains are linked together with the three hydrogen bonds formed among the three second hydroxyl groups of the adjoining glucoses (see subsets (b) and (c) of Fig. 4.24).[117,119] The triple helix of schizophyllan is dissociated to the single chain (s-SPG) by dissolving in dimethylsulfoxide (DMSO).[120] On the other hand, when water is added to the s-SPG/DMSO solution or DMSO is exchanged for water by

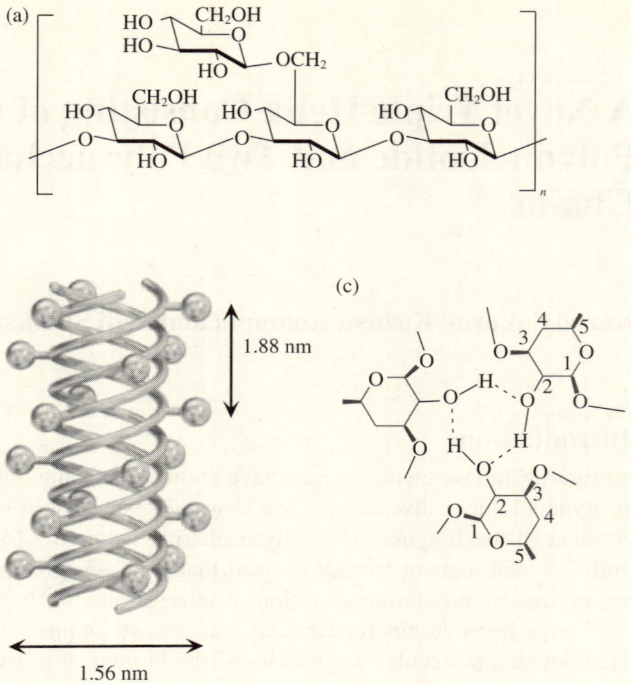


**Fig. 4.24** Molecular structures of β-(1→3)-glucans. The subset (a) shows the repeating unit of schizophyllan, one of the β-(1→3)-glucans, for example. The natural β-(1→3)-glucans take the form of a right-handed triple helix. The subset (b) shows the helix of curdlan; the helix pitch and diameter are 1.88 and 1.56 nm, respectively. Here, the circles represent glucoses. The subset (c) shows how the hydrogen bonds are formed between the adjoining glucoses.

dialysis, s-SPG molecules aggregate owing to the hydrophobic interaction and form both intra- and intermolecular hydrogen bonds.[121)] Although the resultant product is not the same as the original rod-like molecules, the local structure is expected to retrieve the triple helix.[122)]

Recently, we found that when certain single-stranded polynucleotides such as poly(5'-cytidine monophosphate), denoted by poly(C)*, coexists in the renature process of β-(1→3)-glucans mentioned above, the polynucleotide and s-SPG form a macromolecular complex.[123)] Our subsequent studies show that this complexation is a characteristic nature commonly observed for all β-(1→3)-glucans.[124)] As far as we know, this is the first study showing that a neutral polysaccharide can combine with polynucleotides. If s-SPG can enter a cancer cell and combine with

* Hereinafter, poly(X) denotes a homo-polynucleotide and X is the base symbol such as A (adenosine monophosphate), C (cytidine monophosphate), G (guanosine monophosphate), U(uridine monophosphate), I (inosine monophosphate), dA (deoxyadenosine monophosphate), etc.

polynucleotides, the biological system of the cell will be disrupted because polynucleotides are a key molecule for life. Therefore, investigation of the complex may provide new clues to explain the anticancer activity of β-(1→3)-glucans. This section reviews our recent results on these novel polysaccharide/polynucleotide complexes.[125-128]

### 4.4.2 Spectroscopic Changes upon the Complexation[123,126]

Figure 4.25 compares the ultraviolet absorbance (UV) and circular dichroism (CD) spectra between poly(C) itself and the complex made from poly(C) and s-SPG (the complex is denoted as poly(C)/s-SPG). Here the $M_W$ of s-SPG is $1.5 \times 10^5$ and the cytosine base number is about 250. The complexation decreases the absorbance of cytosine at 270 nm by 12% (hypochromism), and increases the CD intensity of the 280 nm positive band and creates a new band at around 240 nm. Other homo-polynucleotides such as poly(A), poly(T) and poly(dA) show similar phenomena in the UV and CD spectra, when they form a complex with s-SPG.[126] Schizophyllan does not have any functional group to absorb the light in this wavelength range. The hypochromism of UV should be related to the decrease in the distance between the stacked bases in the polynucleotides.[129] Furthermore, the increment of CD after the complexation implies that the polynucleotide takes an ordered helical structure even after the complexation, and the helix content (or ordering) is increased.[130] Consequently, the complexation creates a new ordered helical structure of poly(C). This is in strong contrast with polycation/polynucleotide complexes, in which the CD spectra usually considerably depressed due to random aggregation.[126]

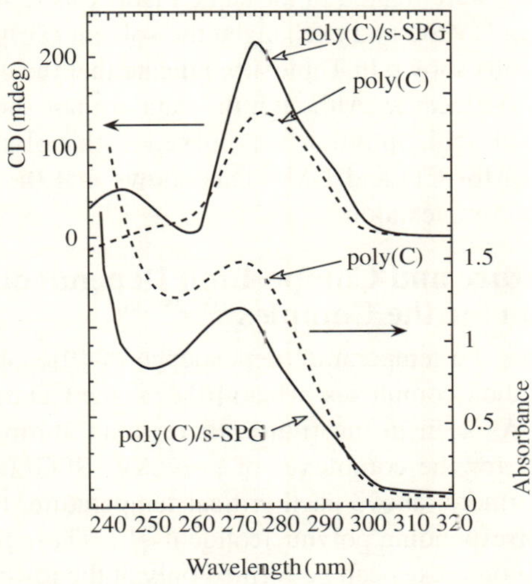


**Fig. 4.25** Comparison of the UV and CD spectra between poly(C) and poly(C) + s-SPG measured at 10 °C for $C_{poly(C)} = 7.16 \times 10^{-3}$ g dL$^{-1}$ ($2.22 \times 10^{-4}$ M) with a 1 cm cell.

### 4.4.3 Molecular Specificity in the Complexation[124,126)]

Table 4.5 summarizes which homo-polynucleotide can induce the complexation. Poly(G), poly(U), poly(I), poly(dG) and poly(dC) show no appreciable change in the UV and CD spectra. However, we found that poly(C), poly(A), poly(dA) and poly(dT) can form a complex with s-SPG. According to a previous work,[129)] the guanines in poly(G) and poly(dG) form a tetramer, the uraciles and cytosines in poly(U) and poly(dC) form a dimer, and the inosines in poly(I) form a tetramer or trimer. For all the cases the hydrogen-bonding sites in the bases are used to form the intramolecular interaction. On the other hand, poly(C), poly(A), poly(dA) and poly(dT) do not form such an intramolecular aggregation; therefore, their hydrogen-bonding sites are unoccupied. There is clear correspondence between complexation ability and the presence of the free hydrogen-bonding sites, i.e., only when the hydrogen bonding-site is available, can the polynucleotide interact with s-SPG, indicating that the hydrogen-bonding interactions are essential to induce the complexation.

The importance of the hydrogen-bonding capability is also proved by the pH dependence of the complextion for the poly(C) + s-SPG system.[131)] The complex forms at pH = 6.5−10 and dissociates at pH = 4−6. The cytosine moiety in poly(C) is protonated at pH = 4−6 and the dimerization of the cytosine occurs with an inter- or intra molecular hydrogen bond, while at pH = 6.5−10 poly(C) maintains a single strand and has unoccupied hydrogen bonding sites. The pH region for the single strand completely agrees with that of complex, confirming that the unoccupied site is necessary for s-SPG to combine with polynucleotides.

We examined whether the other glucans show the same change in the UV or CD spectra when they were treated in the same manner as schizophyllan (i.e., dissolved or dispersed in DMSO and exchanged the solvent for water in the presence of poly(C)). The results shown in Table 4.6 indicate that the complexation is only observed for β-(1→3)-glucans. Although the data are not shown, even when the natural triple helix of schizophyllan was mixed with poly(C), no appreciable change was observed for CD and UV. This shows that the renature process is indispensable for the complexation.

### 4.4.4 Temperature and Composition Dependences of Circular Dichroism for the Complex[123,125,126)]

Figure 4.26 compares the temperature dependence of $[\theta]_{max}$ between two homo-polynucleotides and their complexes, where $[\theta]_{max}$ is the CD intensity at the top of the positive band. As seen in the figure, there is an abrupt decrease in $[\theta]_{max}$ around 32 and 54°C for the complexes of poly(A)/s-SPG and poly(C)/s-SPG,* respectively. Furthermore, above each critical temperature, the $[\theta]_{max}$ merges in the values for the corresponding polynucleotide itself. These features show that s-SPG/polynucleotide complexes can be formed only at the lower temperature range and they dissociate (or melt) upon heating. The melting temperatures ($T_m$) evaluated from the plots are summarized in Table 4.5. It is interesting that the difference in $T_m$ as well as the abrupt decrease is similar to the well-known melting

**Table 4.5** Nucleotide specificity in complex formation in neutral and nonsalt aqueous solution

| | | Complex formation | $T_m$ (°C) | Conformation |
|---|---|---|---|---|
| RNA | poly(C) | Yes | 54 | single chain |
| | poly(A) | Yes | 32 | single chain |
| | poly(U) | No | | intramolecular H-bond (hairpin like) |
| | poly(G) | No | | 4G wire (intramolecular H-bond) |
| | poly(I) | No | | intramolecular H-bond |
| DNA | poly(dC) | No | | intramolecular H-bond |
| | poly(dA) | Yes | (40)* 80 | single chain |
| | poly(dT) | Yes | 10 | single chain |
| | poly(dG) | No | | 4G wire (intramolecular H-bond) |

* Conformational transition temperature, see text for details.

**Table 4.6** Relationship between the capability to form the complex and the glucose linkage of natural glucans

| Name | Units | Linkage | Side chain | |
|---|---|---|---|---|
| Amylose | α-D-Glucose | 1→4 | no | No |
| Carrageenan | β-D-Galactose | alternating 1→4 and 1→3 | sulfonic groups | No |
| Cellulose | β-D-Glucose | 1→4 | no | No |
| Dextran | α-D-Glucose | 1→6 | α(1→3) or α(1→4) | No |
| Pullulan | α-D-Glucose | 1→6 | no | No |
| Lentinan | β-D-Glucose | 1→3 | G G<br>-G-G-G-G-G- | Yes |
| Curdlan | β-D-Glucose | 1→3 | -G-G-G- | Yes |
| Schizophyllan | β-D-Glucose | 1→3 | G<br>-G-G-G- | Yes |

behavior of double-helix DNAs. The poly(C)/s-SPG complex "melts" at a higher temperature than the poly(A)/s-SPG complex, which is similar to the fact that the poly(C)/poly(G) complex melts at higher temperatures (*ca.* 110 °C) than the poly(A)/poly(U) complex (*ca.*, 65 °C). The difference in the melting temperatures of the DNA helix can be ascribed to the difference in the number of the hydrogen bonds (cytosine has 3 hydrogen-bonding sites whereas adenosine has 2 sites). The same discussion can be applied to explain the difference in the melting temperature between the poly(A)/s-SPG and poly(C)/s-SPG complexes, assuming that hydrogen bonding is involved in the complex. In fact, when we evaluated the entropy and enthalpy for the melting of the poly(C)/s-SPG complex, these values are very close to those of DNA.[126)]

As shown in Table 4.5, poly(dA) in the complex shows the conformational

* In this paper, s-SPG + poly(X) stands for a mixture of poly(X) and s-SPG, and does not necessary mean the complex; it can just be a mixture of the two components without any interaction. The complex is explicitly denoted as poly(X)/s-SPG.

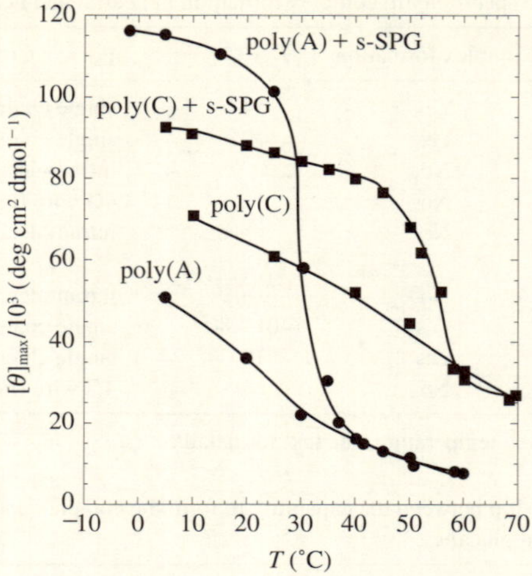


**Fig. 4.26** Temperature dependence of $[\theta]_{max}$ for poly(A) and poly(C) and their mixtures with s-SPG.

transition upon heating at 40 °C, then dissociates at 80 °C. Although not discussed in detail (*cf.* reference[125]), this strange behavior can be ascribed to the fact that the deoxyribose moiety in poly(dA) can take two puckering conformations (C2'-end and C3'-end), while the ribose moiety in poly(A) takes only the C3'-end conformation.[129]

Figure 4.27 plots the relative intensities of CD against $V_W$ for both poly(A) and poly(C) systems. Here, $V_W$ is the water volume fraction in a water/DMSO solution and the relative intensity is defined as the normalized value of $[\theta]_{max}$ by that of poly(A) or poly(C) at $V_W = 1$. In the low $V_W$ region, the individual polynucleotide, i.e., poly(A) or poly(C), exhibits the same values as expected for each mononucleotide. At $V_W = 0.5$ for poly(A) and $V_W = 0.6$ for poly(C), the intensity begins to increase. These increments indicate that the polarity of the solvent becomes significant enough to induce the hydrophilic interaction to lead to stacking between the bases. With increasing $V_W$, the intensity increases essentially linearly with $V_W$. On the other hand, when s-SPG is added, it shows completely different behavior from that of each individual polynucleotide. In the low $V_W$ region (*ca.*, 0−0.45) the CD intensity stays at the same value as that of the mononucleotide. At $V_W = 0.58$ for poly(A) + s-SPG and $V_W = 0.6$ for poly(C) + s-SPG, the intensities increase dramatically and level off at $V_W > 0.7$. The onset composition for the complex formation is close to where the stacking begins in the poly(A) and poly(C) solutions. This coincidence indicates that the hydrophobic interaction between bases is also key to inducing complex formation. Furthermore, the saturation of the CD intensity indicates that the complex formation is completed and most bases are stacked at $V_W > 0.7$.

Careful examination of the difference in the onset composition between

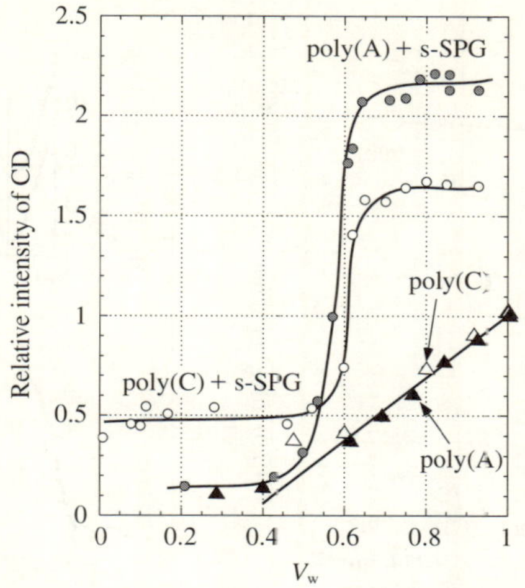


**Fig. 4.27** Composition dependence of the relative intensity of CD for poly(C), poly(C) + s-SPG, poly(A), and poly(A) + s-SPG.

poly(A) + s-SPG and poly(C) + s-SPG reveals that the composition of the poly(A) system is lower than that of the poly(C) system. This can be ascribed to the fact that the adenine moiety is more hydrophobic than the cytosine moiety due to the difference in the number of the aromatic rings; adenine is a purine base and cytosine is a pyrimidine base. This is consistent with our conclusion that the hydrophobic interaction between the bases is important for the complexation.

### 4.4.5 Critical Molecular Weight to Induce the Complexation[125)]

We prepared three s-SPG samples with $M_W$ = 2100, 3800, 7200 with acid hydrolysis from a s-SPG sample ($M_W$ = 15,000), and examined which sample can combine to poly(C) with the base number fixed at 250. Furthermore, we changed the base number in poly(dA) and fixed moleclar weight at 15,000 for s-SPG, and examined how many bases are necessary to induce the complexation. The results are presented in Fig. 4.28(a) and (b). When the number of 5'-cytidine monophosphate is large enough, the critical $M_W$ of s-SPG for the complexation is 3000−5000. This means that, about 16−30 glucose units are necessary for the complexation. On the other hand, 25−30 of deoxyadenosine monophosphate are necessary to combine to the s-SPG with $M_W$= 15,000.

### 4.4.6 Stoichiometry and Molecular Modeling[126,132)]

To determine the stoichiometry of the complex, we measured the CD spectra by changing the molar ratio: $M_{s\text{-}SPG}/(M_{s\text{-}SPG} + M_{poly(C)})$, where $M_{s\text{-}SPG}$ and $M_{poly(C)}$ are the molar concentrations of poly(C) and s-SPG, respectively. Fig. 4.29 plots $[\theta]_{max}$ against the molar ratio. The $[\theta]_{max}$ value increases linearly with increasing molar ratio, following an upward convex, and finally levels off at $[\theta]_{max} = 90 \times 10^3$ deg

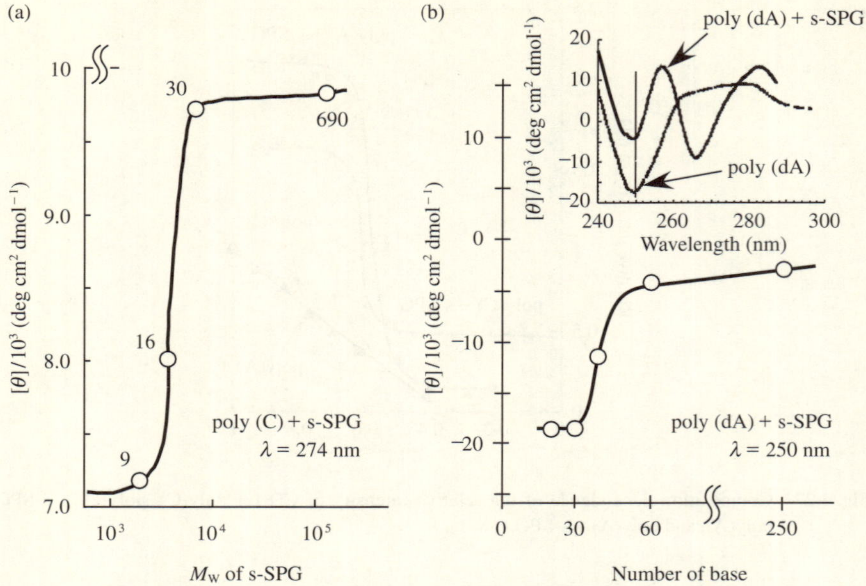


**Fig. 4.28** The critical chain length required to induce complexation. The subset (a) shows the $M_W$ of s-SPG dependence of $[\theta]_{max}$ for the s-SPG + poly(C) system when the base number of poly(C) is fixed at 250. The subset (b) shows that the base number dependence of $[\theta]_{max}$ for the s-SPG + poly(dA) system, fixing $M_w$ = 15000 for s-SPG. The inset of (b) compares the CD spectrum between poly(dA) and the s-SPG/poly(dA) complex.

cm$^2$ dmol$^{-1}$. The cross section of the plateau and increment regions provides the stoichiometric ratio; $n$ is evaluated to be 0.39−0.41. Although the data are not shown, the poly(A) and poly(dA) systems show a similar stoichiometric ratio. These results indicate that two SPG repeating units combine into three base units, namely, 8 glucoses *vs.* 3 bases. When we examined the stoichiometry for the curdlan/poly(C) system,[128)] the results showed 6 glucoses *vs.* 3 bases. As shown in Table 4.6, curdlan has no side chain and schizophyllan has one glucose side chain out of the three main chain glucoses. Therefore, that discrepancy in the stoichiometric number between the curdlan and schizophyllan complexes can be rationalized by assuming that the main chain of the β-(1→3)-glucans participates in the complexation while the side chain of schizophyllan is not involved in the complexation.

We can speculate a few possible structures based on the following facts or assumptions: (1) since the second hydroxyl groups of the main chain, which are located inside the triple helix column, participate in the hydrogen-bonding interaction, the same hydroxyl groups should be involved in the complex formation; (2) the side glucose chain only provides solubility and is not involved in the complex formation; and (3) the CD and UV data suggest the formation of a helical structure in the complex. Fig. 4.30 illustrates the two possible structures for the poly(C)/s-

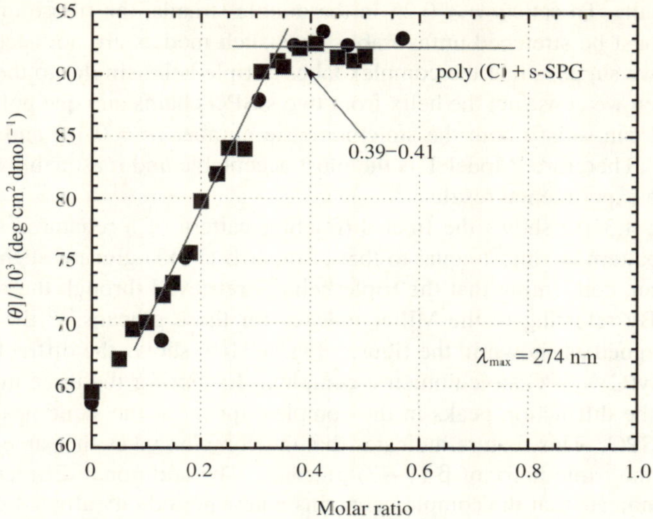


**Fig. 4.29** The Job plot for the poly(C) + s-SPG system. The molar ratio is defined by $M_{s\text{-SPG}}/(M_{s\text{-SPG}} + M_{poly(C)})$, where $M_{s\text{-SPG}}$ and $M_{poly(C)}$ are the molar concentrations of poly(C) and s-SPG, respectively. The cross section of the plateau and increment regions shows that the stoichiometric ratio = 0.39−0.41.

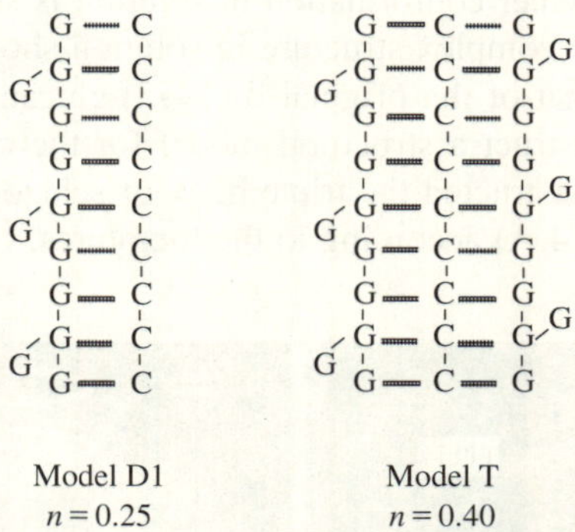


**Fig. 4.30** Two possible structures for the poly(C)/s-SPG complex. G and C denote the glucose and cytosine in the chains, respectively. In Model D1, every glucose residue in the main chain interacts with the cytosine. Model T supposes that the complex adopts a triple helix similar to that existing in the β-(1→3)-glucans. The numbers under the models indicate the stoichiometric numbers for each model.

SPG complex. In Model D1, every glucose residue in the main chain interacts with the cytosine in a DNA-like double helix. Since the distance between the neighboring C-2 carbon's OH is about 4 Å and the stacking distance between cytosine moieties is 3.0−3.6 Å, this model is reasonable from the stereochemistry. However, the stoichiometric number ($n$ = 0.25) does not agree with the experimental result. To satisfy $n = 0.25$ in the double strands, the ribose or phosphate moieties must be stretched unfavorably, thus such models are not acceptable. In Model T, we suppose that the complex takes a triple helix similar to the original t-SPG. When we construct the helix from two s-SPG chains and one poly(C) chain, this model can satisfy both the stoichiometric number ($n = 0.40$) and the stereochemistry. Therefore, Model T is the most acceptable and reasonable structure to explain the experimental results.

Figure 4.31(a) shows the fiber-diffraction pattern of a renatured schizophyllan. The pattern is almost equal to that of other β-(1→3)-glucans such as curdlan and lentinan, confirming that the triple helix is retrieved through the renaturation process. By referring to the Miller indexes for these glucans,[119] the reflections can be assigned as shown in the figure. Fig. 4.31(b) shows the diffraction pattern from a poly(C)/s-SPG stoichiometric complex. Except for the three meridian diffractions, the diffraction peaks in the complex appear at the same position as the renatured SPG. This feature indicates that the complex takes the same helix pitch as that in the triple helix of β-(1→3)-glucans. The additional diffractions in the meridian indicate that the complex contains a new periodicity aligned perpendicular to the fiber axis. This meridian diffraction can be associated to the stacked cytosine base. However, the crystallinity and crystalline orientation are not good enough to determine the structure. The important conclusion suggested by the X-ray crystallography is that the poly(C)/s-SPG complex takes the helix similar to that of the natural triple helix of β-(1→3)-glucans.

Since the local polymer conformation in solution is sometimes related to the crystalline structure, the complex structure in solution should be represented by a triple helix similar to that of the original β-(1→3)-glucans. Based on the above discussion, we can construct a structural model for the complex with computer chemistry. First, we constructed the triple helix of schizophyllan (a right-handed $6_1$ triple helix with a 17.4 Å) according to the literatures,[117,118] and took one chain

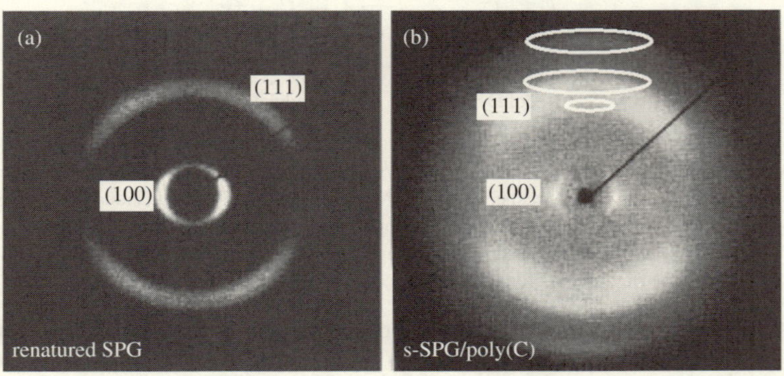


**Fig. 4.31** Comparison of the X-ray diffraction patterns between the renatured s-SPG (a) and the s-SPG/poly(C) complex (b). The fiber axis is vertical and the Miller indexes are assigned by referring to lentinan and curdlan.

out of the helix, thus forming a hypothetical double helix. We created a poly(C) single helix referring to the crystal structure,[129] i.e., a right-handed $6_1$ helix with an 18.6 Å pitch and the ribose has the A-type ($C_3$'-end) configuration. Next, we tried to merge the schizophyllan double helix and the poly(C) single helix. The poly(C) chain can fit the groove that has been created after one SPG chain is taken out because of the similarity in the helix parameters between poly(C) and schizophyllan. Both schizophyllan and poly(C) adopt a right-handed $6_1$ helix with similar pitch. In the third step, using the Discover 3 program and the Amber force field, we calculated the most stable conformation of poly(C) confined in the groove, fixing all atom positions in schizophyllan. The calculation was converged

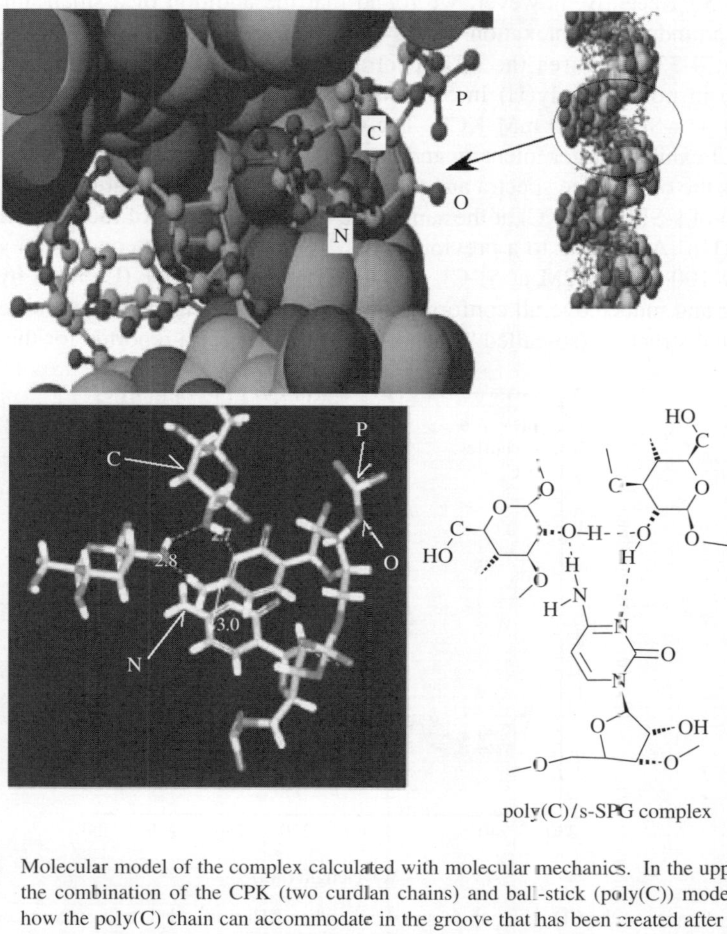


**Fig. 4.32** Molecular model of the complex calculated with molecular mechanics. In the upper panel, the combination of the CPK (two curdlan chains) and ball-stick (poly(C)) models shows how the poly(C) chain can accommodate in the groove that has been created after one SPG chain is taken out. The lower panel shows the hydrogen bonding site speculated from the calculation. The numbers in the figure show the distance between the atoms in the hydrogen bonds in Å.

and indicated that poly(C) can take a conformation that is very close to the original one without creating any steric hindrance, as shown in Fig. 4.32. Furthermore, according to the calculated model of the complex, the cytosine can be located close to the second hydroxyl groups to form hydrogen bonding (see Fig. 4.32). Therefore, at present, we believe that the amino group attached to the fourth carbon of cytosine donates its proton to the oxygen atom of the second hydroxyl group of the adjacent glucose, and the hydroxyl group of the other adjacent glucose donates the proton to the third nitrogen of the cytosine, as shown in Fig. 4.32.

### 4.4.7 Cation-induced Complexation[127)]

Poly(U) cannot combine with s-SPG in nonsalt or neutral solutions, as seen in Table 4.5. Recently, however, we found that the addition of a small amount of cation can induce complexation.

Fig. 4.33 compares the CD spectra among the following four solutions: poly(U) in no salt, poly(U) in 50 mM KCl, poly(U) + s-SPG in nonsalt, and poly(U) + s-SPG in 50 mM KCl. The CD spectrum of poly(U) + s-SPG in 50 mM KCl exhibits larger intensity and the peak top emerges at a lower wavelength, whereas the other three spectra are essentially identical. This feature indicates that addition of s-SPG and KCl at the same time induces stacking of the uracil moieties in poly(U). According to a previous work,[129)] in a given salt concentration range (usually 100 mM − 2M at 5 °C), two uracil residues of poly(U) form hydrogen bonding and induce overall conformational transition from a random coil to a double helical structure (so-called "hairpin"). The CD spectra reported for the hairpin

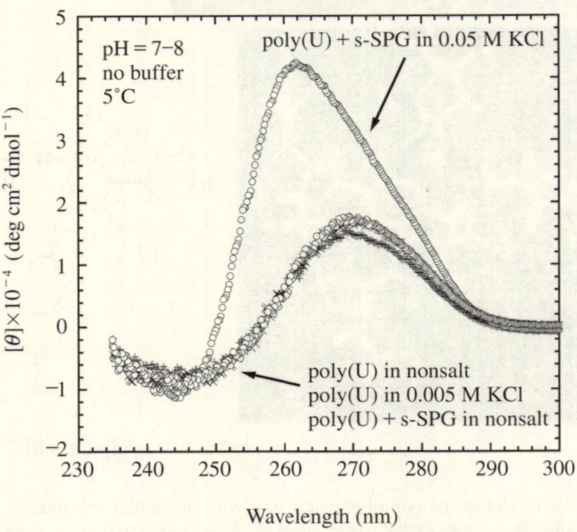


**Fig. 4.33** Change in the CD spectrum occurred only when KCl was added to the mixture of poly(U) and s-SPG (○). Poly(U) in nonsalt (×), poly(U) in 50 mM KCl (×), poly(U) and s-SPG in nonsalt(○), and poly(U) and s-SPG in 50 mM KCl (○). The poly(U) and s-SPG molar concentrations were $2.6 \times 10^{-4}$ and $6.4 \times 10^{-4}$ M, respectively.

is quite similar to that for poly(U) + s-SPG in 50 mM KCl in the figure, however, the salt concentration of 50 mM is not high enough to induce the hairpin conformation. In fact, the CD spectrum for poly(U) in 50 mM KCl is essentially identical to that of poly(U) in nonsalt. Therefore, the addition of some alkaline cations to the mixture of poly(U) and s-SPG can induce complexation between poly(U) and s-SPG, otherwise nothing happens between poly(U) and s-SPG. We found that the complex stability strongly depends on both the cation species and the salt concentration. The magnitude of CD spectral change follows the order $Rb^+ \gg K^+ > Na^+ \approx Cs^+$, whereas $Li^+$ does not induce any change. This stability order is confirmed by the salt concentration dependence of the fluorescence polarization. Furthermore, $^{23}$Na NMR spectroscopy reveals that the product system is stabilized by the addition of cations, suggesting that the OH group in schizophyllan and the phosphate anion in poly(U) synergistically form a specific ligand system for the cations.

### 4.4.8 Switching the Complexation On/Off

Since the complexation is sensitive to the cation concentration, it can be switched on and off by changing the KCl concentration. Cryptand [2,2,2] is a well-known strong ligand for potassium, the effective (or free) potassium cation concentration in the system can be controlled by adding cryptand.[124] Fig. 4.34 demonstrates that the complexation is almost reversibly switched on and off by alternately adding

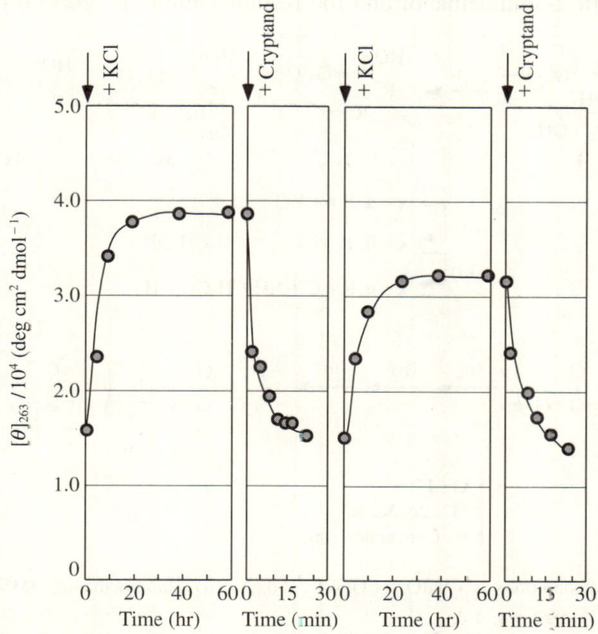


**Fig. 4.34** Switching the complexation by adding KCl and cryptand alternately. The poly(U) and s-SPG molar concentrations were $2.6 \times 10^{-4}$ and $6.4 \times 10^{-4}$ M, respectively.

KCl and cryptand. The potassium concentration inside cells (*ca.* 160 mM) is higher than that in blood plasma (*ca.* 4 mM), so using this difference, it is possible to construct an oregano-position sensitized drug carrier, where the drug has a poly(U) tail as a s-SPG binding site and s-SPG acts as a drug protector as well as a targeting carrier (SPG is known to specifically accumulate in tumor cells, lymph nodes, and bone marrow). Once this complex enters the cell, the complex dissociates and releases the drug.

### 4.4.9 Chemical Modification to Enhance Affinity[128)]

To utilize the complex as a gene carrier for antisense oligonucleotides or an affinity column to separate a certain polynucleotide from others, it would be a great advantage to develop a method for tuning the affinity between the glucans and particular polynucleotides. We found that such a tunable affinity can be achieved when we site-specifically introduce a functional group to the side chain of the β-(1→3)-glucans.[128,133)] Since the molecular model of the complex indicates that the β-(1,3)-glucan main chain is essential for the complexation, the main chain should be kept intact throughout the chemical reaction. Fig. 4.35 schematizes the reaction scheme used when aminoethanol is introduced. In the first step, the side chain is oxidized by periodate oxidation to yield aldehyde. Since the periodate only cleaves a 1,2 diol and there is no such diol in the β-(1,3)-glucan main chain, the main chain cannot react with periodate. In the second step, the aldehyde group is reacted with 2-aminoethanol and the resultant imine is reduced to give a sec-

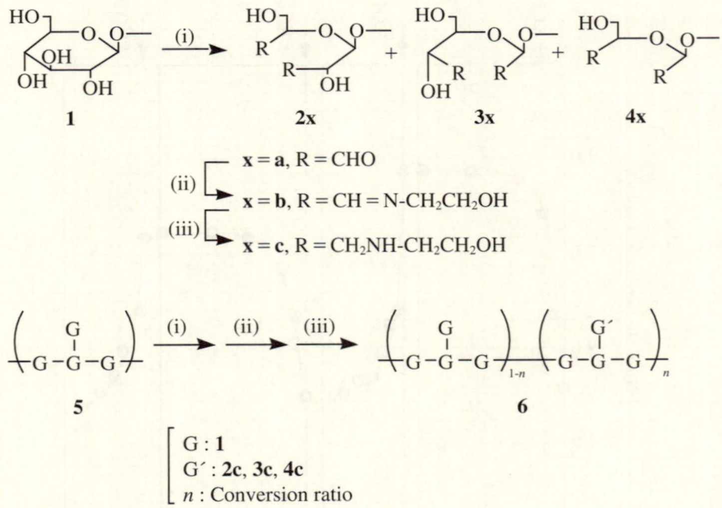


Reagents and conditions : (i) $NaIO_4$, $H_2O$, 4°C, 2 days ; (ii) 2-aminoethanol, DMSO, r. t., 2 days ; (iii) $NaBH_4$, DMSO, r. t., 1 day.

**Fig. 4.35** Reaction scheme for introducing aminoethanol to the schizophyllan side chain through periodate oxidation and reductive amination.

ondary amine through reductive amination. The modification level (the modified side chain molar percent) can be controlled in the range of 1−30 %.

We prepared a 2-aminoethanol-appended schizophyllan with a modification level of 2.4% with the method presented in Fig. 4.35, and hereinafter designated as N-s-SPG. When N-s-SPG is mixed with poly(C), it forms a complex that exhibits the same spectroscopic behavior as the poly(C)/s-SPG complex.[128] This indicates that the presence of the anion group in the side chain does not obstruct the complexation at all. Furthermore, the coincidence of the CD spectra implies that the complex structure is almost identical between the poly(C)/s-SPG and poly(C)/N-s-SPG complexes. In fact, the Job plot for the poly(C)/N-s-SPG system shows that the stoichiometric number $n$ is close to 0.40, similar to that of poly(C)/s-SPG.

Figure 4.36 compares the temperature dependence of $[\theta]_{max}$ between the poly(C) + s-SPG and poly(C) + N-s-SPG systems. The poly(C)/N-s-SPG complex shows dissociation cooperativity (sharpness of the melting curve) similar to that of the poly(C)/s-SPG complex. However, the melting temperature of the poly(C)/N-s-SPG complex increases by 8 °C, which is higher than that of unmodified s-SPG. It is noteworthy that only 2.4 mol% of modification can induce such high thermal stabilization. Our study showed that when we added the Tris to the poly(C)/N-s-SPG complex, the melting temperature decreased with increasing cation concentration, suggesting that the stabilization, i.e., increment of $T_m$, in the poly(C)/N-s-SPG is related to the ion-pair formation by the introduced amino groups. In general, the electrostatic interactions are relatively strong and are classified as long-range interactions (longer than the hydrogen-bonding interactions),

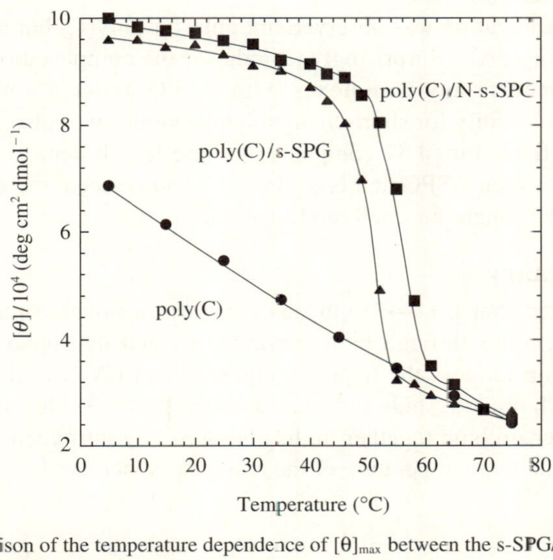


**Fig. 4.36** Comparison of the temperature dependence of $[\theta]_{max}$ between the s-SPG/poly(C) and N-s-SPG/poly(C) complexes.

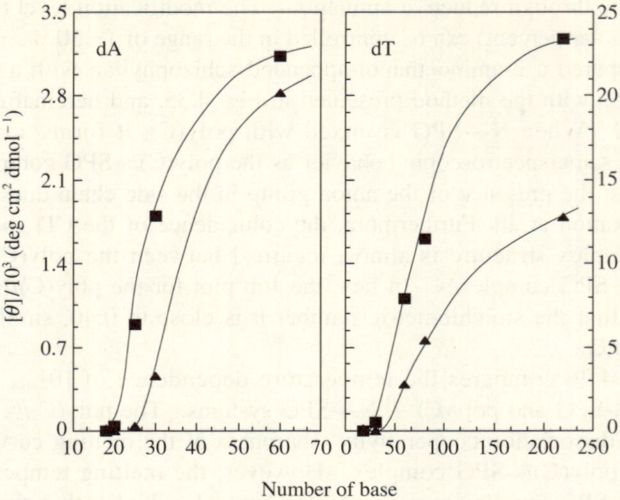


**Fig. 4.37** Comparison of the $[\theta]_{max}$ *vs.* base number plot between the s-SPG and N-s-SPG complexes.

so that the ion-pair formation between the polymer chains tends to suppress the cooperative dissociation of the multivalent hydrogen bonding. In other words, the ion pair has the ability to pin the polymer chains together to prevent them from dissociating. There is no doubt that this "pinning effect" contributes to enhance the melting temperature.

Thermal stabilization was observed not only for poly(C) but also for poly(A), poly(dT) and poly(dA). Surprisingly, it induced the complexation with poly(U), which does not form any complexes with s-SPG under a nonsalt condition. Furthermore, the affinity for short-chain polynucleotides was also enhanced by the amino modification. Fig. 4.37 compares the base length dependence for dA and dT oligomers between s-SPG and N-s-SPG. As can be seen, the critical base pair length is shifted through the small modification.

## 4.4.10 Summary

We demonstrated that β-(1→3)-glucans can form a novel complex with some homo-polynucleotides through hydrogen-bonding and hydrophobic interactions. The complexation induces the hypochromic effect in UV and creates a new CD spectrum. Stoichiometric study indicates that two β-(1→3)-glucan chains and one polynucleotide chain bind together in the complex. By introducing a cation to the side chain, the melting temperature of the complex is increased.

# 4.5 Physicochemical Properties of Spider Silk–An Approach to Nanostructure

**Shigeyoshi Osaki**

## 4.5.1 Introduction

There is much interest in spider silks because spiders are able to capture insect prey skillfully using their orb-web and to fall quickly using their lifeline to escape danger from invaders. Spider silks may also be useful as a new material in the next generation because of their physicochemical properties. Recently, a number of researchers have focused on spider silks because of the development of the genetic technology, which provides the possibility of mass producing spider silks[134-136] for use in bullet-proof wears, stockings, suture and other purposes. Thus, it is very important to determine whether the physicochemical properties of spider silks are truly excellent, and to investigate such properties from the viewpoint of nanostructure.

Many properties of spider silks have been studied including the physicochemistry, mechanical stress, elastic modulus and strain energy.[137-145] The present author has investigated the thermal, optical, molecular, aging and mechanical properties of spider silks.[146-154]

It is meaningful to study the thermal properties of spider's silk thread from the heat resistant point of view since spider silks adhere to some objects such as rock and metal and may be heat-stable even at high temperatures between 60 °C and 100 °C. However, the thermal properties have not previously been studied, probably due to the difficulty in collecting draglines. Spider threads have been studied mainly from the morphological point of view.[155]

The orb-web is irradiated outdoors by ultraviolet (UV) rays. It is interesting to ask from the aging point of view how UV-irradiation affects spider silks at different stages of life and to study the effects of UV-irradiation. However, there have been no reports on the effects of UV rays upon draglines.

Draglines act as a mechanical lifeline supporting their weight (Fig. 4.38) as spiders move and escape from various dangers. It is intriguing to examine whether the dragline is mechanically strong and whether the quality of mechanical safety changes over the spider's life time because the quality of danger changes as they grow. The elastic modulus of draglines gives an intrinsic value reflecting the mechanical strength in the region where the stress-strain behavior is linear. The elastic modulus has been estimated by the stress-strain curve and X-ray diffraction

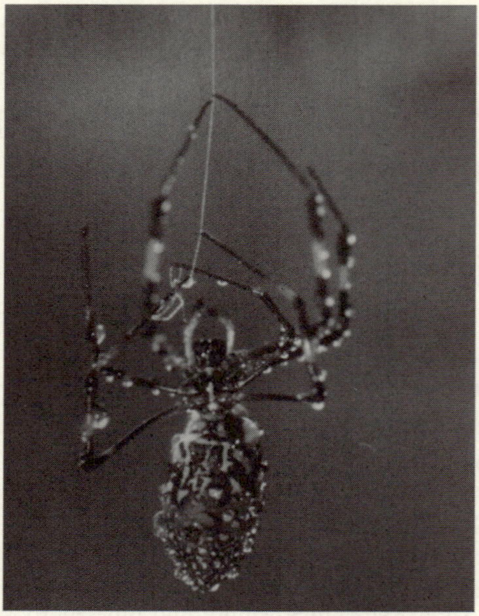

**Fig. 4.38** A *Nephila clavata* spider hanging from its thin dragline.

methods.[142,143,156] However, there is a question as to whether the elastic modulus had been accurately determined because of the following four problems: sample preparation, mechanical hysteresis, recognition of true draglines and measurement of the cross section. These problems have caused difficulties in determining the elastic modulus accurately. On the other hand, there is also a report of a preliminary synchrotron X-ray study of the spider dragline silks.[157]

It is uncertain whether spider silks show actual characteristic properties even though many researchers have been interested in the physicochemical properties of spider silks.[137-145] This is mainly due to the difficulty in obtaining sufficient numbers of dragline samples appropriate for thermal, optical and aging experiments, in preparing the samples with no mechanical hysteresis and in measuring the mechanical properties because the silk samples are very thin.

Recently, the author overcame such difficulties and was able to measure the physical properties of spider silks. As a result new information was obtained from the measurements of spider silks. An approach to the mass production of spider silks actually became meaningful due to the development of genetic technology.[134-136] It is very important to study the characteristics of the physicochemical properties of spider silks and to explain such properties from the nanostructural point of view.

The mechanical strength of draglines as a lifeline for spiders may be related to the spider's weight, but no studies have yet been reported regarding the relationship. Thus, the safety coefficient for the lifeline should be considered for the mechanical strength of draglines. The spider's 400 million years of evolution should have allowed the mechanical properties of draglines to reach maximum efficiency.

This section describes the physicochemical properties of spider silks such as thermal, aging, mechanical and infrared properties and an approach to explaining such properties from the nanostructural point of view.

## 4.5.2 Materials and Methods

### A. Spider Silks

Spiders are found all over the world and there are 40,000 different species. About half these spiders build orb-webs. Ecologically one can divide spiders into two groups: web spiders and wandering spiders. All spiders have the ability to produce silken threads from spinning glands, and use the spinning glands for constructing snares and making egg cocoons and draglines.[155)]

The different spinning glands in the spider's abdomen produce different kinds of silks, such as draglines, frame threads, mooring radial threads, spiral threads, sticky material, egg cocoon, swathing bands and attachment discs.

A typical web spider secretes different threads such as frame threads, mooring threads, radial threads, spiral threads and a hub for constructing orb-web. They also secrete threads such as attachment discs and draglines, as shown in Fig. 4.39. The radial threads converge in a central spot called the hub, which consists of

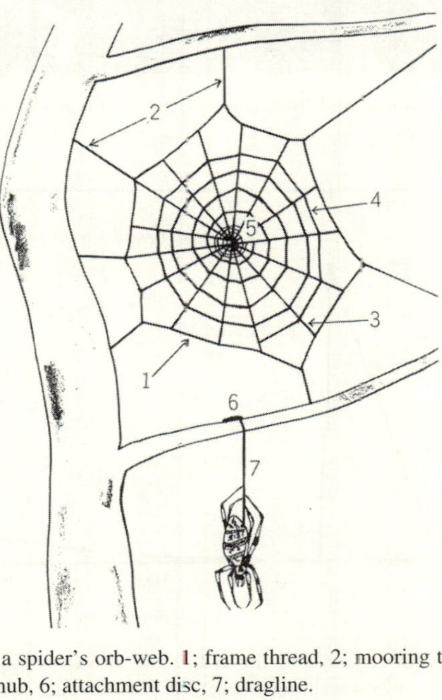


**Fig. 4.39** Schematic view of a spider's orb-web. 1; frame thread, 2; mooring thread, 3; radial thread, 4; spiral thread, 5; hub, 6; attachment disc, 7; dragline.

irregularly interconnected threads and mechanically supports the orb-web. The spiral threads are covered with sticky material for capturing prey (Fig. 4.40). The frame threads serve as insertion sites for the radial threads and for constructing the frame of the orb-web. Mooring threads are for mooring the orb-web and capturing threads are for wrapping insects. The dragline acts as a mechanical lifeline. The orb-web is mechanically supported by a main frame of radial threads, which are similar to draglines because they are both secreted from the ampullate gland.

The force-elongation curves of the radial and spiral threads prepared from

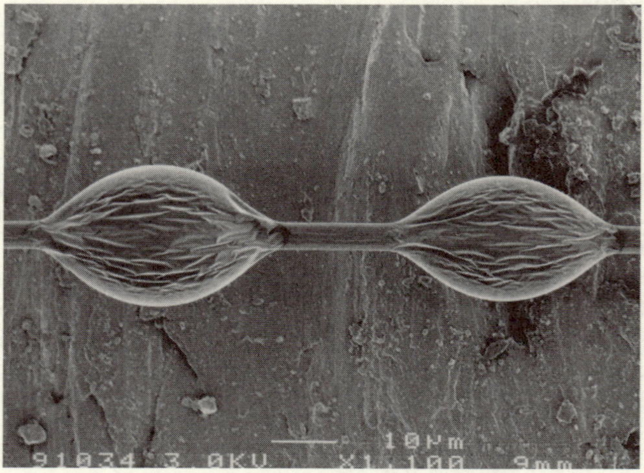


**Fig. 4.40** Spiral thread observed by scanning electron microscope. Sticky materials constitute globular shapes.

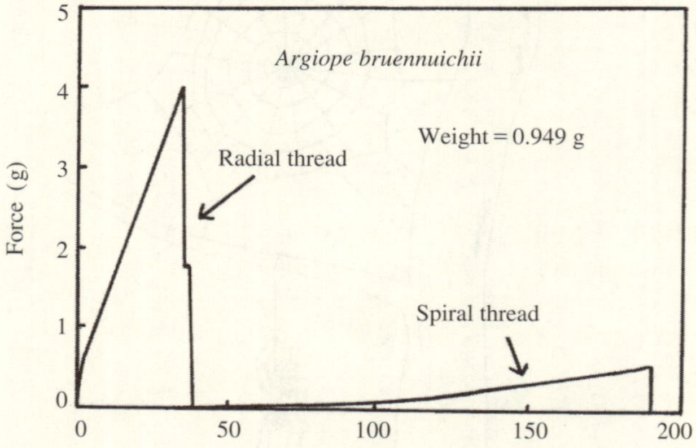


**Fig. 4.41** Force-elongation curves of the radial and spiral threads prepared from an *Argiope bruennuichii* spider with a weight of 0.949 g. Stretching velocity; $3.3\times10^{-4}$ m/s.

*Argiope bruennuichii* spider are described in Fig. 4.41.[158] The radial threads showed high mechanical breaking strength. They are used to construct the main frame of the orb-web. On the other hand, the spiral threads showed high elastic property, effective for capturing flying insects, given its low mechanical breaking strength. The orb-web is constructed so that only the connecting part of the spiral threads with the radial threads will break so as to avoid breaking the entire orb-web when a captured insect behaves violently.

### B. Samples for Thermal, Optical and Aging Measurements

The samples used here were mainly draglines secreted from *Nephila clavata* spiders (Japanese golden web spider). The samples were automatically wound with a 5 cm × 10 cm rectangular frame covered with plastic tube, as shown in Fig. 4.42. The samples were then used for thermal, optical and aging measurements.[160]

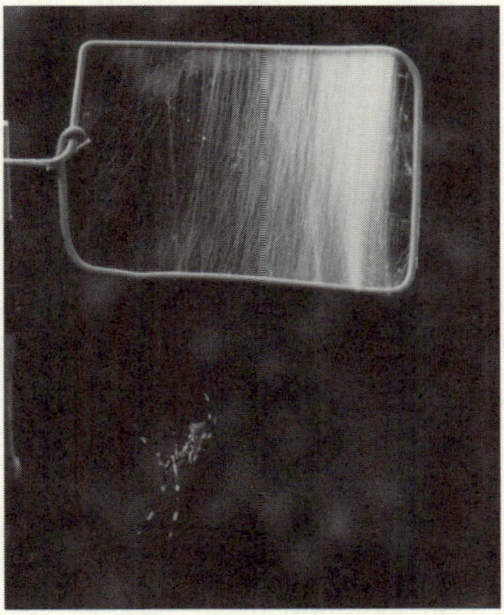

**Fig. 4.42** Draglines are automatically wound by a 5 cm × 10 cm rectangular frame covered with plastic tube when the *Nephila clavata* spider is falling from a dragline

### C. Preparation of Samples for Mechanical Measurements

The dragline samples for mechanical measurements and scanning electron microscopic (SEM) observation were prepared from a spider falling naturally from a wooden bar in order to avoid mechanical hysteresis, as shown in Fig. 4.43.

### D. Weight of *Nephila clavata* Spiders

In Japan, *Nephila clavata* spiders are born from an egg cocoon in late spring and reach divergence of sex in late summer, accompanied by the final ecdysis in summer. Male spiders visit the orb-webs of female spiders for courtship in autumn. After mating, pregnant female spiders lay their eggs inside silk cocoons in late

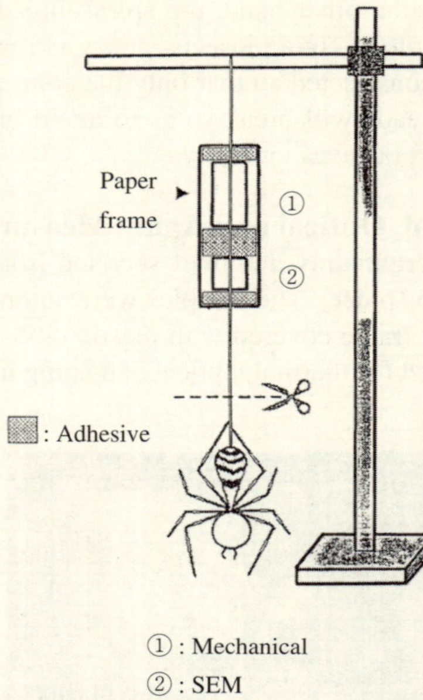


**Fig. 4.43** Schematic figure for preparing sample draglines for mechanical measurement and for scanning electron microscopic observation from *Nephila clavata* spider (Japanese golden web spider) falling from a wooden bar.

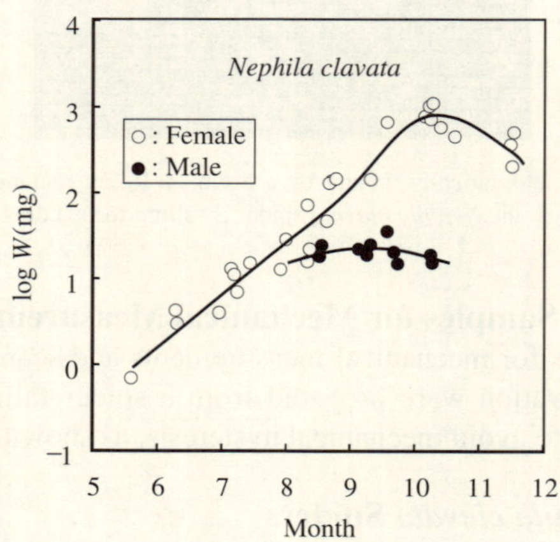


**Fig. 4.44** Weight ($W$) of *Nephila clavata* spiders in different months (female and male).

autumn and die soon afterward. The weight ($W$) of the female *N. clavata* spider in Japan increases exponentially from spring to autumn, when it reaches maturity. For example, a female spider's weight increases from about 1 mg in spring, to about 50 mg in summer, and up to about 1 g in autumn, as shown in Fig. 4.44. It is very difficult to differentiate the spider's sex before late summer.

## E. Thermal Measurements

Differential scanning calorimetry (DSC) and thermogravimetry (TG) were carried out at a heating rate of 20 °C/min using a micro TG-DSC Thermoflex 8100 manufactured by Rigaku Electric Co., Ltd. DSC aids in measuring change in heat by automatically compensating the amount of heat so that the temperature of the sample is always kept equal to that of the standard sample. TG assists in the measurement of change in sample weight with the use of a thermobalance.

## F. Density of Draglines

The average density of draglines at 20 °C was determined to be 1.13 $g/cm^3$ for juvenile *N. clavata* in July and 1.29 $g/cm^3$ for mature female *N. clavata* in October by a flotation method using a mixture of benzene and chloroform.[148]

## G. Scanning Electron Microscopy

Figure 4.45 is an electron microphotograph of a dragline, showing that the dragline consists of a double filament even though it appears as a single filament to the naked eye.[153] Double filaments usually adhere together and align in parallel. Each of the filaments has a nearly circular cross section when examined by a scanning electron microscope. A cylindrical filament has a minimum surface area, which is the most efficient structure for the spider to spin from its glands.

(a) (b)

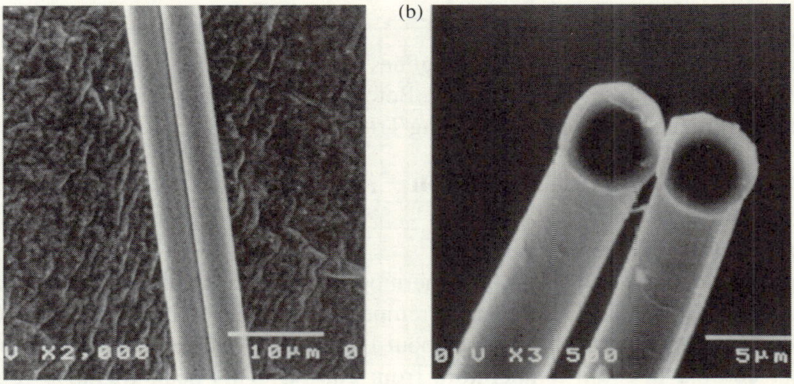


**Fig. 4.45** Electron microphotograph of a dragline from a *Nephila clavata* spider. The dragline consists of double filaments. (a) side view, (b) cross section view.

**H. Elastic Modulus**

The elastic modulus ($E$) is defined by

$$E = (F/S)/(\Delta L/L) \tag{4.1}$$

Here, $F$ is the force applied to the dragline, $S$ is the cross sectional area of the double filament, $L$ is the initial length of the dragline, and $\Delta L$ is the elongation of the dragline when the $F$ is applied. The $F$ and $\Delta L$ are determined from the force-elongation curve within the elastic limit point and the S from the electron scanning microscopy.

**I. ESR Measurements**

Electron spin resonance (ESR) spectra of spider silks inserted into a quartz tube of diameter of 4.8 mm were measured by an ESR spectrometer (JES-RE2X, JEOL). Irradiation of UV-rays upon the silks was carried out with a mercury lamp with a maximum wavelength of 250−365nm.

**J. Ultraviolet Rays**

UV-A and UV-B rays are defined as wavelengths between 320−400 nm and 290−320 nm, respectively. Rays with wavelengths longer than 270 nm and an intensity distribution similar to UV-A and UV-B rays were produced by a 660 watt xenon-arc lamp (Suntester XF-180, Shimadzu, Japan). The UV-A* and UV-B* were prepared by excluding wavelengths shorter than 320 nm and 290 nm, respectively, from the xenon-arc lamp using special filters. The UV-B* contains both UV-B and UV-A rays while UV-A* contains only UV-A rays. The intensity of UV-B* between 290−320 nm was approximately double the intensity of actual UV-B rays and the intensity of UV-A* between 320−400 nm was identical to that of actual UV-A.

**K. Amino Acid Composition**

The amino acid composition of draglines was analyzed by a Hitachi High Speed Amino Acid Analyzer (835-50 type). Before the amino acid analysis, the silk sample was purified with water containing Triton X100.

### 4.5.3 Results and Discussion

**A. Thermal Properties**

Figure 4.46 shows DSC and TG thermograms of the draglines from female *N. clavata* collected in October in a temperature range of 20 °C to 200 °C. An endothermic peak was observed at about 100 °C and disappeared in the second run after cooling to room temperature from 200 °C. This peak was also found to reduce in height after decompressing at a pressure of 1.33 Pa for 3 h at room temperature. This suggested that the peak may be due to the desorption of water molecules adsorbed on silk fibers. The adsorbed water was estimated to be about 10% of the sample weight from the TG curve. The broadness of this endothermic peak

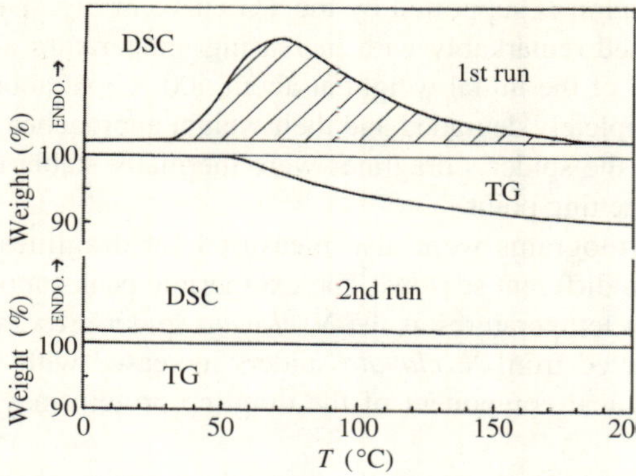


**Fig. 4.46** DSC and TG thermograms for draglines prepared from female *N. clavata* collected in October. Temperature range: room temperature to 200 °C; ENDO, endothermic; *T*, temperature.

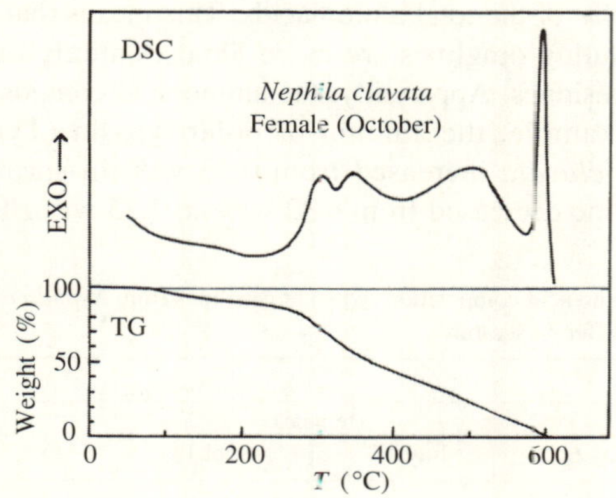


**Fig. 4.47** DSC and TG thermogram for draglines prepared from female *N. clavata* collected in October. Temperature range: room temperature to 600 °C; EXO, exothermic; *T*, temperature.

reflects the distribution of adsorption sites of water molecules contained in draglines.[148)]

Figure 4.47 shows the DSC and TG thermograms of draglines from female *N. clavata* collected in October in a range of up to 600 °C. The color of the draglines changed to deep yellow at about 200 °C and to black at about 400 °C, suggesting that the draglines began to decompose above 200 °C. Exothermic peaks can be observed at about 300 °C, 340 °C, 500 °C and 600 °C. The first three peaks are due to the degradation of macromolecular proteins, while the last is ascribed to the evaporation of all chemical species.[148)]

Such consideration is supported by the TG curve in Fig. 4.47. The weight of the sample decreased remarkably with increasing temperature above 200 °C and reached about 40% of the initial weight at about 400 °C. At about 600 °C, all the draglines were completely degraded and their weight approached zero. These findings indicated that the spider's draglines were thermally stable up to 200 °C and did not show any melting point.

The DSC thermograms were also measured for draglines from female *N. clavata* collected in different seasons. The exothermic peaks appearing above 280 °C shifted to higher temperatures as the *N. clavata* spider grew, and the density of the draglines collected from *N. clavata* spiders increased with season. This suggests that the chemical component of the dragline protein may change with the season.

In order to clarify the change in the thermal properties of spider draglines by season and sex, the author studied the amino acid composition of draglines consisting of proteins. Table 4.7 shows the amino acid composition of draglines at different seasons. The major constituents of amino acid of the draglines from *N. clavata* spider are glycine (Gly), alanine (Ala) and glutamic acid (Glu) residues, and occupy 70 – 80% of the total amino acids. This means that the macromolecular proteins constituting draglines are made almost entirely of a repeat of three major amino acid residues. Apparently, the amino acid composition changes with the season. For example, the amount of polar tyrosin (Tyr) residues in the draglines from *N. clavata* increased from 0.53 wt% in spring to 3.86 wt% in autumn, while proline decreased from 6.32 wt% to 3.35 wt%. The amount of pro-

**Table 4.7** Amino acid composition (wt%) of draglines from *Nephila clavata* collected in different seasons

| Amino acid | *Nephila clavata* | | | | |
|---|---|---|---|---|---|
| | (female) | | | | (male) |
| | May 6 | July 15 | Sept 1 | Oct 11 | Oct 1 |
| Asp | 2.33 | 1.74 | 1.52 | 1.45 | 1.92 |
| Thr | 1.04 | 0.86 | 0.91 | 0.97 | 1.09 |
| Ser | 4.67 | 3.82 | 3.69 | 3.60 | 3.68 |
| Glu | 10.06 | 9.95 | 9.72 | 9.24 | 9.31 |
| Gly | 38.50 | 40.73 | 40.57 | 40.05 | 41.98 |
| Ala | 27.46 | 28.25 | 28.17 | 29.88 | 30.47 |
| Cys | 0.16 | 0.30 | 0.28 | 0.22 | 0.47 |
| Val | 1.31 | 1.08 | 1.11 | 1.15 | 1.25 |
| Met | 0.16 | 0.08 | 0.27 | 0.17 | 0.09 |
| Ile | 0.74 | 0.56 | 0.54 | 0.56 | 0.66 |
| Leu | 2.97 | 3.18 | 3.16 | 3.02 | 3.32 |
| Tyr | 0.53 | 2.88 | 3.69 | 3.86 | 2.66 |
| Phe | 0.63 | 0.43 | 0.41 | 0.40 | 0.52 |
| Lys | 0.44 | 0.32 | 0.29 | 0.28 | - |
| His | 0.40 | 0.09 | - | 0.08 | - |
| Arg | 1.79 | 1.65 | 1.62 | 1.73 | - |
| Pro | 6.32 | 4.07 | 4.04 | 3.35 | 2.57 |

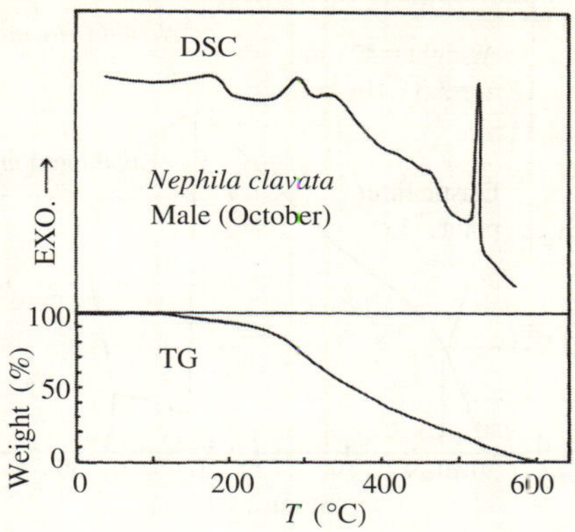


**Fig. 4.48** DSC and TG thermograms for draglines prepared from male *N. clavata* collected in October: EXO, exothermic; *T*, temperature.

line in the draglines from male *N. clavata* in autumn was smaller than that from female *N. clavata* in the same season. The DSC and TG thermograms taken for the draglines from male *N. clavata* in autumn are also shown in Fig. 4.48. The thermal degradation of the draglines from male *N. clavata* occurred at temperatures lower than that from females (see Fig. 4.47). Thus, this may be caused by a difference not only in the physical properties such as the morphology of silks, the molecular weight and the crystallinity, but also in the chemical structure ascribed to the amino acid composition.

## B. Mechanical Lifeline of Draglines

Draglines act as a lifeline for spiders when moving and falling from high places. The mechanical strength of the draglines may be related to the spider's weight since spiders hang from them.

Figure 4.49 shows the stress-strain curve of the draglines secreted by a *Nephila clavata* spider weighing 471 mg, measured at a stretching velocity of 3.3 $\times 10^{-4}$ m/s. The elastic limit stress (ELS) was defined as the stress at which the stress-strain behavior changes from linear to nonlinear, and the mechanical breaking stress (MBS) was defined as the stress at the breaking point. The elastic limit strain (ELN) and breaking strain corresponded to the strain at the elastic limit point and breaking point, respectively.[152,153]

Figure 4.50 shows the relationship between the elastic limit strength (*ES*) and the stretching velocity (*V*) for draglines secreted from *N. clavata* spiders of different weights. The elastic limit strength increased slightly with increasing stretching velocity and was definitely larger for heavier spiders. Reliable data at a fixed velocity can be evaluated in comparison with the data on velocity dependence. Here, the *ES* is given by the ELS times the cross sectional area (*S*) of the dragline.

The *ES* for draglines of *N. clavata* spider weighing more than 50 mg at a

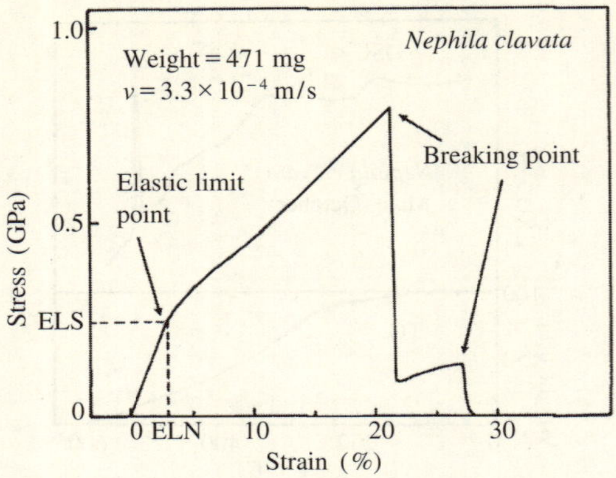


**Fig. 4.49** Stress-strain curve of a dragline secreted from a *N. clavata* spider with a weight of 471 mg, at a stretching velocity of $3.3 \times 10^{-4}$ m/s. Elastic limit stress (ELS) and elastic limit strain (ELN) are defined, respectively, as the stress and the strain at which the stress-strain curve shifts from the linear to the nonlinear region.

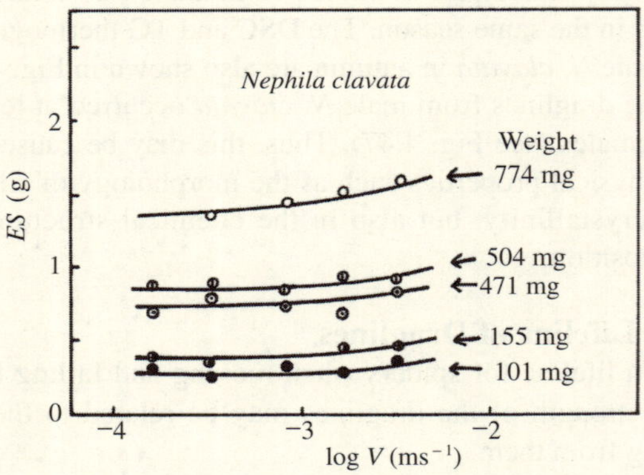


**Fig. 4.50** Relationship between the elastic limit strength (*ES*) and stretching velocity (*V*) for draglines secreted from *N. clavata* spiders of different weights.

stretching velocity of $3.3 \times 10^{-4}$ m/s was plotted against spider weight $W$ in Fig. 4.51. The $ES$, defined as the mechanical strength at the elastic limit point, increased linearly with increasing $W$. Experimentally, we may express $ES$ as

$$ES = 6.76 + 1.98\,W \tag{4.2}$$

with a correlation coefficient of 0.962. Here, the slope was estimated to be near 2.0 for spiders more than 50 mg. The stress-strain curve of the dragline enters a nonlinear region when the stress exceeds about twice the spider's weight. In other words, a spider safely captures and pulls an insect using its dragline if the insect's weight is less than the spider's weight.[153)]

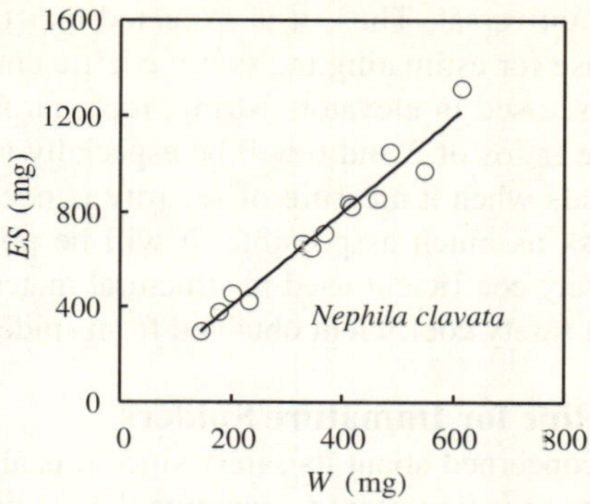


**Fig. 4.51** *ES* at a stretching velocity of $3.3 \times 10^{-4}$ m/s for a *N. clavata* spider's dragline is plotted against the spider's weight (*W*). The spider's weight is more than 50 mg.

The SEM observation demonstrated that the dragline consisted of a double filament (see Fig. 4.45). As shown in Fig. 4.49, a second breaking point in the stress-strain curve is often observed except when the double filaments break perfectly at the same time. The magnitude of the second breaking stress may depend on the degree of imperfect breaking. Even though one filament of the dragline is broken, the other filament will be available for supporting the spider's weight. The spider's weight corresponds to the elastic strength for the single filament. Thus, the elastic limit strength gives minimun safety and maximum efficiency for supporting the spider's weight.

The relationship between the breaking strength (*BS*) and the stretching velocity (*V*) was also examined for draglines secreted from *N. clavata* spiders of different weights. The breaking strength was larger for the heavier spiders. The deviation in the data of the breaking strength was much larger than that of the elastic limit strength. This may come from the nonlinearity of the breaking strength.

The relationship between *BS* and *W* was examined for *N. clavata* spider draglines at a stretching velocity of $3.3 \times 10^{-4}$ m/s. The breaking strength also increased linearly with increasing weight, with the slope determined to be near 6. The relationship between the *BS* and *W* was found to be expressed by the following empirical equation:

$$BS = 0.236 + 6.352\,W \tag{4.3}$$

with a correlation coefficient of 0.965. Thus, draglines should break at a stress of about six times the spider's weight. The probability of the breaking of draglines will be induced because of the nonlinearity of stress-strain curve just after the mechanical stress exceeds the value of the *ES*.

Since spiders move, jump and fall rapidly, maximal efficiency is needed for the mechanical strength of draglines from the viewpoint of safety, and is likely the result of 400 million years of evolution.

The safety coefficient should be based on scientific law from the mechanical point of view and be universal. Thus, it is expected that the findings described here may provide a base for estimating the safety coefficient for various structural materials, such as those used in elevators, string, rope, airplanes, bridges, pillars, wheels and fibers. The ratios of 2 and 6 will be especially useful as safety coefficients for such materials when a measure of security is necessary. When humans desire to eliminate risk as much as possible, it will be possible to estimate the actual value of the safety coefficient used in structural material to a higher degree than the most efficient safety coefficient obtained from spider draglines.

## C. Mechanical Lifeline for Immature Spiders

Every living thing is concerned about its safety since it is always surrounded with various dangers. Hence, it is important to examine the quality of safety during the spider's entire life as it relates to preventing spiders from having dangerous accidents depending on the living environments which change with growth. If the mechanical properties of draglines are measured at all stages of a spider's growth, it will be possible to understand the quality of safety throughout the spider's lifetime.

It is important to measure the degree of safety since a spider's life depends on the draglines.

Figure 4.52 shows the relationship of the elastic limit strength *ES vs.* spider's weight *W* for spiders weighing less than 50 mg. The *ES* increased linearly with increasing *W*. Experimentally, we may express *ES* as

$$ES = 0.95 + 3.00\,W \tag{4.4}$$

with a correlation coefficient of 0.984. The slope was determined to be about 3, which was larger than that obtained in Fig. 4.51.

Figure 4.53 shows the safety coefficient (*SC*) plotted against the spider's weight (*W*). Here, the ratio of the elastic limit strength of dragline to the spider's weight is defined as the *SC*. A larger *SC* means a higher degree of safety. The *SC*

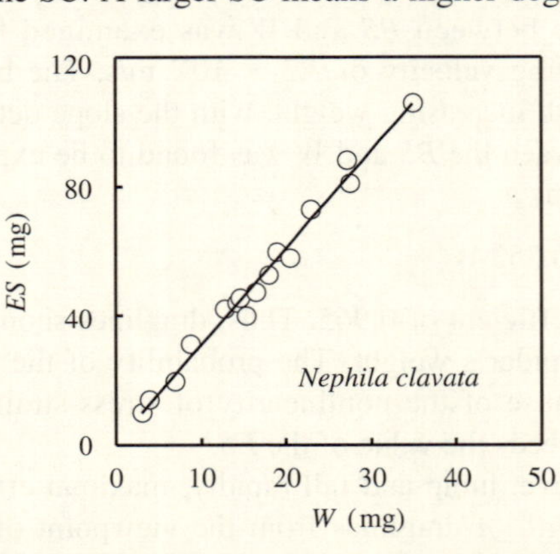


**Fig. 4.52** Elastic limit strength ES is plotted against the spider's weight (*W*). The spider's weight is less than 50 mg.

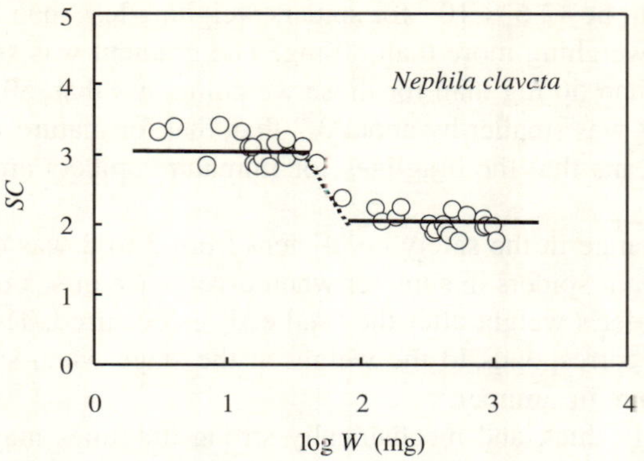


**Fig. 4.53** Spider's weight (*W*) dependence of safety coefficient (*SC*) for the draglines secreted from *N. clavata* spiders. The *SC* is defined as the ratio of the elastic limit strength of the dragline to the spider's weight. The standard error in the observed values of elastic limit strength and cross section area was 5%.

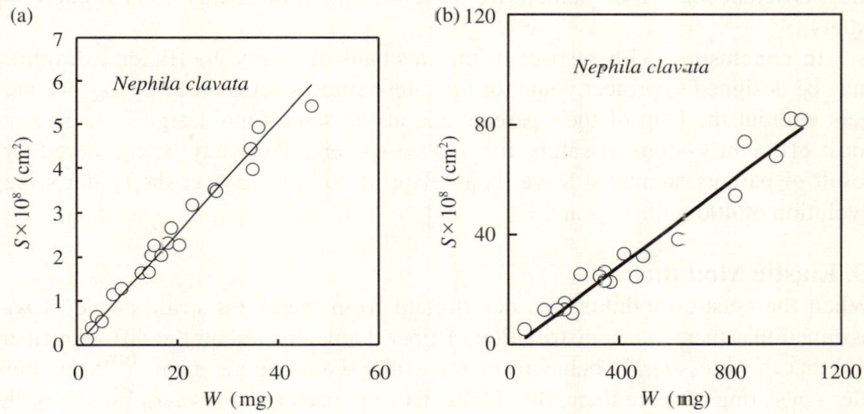


**Fig. 4.54** (a) A cross section area (*S*) of draglines plotted against (*W*) for immature spiders (*N. clavata*) weighing less than 50 mg, (b) The *S* of draglines plotted against the *W* for mature spiders (*N. clavata*) weighing more than 50 mg.

was determined to be about 2 for mature spiders and about 3 for immature spiders. The *SC* changes drastically from 3 to 2 at about 50 mg of the spider's weight. The higher *SC* reguires mechanically stronger draglines for the spider to hang safely.[159)]

Here, in order to consider the draglines from the structural point of view, it is important to estimate the value (*S*/*W*) of the cross section area S of the dragline per spider's weight *W*.

Figures 4.54(a) and (b) show a cross section (S) of a dragline plotted against the spider's weight *W* for spiders of less than 50 mg and more than 50 mg. *S* increases with increasing *W*. We can observe a break at about 50 mg. The slope was determined to be $12.6 \times 10^{-6}$ for spiders weighing less than 50 mg and $7.6 \times 10^{-6}$ for spiders weighing more than 50 mg. The gradient was smaller for spiders weighing more than 50 mg than for those weighing less than 50 mg. The *S/W* for immature spiders was smaller by about 3/2 than that for mature spiders. The large value of *S/W* means that the draglines for immature spiders are thicker than for mature ones.

A drastic change in the safety coefficient from 3 to 2 was observed at about 50 mg of *N. clavata* spiders in summer when divergence of sex and rapid increase in the female spider's weight after the final ecdysis occurred. The spider's weight of about 50 mg corresponds to the weight at the stage when spiders grow from immature to mature in summer.

The relatively thick and mechanically strong draglines may be required for immature spiders to survive without fear of breaking. Actually, the spider's weight increases exponentially by month even though the cross section area of draglines related to the size of secreting gland increases gradually by month.

It was found that the safety coefficient of about 3 is based on relatively thick draglines for young or immature spiders while the safety coefficient of about 2 is most efficient for mature spiders from the viewpoint of energy consumption and safety.

In conclusion, with respect to the mechanical safety coefficient, draglines may be designed to protect young or immature spiders surrounded by various dangers without the help of their parents and at the same time designed to give the most efficient system of safety for mature spiders. This may be the hereditary result of parents having to leave their offspring to survive over the spider's long evolution of 400 million years.

## D. Elastic Modulus

When the elastic modulus was determined from the stress-strain curve, it was assumed that there was a distribution of fiber diameters among the different fibers and that the stress-strain behavior of each fiber would be identical.[143)] When bundles consisting of more than 5000 filaments were used for the determination of the X-ray modulus,[142)] it was difficult to count the number of filaments, to determine the total area by accurately measuring the cross sectional area of each filament and to evaluate the stress applied only to the crystalline region of the dragline consisting of an almost amorphous region.

The question was whether the elastic modulus had been accurately determined because of the following four problems. First, there was a difficulty in preparing samples for the mechanical measurements of the draglines. If a force greater than the elastic limit strength was applied to a dragline, the dragline could not be used as a sample because of the mechanical hysteresis. Secondly, the accuracy of the data in the mechanical measurements was very poor because of difficulties in measuring the markedly low stress and strain. Thirdly, it was very difficult to obtain only true draglines as samples from the spiders because they secrete seven kinds of threads. Fourthly, it was difficult to measure the cross sectional

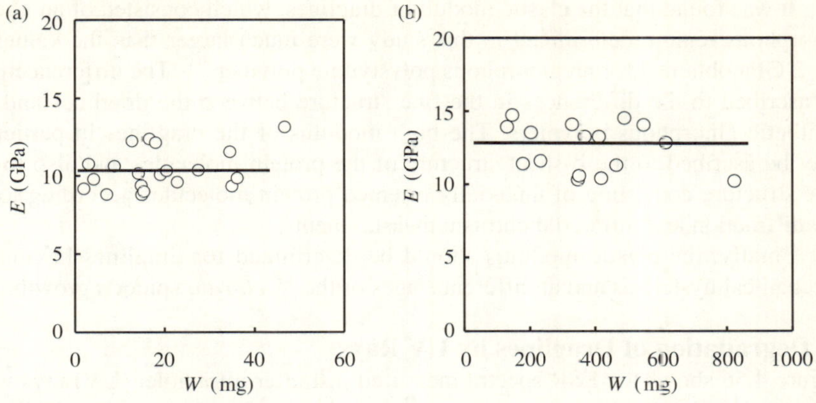


**Fig. 4.55** (a) Elastic modulus *E* of draglines of spiders weighing less than 50 mg, (b) Elastic modulus *E* of draglines of spiders weighing more than 50 mg.

area using the same dragline used for the mechanical measurements. These problems have caused difficulties in estimating the elastic modulus accurately.

However, we overcame these problems for the mechanical measurements and were able to accurately determine the elastic moduli of the draglines.

The relationship between the elastic limit strain ELN of a dragline and spider weight W was examined. The ELN was determined to be 2.31 ± 0.25% for spiders weighing less than 50 mg and 2.6 ± 0.14% for spiders weighing more than 50 mg.

Using the data from *ES*, ELN and *S*, the elastic modulus (*E*) was determined for draglines of spiders of different weights. Fig. 4.55(a) shows the elastic modulus for spiders weighing less than 50 mg. The elastic modulus was determined to be 10.3 ± 1.2 GPa. Fig. 4.55(b) also shows the elastic modulus for dragline of spiders weighing more than 50 mg. The moduli were determined to be 12.9 ± 1.7 GPa for spiders weighing more than 50 mg. The relatively small values of *ca.* 10 GPa for immature spiders and old spiders may come from the relatively large cross sectional area of the dragline and from the relatively small value of elastic limit strength, respectively. Thus, it was found that the elastic modulus closely depended on the stage of spider growth.[154)]

In this study, we observed two breaks in the relationship of elastic modulus *vs.* spider weight. The first break at about 50 mg corresponded to the season when the difference in the sex of *N. clavata* spiders is clearly recognized.

On the other hand, Becker *et al.*[142)] reported that the X-ray modulus of draglines out of *N. clavipes* was *ca.* 17 GPa, although some problems were identified with the measurements as described in the Introduction. This value was comparable to the elastic moduli determined in the present anthor's study.

The most important issue when determining an accurate elastic modulus is to obtain true draglines from the spiders and then measure the cross sectional area of the same draglines used to measure the stress-strain curve.

It was found that the elastic moduli of draglines, which consisted of an almost amorphous region, determined in this study were much larger than the values of *ca.* 2 GPa obtained for an amorphous polystyrene polymer.[160] The difference may be ascribed to the differences in the fine structure between the draglines and the synthetic amorphous polymers. The high modulus of the draglines in particular may be ascribed to the β-sheet structure of the protein molecules and also to the fine structure consisting of uniaxially oriented protein molecules according to X-ray diffraction and infrared dichroism measurements.

Finally, the elastic modulus should be determined for draglines having no mechanical hysteresis and at different stages of the *N. clavata* spider's growth.

## E. Degradation of Draglines by UV Rays

Figure 4.56 shows the ESR spectra measured just after ultraviolet (UV) rays were irradiated to the draglines, which were prepared from female *N. clavata* collected in September, for different periods of time (original, 5, 20, and 60 min). The magnetic field applied to the sample was 3,284 gauss ± 2,500 gauss and the *g* value was determined to be 2.0057 for the observed radical. The spectrum changed and the signal intensity increased with increase in the time of UV-irradiation.[161] It is well known that ultraviolet rays induce the breaking of chemical bonds in protein molecules. Since the signal intensity in the ESR spectrum expressed the quantity of radicals due to the chemical decomposition of silk protein, the increase in the signal intensity may correspond to that in the degradation of the draglines. The results showed that UV rays induced the degradation of the protein molecules due to the breaking of chemical bonds.[150]

Figure 4.57 shows the radical intensity, which was defined as the signal per sample weight at a given time, for draglines of the *N. clavata* appearing in Fig. 4.56 plotted against the irradiation time for UV rays. The radical intensity for draglines of female *N. clavata* initially increased rapidly with irradiation time then

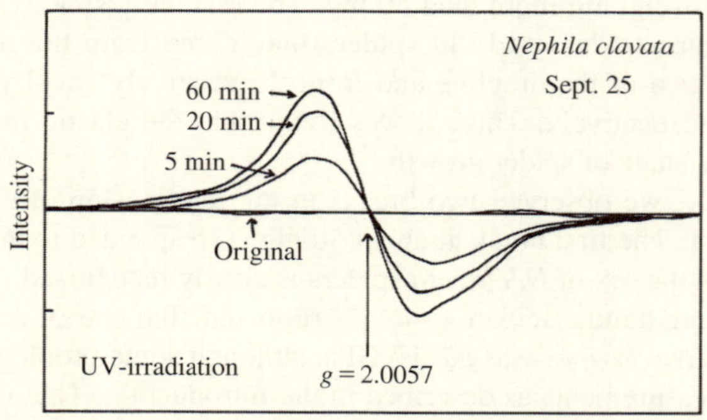


**Fig. 4.56** Spectra of ESR for draglines of *N. clavata* at varying times (original, 5, 20 and 60 minutes) of ultraviolet irradiation when ultraviolet rays are irradiated to the draglines secreted from *N. clavata* spiders collected in August. The *g*-value was determined to be 2.0057.

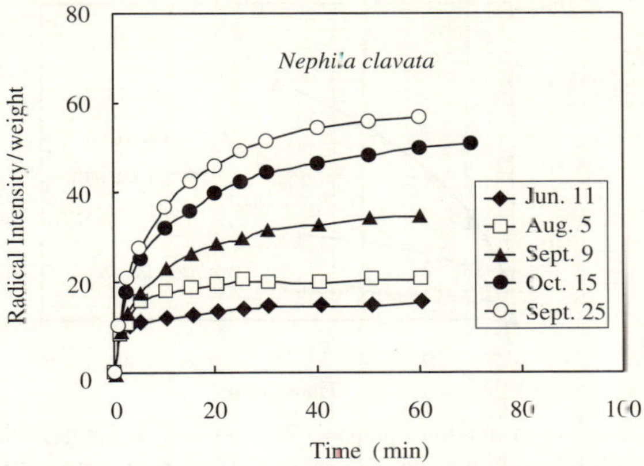


**Fig. 4.57** Time dependence of UV-irradiation upon the radical intensity for draglines from *N. clavata* collected in different months.

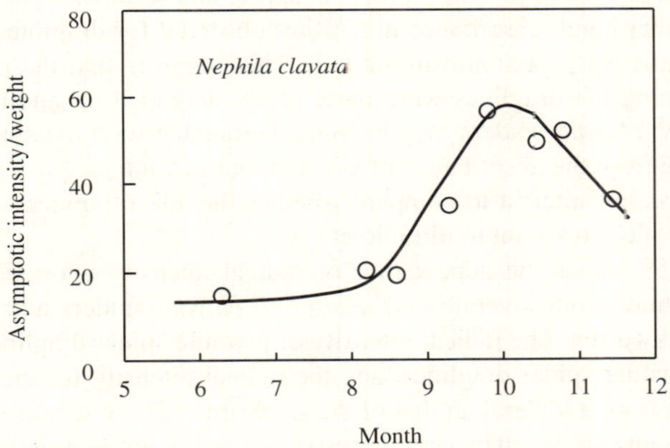


**Fig. 4.58** The asymptotic value of the relative radical intensity for draglines from *N. clavata* collected at different stages of growth.

gradually approached an asymptotic value, as shown in Fig. 4.57.[161] This suggested that the proteins constituting the dragline decomposed chemically from the UV irradiation, and that the relative signal intensity for mature spiders was larger than that for juvenile spiders. Thus, the draglines of the mature spiders have many radicals in the protein molecules, suggesting that the draglines of the mature spiders degrade more easily in autumn than those of the juvenile spiders in spring. This finding indicates that aging should be a consideration when obtaining biomaterials, such as the silk secreted from an old spider.[150]

Figure 4.58 shows the asymptotic values of relative radical intensity for draglines collected at different stages of growth. The radical intensity started to increase in late August, reached maximum in October, after which it decreased.

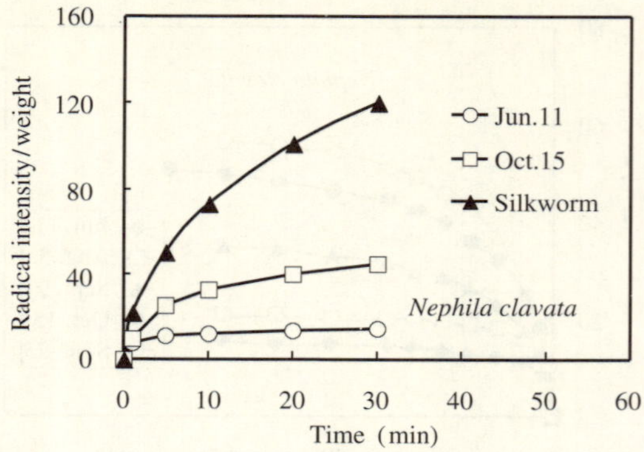


**Fig. 4.59** Radical intensity of draglines prepared from *N. clavata* spiders plotted against UV-irradiation time for silks from juvenile and mature spiders and out of silkworm.

This supported the finding that draglines of mature spiders in autumn degrade more easily than those of juvenile spiders in spring and summer.[161)]

On the other hand, absorbance at 275 nm observed for draglines collected in different months was maximum in autumn. This suggests that the protein molecules constituting the draglines were more easily degraded in autumn. Thus, the draglines of *N. clavata* spiders may become degradable with the spider's growth and may come from the absortbance of UV at about 275 nm.

Here, it was of interest to compare whether the silk of spiders or the silk of silkworms are more resistant to ultraviolet rays.

Figure 4.59 shows the dependence of radical intensity upon UV-irradiation time for draglines from juvenile and mature *N. clavata* spiders and for silk prepared from silkworms. The radical intensity of juvenile spider draglines was lower than that of mature spider draglines, and the radical intensity for the draglines of the spiders was much lower than that of the silkworms. These results show that the draglines of young spiders had a strong resistance to UV-irradiation in comparison with the draglines from old spiders and also that the spider draglines were more resistant to ultraviolet rays than the silk of silkworms. Thus, the draglines from spiders will be useful as a new material for textiles from the viewpoint of UV-resistance.

Finally, it was suggested that an absorbance at about 275 nm may be the trigger for the protein molecules of draglines to decompose easily. However, such a small absorbance at about 275 nm for draglines from juvenile spiders may result in strong UV-resistance. On the other hand, the absorbance at about 275 nm of spider silk in the UV region was much smaller than that of the silkworm, suggesting that spider draglines have a strong chemical structure to resist UV-irradiation.

### F. Strengthening of Draglines by UV Rays

UV rays from sunlight are harmful to a variety of living organisms including animals and humans, as well as to biomaterials such as silkworm silk. Spiders fre-

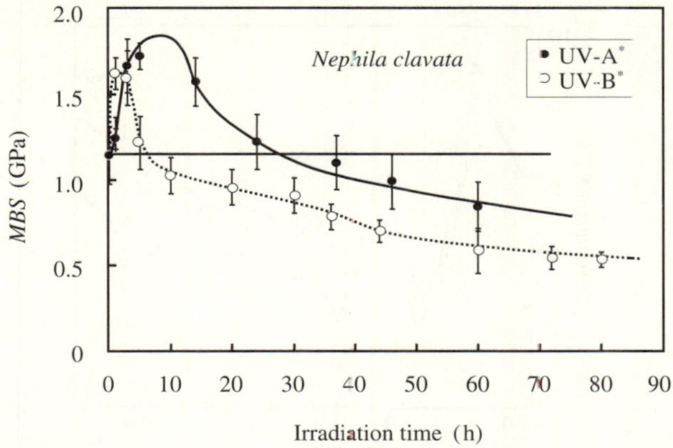


**Fig. 4.60** Time dependencies of the mechanical breaking stress (*MBS*) of draglines (mean value ± standard error) secreted from a *N. clavata* spider weighing 540 mg. Draglines were irradiated with UV rays of UV-A* or UV-B* , respectively. The horizontal line is drawn using a value of 1.2 GPa as the starting point. ● ; UV-A*, ○ ; UV-B*.

quently rebuild new orb-webs, although the reasoning behind this is unknown. Such rebuilding may be related to the decomposition of the orb-web induced by UV rays from sunlight. Consequently, the effects of UV rays on the mechanical breaking stress (MBS) of spider silk are of great interest.

Figure 4.60 shows the time dependence of UV-irradiation upon the MBS of draglines secreted from a *N. clavata* spider weighing 540 mg. MBS is defined as the mechanical breaking strength (BS) divided by the cross sectional area of the draglines.[162] The MBS increased rapidly shortly after commencing irradiation with UV-A* or UV-B*, peaked fairly quickly, then decreased gradually as it approached the respective asymptotic value with time of UV-irradiation.

Compared to the MBS at the starting point of UV-irradiation, the BS was larger by *ca.* 60% for UV-A* and *ca.* 40% for UV-B* irradiation. Thus, irradiation with UV-A* showed better improvement of BS. The peaks in the MBS may come about as a consequence of two factors, an increase in molecular weight due to cross-linking of protein molecules and a decrease in molecular weight due to decomposition.[162]

Consequently, spiders effectively utilize the UV-A rays to mechanically strengthen the orb-web. The study of spider silk provides valuable information regarding the benefits of UV-rays upon essential activities of living organisms, the characteristics of biomaterials and development of industrial materials.

## G. Infrared Dichroism

Figure 4.61 shows the infrared absorption spectra of draglines measured after polarized infrared rays were irradiated parallel and perpendicular to the draglines which were aligned in parallel. The spectra shows three different absorptions of 1510 $cm^{-1}$, 1630 $cm^{-1}$ and 3200 $cm^{-1}$, suggesting the presence of amino acid

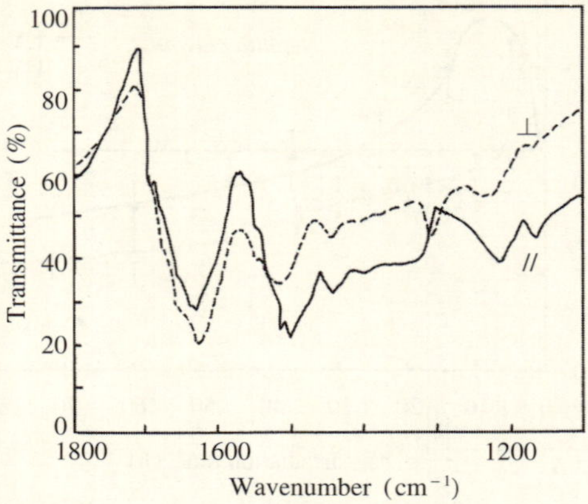


**Fig. 4.61** Dichroism of infrared spectra of draglines when polarized infrared rays are irradiated to the draglines secreted from *N. clavata* spiders in two different perpendicular directions. —— ; parallel, ----- ; perpendicular.

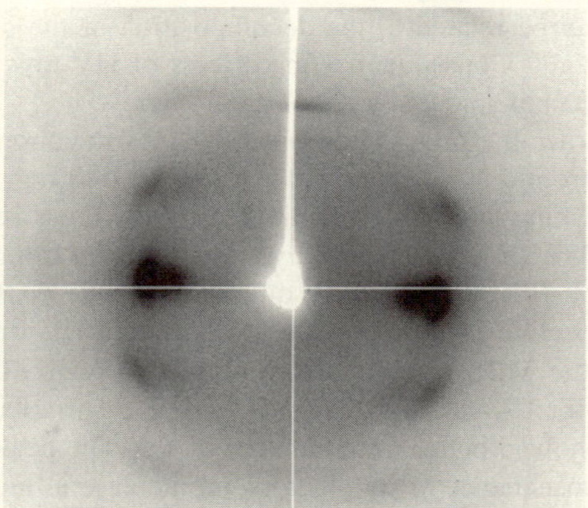

**Fig. 4.62** X-ray diffraction photograph of draglines secreted from a *N. clavata* spider.

residues. The infrared dichroism shows that the protein molecules constituting the draglines were on average arranged parallel along the long axis of the draglines.[158] The microwave method also shows that the protein molecules are aligned parallel to the long chains of draglines.[158,163] Such uniaxially oriented protein molecules give mechanically strong silks.

### H. Fine Structure of Dragline

Protein materials constituting the draglines are liquid fibroins inside the spinning gland of the spider, and the stretching force causes them to change into a solid dragline just after secretion from the gland. The molecular weight was determined to be about 700,000 for the liquid silk inside the gland, about 520,000 for solidified draglines just after collection, and about 340,000 for solidified draglines several weeks after collection.[164)] Recently, however, molecular weight was determined to be about 270,000 for liquid silk inside the spinning glands of a *N. clavata*[165)] and the molecular distribution was very sharp.

Such protein molecules show an α-configuration with predominantly intramolecular hydrogen bonds for the liquid fibroin, but a β-configuration in a pleated sheet form with intermolecular hydrogen bonds for the solidified dragline. It is estimated that pulling the liquid fibroins changes the α-configuration into a β-configuration.[155)] The crystallinity is very low for spider draglines compared with silkworm silks. At present, it is very difficult to determine the crystal structure of proteins constituting a dragline because of its very low crystallinity (Fig. 4.62).[165)]

However, from experimental results, the magnitude of the plane distance in the X-ray diffraction pattern suggests that the amino acid residue has large side chains and that the proteins have an anti-parallel structure in the crystalline region.

In any case, uniaxially oriented molecules and the β-configuration having a pleated sheet form with intermolecular hydrogen bonds provide the characteristic properties of spider silks.

## 4.5.4 Conclusion

Spider silks can be expected to be utilized as a new material instead of silkworm silks, since the physicochemical properties of spider silks, such as heat-resistance,[148)] UV-resistance,[150,151)] mechanical strength,[152-154)] and flexibility are better.

Draglines composed of repeating Gly, Ala and Glu amino acid residues may give heat-resistant[148)] and UV-resistant[150,151)] properties. Spider silks composed of an almost amorphous and β-sheet structure[155,165)] give flexible and stretch[153)] properties for spiders to move easily and quickly. Uniaxially stretched silks especially[146)] give mechanically strong properties and a high modulus.[154)] These excellent properties have developed over 400 millions years of evolution It will be necessary to further examine spider silks from the viewpoint of nanostructure.

Genetic technology revealed the sequence of the amino acids of spider fibroins.[134,135)] Draglines consist of two kinds of fibroins (spidroin 1 and spidroin 2) containing a polyalanine sequence. Spidroin 1 consists of a repeat of three kinds of segments and the polyalanine region consists of a β-sheet structure. Spidroin 2 consists of a β-turn structure and a β-sheet structure. Thus, it is estimated that the stretching force destroys the β-turn structure and produces a β-sheet structure. Such a structure may explain the high elastic modulus of spider silks.[134,135)]

Even though we live in a world of high technology, it is necessary to study the science and technology of natural phenomena. The findings of the most efficient safety[152,153)] obtained for spider draglines is a typical example from natural phenomena. Recently, the author obtained the interesting result that ultraviolet rays remarkably strengthened the orb-web of spiders,[162] and that the mechanical properties of spider draglines give an origin for evaluating the degree of trust.[166,167] Such natural phenomena should be examined from the viewpoint of nanostructure.

Currently, it is believed that a new type of material having properties similar to those of spider silk should be produced using a genetic method. To do so it would be beneficial to study the physicochemical properties of spider silk in detail from the viewpoint of nanostructure.

# References


< Section 4.1 >

1. Fraústo da Silva, J. J. R., Williams, R. J. P., *The Biological Chemistry of the Elements*, Oxford University Press, Oxford, 1991.
2. Lippard, S. J., Berg, J. M., *Principles of Bioinorganic Chemistry*, University Science Books, 1994.
3. Lippard, S. J., *Science*, 1993, **261**, 699.
4. Reedijk, J., *Macromolecule-Metal Complexes* (Ciardelli, F., Tsuchida, E., Wöhrle, D. eds.), p.131, Springer-Verlag, Berlin and Heidelberg, 1995.
5. Reedijk, J., *Makromol. Chem. Macromol. Symp.*, 1994, **80**, 95.
6. Housecroft, C. E., Sharpe, A. G., *Inorganic Chemistry*, Prentice Hall, Harlow, 2001.
7. Pearson, R. G., *J. Am. Chem. Soc.*, 1963, **85**, 3533.
8. Taube, H., *Chem. Rev.*, 1952, **50**, 69.
9. Shenton, W., Mann, S., Colfen, H., Bacher, A., Fischer, M., *Angew. Chem. Int. Ed.*, 2001, **40**, 442.
10. Wolfe, S. A., Grant, R. A., Elrod-Erickson, M., Pabo, C. O., *Structure*, 2001, **9**, 717.
11. Walkup, G. K., Imperiali, B., *J. Am. Chem. Soc.*, 1997, **119**, 3443
12. Pierre, J. L., Fontecave, M., Crichton, R. R., *Biometals*, 2002, **15**, 341.
13. Crichton, R. R., Pierre, J. L., *Biometals*, 2001, **14**, 99.
14. Zangger, K., Oz, G., Otvos, J. D., Armitage, I. M., *Protein Sci.*, 1999, **8**, 2630.
15. Webb, E., *Enzyme Nomenclature, IUB Recommendations 1984*, Academic Press, New York, 1984.
16. Reedijk, J., Bouwman, E., *Bioinorganic Catalysis,* Marcel Dekker., New York, 1999.
17. Guo, Z. J., Sadler, P. J., *Angew. Chem. Int. Ed.*, 1999, **38**, 1513.
18. Reedijk, J., *Curr. Opin. Chem. Biol.*, 1999, **3**, 236.
19. Guo, Z. J., Sadler, P. J., *Advances in Inorganic Chemistry*, Vol. 49, p.183, 2000.
20. Reedijk, J., *Chem. Rev.*, 1999, **99**, 2499.
21. Reedijk, J., *Chem. Commun.*, **1996**, 801.
22. Bloemink, M. J., Reedijk, J., Farrell, N., Qu, Y., Stetsenko A. I., *J. Chem. Soc., Chem. Commun.*, **1992**, 1002.
23. Brabec, V., Kasparkova, J., Vrana, O., Novakova, O., Cox, J. W., Qu, Y., Farrell, N., *Biochemistry*, 1999, **38**, 6781.
24. Komeda, S., Lutz, M., Spek, A. L., Chikuma, M., Reedijk, J., *Inorg. Chem.*, 2000, **39**, 4230.
25. Teuben, J. M., Ph. D. thesis, Leiden University, 2000.
26. Ohndorf, U. M., Rould, M. A., He, Q., Pabo, C. O., Lippard, S. J., *Nature*, 1999, **399**, 708.
27. Sahni, S. K., Reedijk, J., *Inorg. Chim. Acta*, 1988, **150**, 169.
28. Hoorn, H. J., de Joode, P., Dijkstra, D. J., Driessen, W. L., Kooijman, H., Veldman, N., Spek, A. L., Reedijk, J., *J. Mater. Chem.*, 1997, **7**, 1747.
29. van de Water, L. G. A., ten Hoonte, F., Driessen, W. L., Reedijk, J., Sherrington, D. C., *Inorg. Chim. Acta*, 2000, **303**, 77.
30. van Pouderoyen, G., den Blaauwen, T., Reedijk, J., Canters, G. W., *Biochemistry*, 1996, **35**, 13205.
31. Challa, G., Reedijk, J., van Leeuwen, P. W. N. M., *Polym. Adv. Techn.*, 1996, **7**, 625.
32. Sigel, H., *Metal Ions in Biological Systems*, Marcel Dekker, New York, 1980.

< Section 4.2 >

33. Harada, A., Okamoto, K., Kamachi, M., Honda, T., Miwatani, T., *Chem. Lett.*, **1990**, 917.
34. Harada, A., Okamoto, K., Kamachi, M., *Chem. Lett.*, **1991**, 953.
35. Harada, A., Fukushima,, H., Shiotsuki, K., Kamachi, M., *Supramol. Chem.*, 1993, **2**, 153.
36. Harada, A., Shiotsuki, K., Fukushima, H., Yamaguchi, H., Kamachi, M., *Inorg. Chem.*, 1995, **34**, 1070.
37. Harada, A., Yamaguchi, H., Okamoto, K., Fukushima H., Shiotuki, K., Kamachi, M., *Photochem. Photobiol.*, 1999, **70**, 298.
38. Yamaguchi, H., Kamachi, M., Harada, A., *Angew. Chem. Int. Ed.*, 2000, **39**, 3829.

39. Harada, A., Fukushima, H., Shiotsuki, K., Yamaguchi, H., Oka, F., and Kamachi, M., *Inorg. Chem.*, 1997, **36**, 6099.
40. Yamaguchi, H., Harada, A., *Biomacromolecules*, 2002, **3**, 1163.
41. Lehn, J. -M., *Angew. Chem. Int. Ed. Engl.*, 1990, **29**, 1304.
42. Harada, A., Yamaguchi, H., Kamachi, M., *Chem. Lett.*, **1997**, 1141.
43. Harada, A., Yamaguchi, H., Oka, F., Kamachi, M., *Proc. SPIE Int. Soc. Opt. Eng.*, 1999, **3607**, 126.
44. Harada, A., Yamaguchi, H., Tsubouchi, K., Horita, E., *Chem. Lett.*, 2003, **32**, 18.
45. Davies, D. R., Padlan, E. A., Segal, D. M., *Ann. Rev. Biochem.*, 1975, **44**, 639.
46. Silverton, E. W., Navia, M. A., Davies, D. R., *Proc. Natl. Acad. Sci. U.S.A.*, 1977, **74**, 5140.
47. Tomalia, D. A., Naylor, A. M., Goggard III, W. A., *Angew. Chem. Int. Ed. Engl.*, 1990, **29**, 138.
48. Yamaguchi, H., Kubota, K., Harada, A., *Proc. SPIE Int. Soc. Opt. Eng.*, 2000, **3922**, 228.
49. Yamaguchi, H., Kubota, K., Harada, A., *Chem. Lett.*, **2000**, 384.
50. Yamaguchi, H., Kubota, K., Harada, A., *Nucleic Acids Symp. Ser.*, 2000, **44**, 229.
51. Brown P. O., Cozzarelli, N. R., *Proc. Natl. Acad. Sci. U.S.A.*, 1981, **78**, 843.
52. Krasnow, M. A., Cozzarelli, N. R., *J. Biol. Chem.*, 1982, **257**, 2687.
53. Krasnow, M. A., Stasiak, A., Spengler, S. J., Dean, F., Koller, T., Cozzarelli, N. R., *Nature*, 1983, **304**, 559.
54. Droge, P. D., Cozzarelli, N. R., *Methods Enzymol.*, 1992, **212**, 120.
55. Levene, S. D., Donahue, C., Boles, T. C., Cozzarelli, N. R., *Biophys. J.*, 1995, **69**, 1036.
56. Wang, J. C., *Ann. Rev. Biochem.*, 1985, **54**, 665.
57. Binnig, G., Quate, C. F., Gerber, Ch., *Phys. Rev. Lett.*, 1986, **56**, 930.
58. Blackford, B. L., Jericho, M. H., Mulhern, P. J., *Scanning Microsc.*, 1991, **5**, 907.
59. Butt, H. -J., Guckenberger, R., Rabe, J. P., *Ultramicroscopy*, 1992, **46**, 375.
60. Hoh, J. H., Hansma, P. K., *Trends Cell Biol.*, 1992, **7**, 208.
61. Yang, J., Tamm, L. K., Somlyo, A. P., Shao, Z., *J. Microsc.*, 1993, **171**, 183.
62. Smith, B. L., *Prog. Biophys. Mol. Biol.*, 2000, **74**, 93.
63. Engel, A., Müller, D. J., *Nat. Struct. Biol.*, 2000, **7**, 715.
64. Pereira, R. S., *Biochem. Pharmacol.*, 2001, **62**, 975.
65. Hansma, H. G., *Ann. Rev. Phys. Chem.*, 2001, **52**, 71.
66. Allison, D. P., Hinterdorfer, P., Han, W., *Curr. Opin. Biotechnol.*, 2002, **13**, 47.
67. Fotiadis, D., Scheuring, S., Müler, S. A., Engel, A., Müller, D. J., *Micron*, 2002, **33**, 385.
68. Lindsay, S. M., Nagahara, L. A., Thundat, T., Knipping, U., Rill, R. L., *J. Biomol. Struct. Dyn.*, 1989, **7**, 279.
69. Weisenhorn, A. L., Gaub, H. E., Hansma, H. G., Sinsheimer, R. L., Kelderman, G. L., Hansma, P. K., *Scanning Microsc.*, 1990, **4**, 511.
70. Weisenhorn, A. L., Egger, M., Ohnesorge, F., Gould, S. A. C., Heyn, S. -P., *Langmuir*, 1991, **7**, 8.
71. Hansma, H. G., Weisenhorn, A. L., Gould, S. A. C., Sinsheimer, R. L., Gaub, H. E., *J. Vac. Sci. Technol. B*, 1991, **9**, 1282.
72. Bustamante, C., Vesenka, J., Tang, C. L., Rees, W., Guthold, M., Keller, R., *Biochemistry*, 1992, **31**, 22.
73. Vesenka, J., Guthold, M., Tang, C. L., Keller, D., Delaine, E., Bustamante, C., *Ultramicroscopy*, 1992, **42-44**, 1243.
74. Hansma, H. G., Vesenka, J., Siegerist, C., Kelderman, G., Morrett, H., *Science*, 1992, **256**, 1180.
75. Hansma, H. G., Hansma, P. K., *Proc. SPIE Int. Soc. Opt. Eng.*, 1993, **1891**, 66.
76. Hansma, H. G., Bezanilla, M., Zenhausern, F., Adrian, M., Sinsheimer, R. L., *Nucleic Acids Res.*, 1993, **21**, 505.
77. Lyubchenko, Y. L., Shlyakhtenko, L. S., Harrington, R. E., Oden, P. I., Lindsay, S. M., *Proc. Natl. Acad. Sci. U.S.A.*, 1993, **90**, 2137.
78. Samori, B., Siligardi, G., Quagliariello, C., Weisenhorn, A. L., Vesenka, J., Bustamante, C. J., *Proc. Natl. Acad. Sci. U.S.A.*, 1993, **90**, 3598.
79. Hansma, H. G., Laney, D. E., Bezanilla, M., Sinsheimer, R. L., Hansma, P. K., *Biophys. J.*, 1995, **68**, 1672.
80. Leatherbarrow, R. J., Stedman, M., Wells, T. N. C., *J. Mol. Biol.*, 1991, **221**, 361.
81. Mazeran, P. -E., Loubet, J. -L., Martelet, C., Theretz, A., *Ultramicroscopy*, 1995, **60**, 33.
82. Han, W., Mou, J., Sheng, J., Yang, J., Shao, Z., *Biochemistry*, 1995, **34**, 8215.
83. Zhang, Y., Sheng, S., Shao, Z., *Biophys. J.*, 1996, **71**, 2168.
84. Yamaguchi, H., Harada, A., Kamachi, M., *React. Polym.*, 1998, **37**, 245.

85. Yamaguchi, H., Harada, A., *Chem. Lett.*, **2002**, 382.

< Section 4.3 >

86. Frankel, R. B., Mann, S., *ENCYLCOPEDIA of Inorganic Chemistry* (King, R. B. ed.), p.269, John Wiley & Sons, Chichester, New York, Brisbane, Toronto, Singapore, 1994.
87. Wada, K., Kobayashi, I., *Biomineralization and Hard Tissue of Marine Organisms* (Wada, K. and Kobayashi, I. eds.), Tokaidaigaku shuppankai, 1996.
88. a)Mann, S., *Nature*, 1988, **332**, 119.
b)Takahashi, K., Yamamoto, H., Onoda, A., Doi, M., Inaba, T., Chiba, M., Kobayashi, A., Taguchi, T., Okamura, T., Ueyama, N., *Chem. Commun.*, **2004**, 996.
89. Berman, A. *et al.*, *J. Phys. Chem.,* 1993, **97**, 5162.
90. Addadi, L., Weiner, S. *Proc. Natl. Acad. Sci. U. S. A.,* 1985, **82**, 4110.
91. Berman, A., Addadi, L., Weiner, S., *Nature*, 1988, **331**, 546.
92. Aizenberg, J., Albeck, S., Weiner, S., Addadi, L., *J. Cryst. Growth*, 1994, **142**, 156.
93. Albeck, S., Aizenberg, J., Addadi, L., Weiner, S., *J. Am. Chem. Soc.,*1993, **115**, 11691.
94. Albeck, S., Weiner, S., Addadi, L., *Chem. Eur. J.,* 1996, **2**, 278.
95. Belcher, A. M., *Nature,* 1996, **381**, 56.
96. Falini, G., Albeck, S., Weiner, S., Addadi, L., *Science,* 1996, **271**, 67.
97. Brooks, R., Clark, L. M., Thurston, E. F., *Phil. Trans. A,* 1950, **243**, 145.
98. Mann, S., Heywood, B. R., Rajam, S., Birchall, J. D., *Nature*, 1988, **334**, 692.
99. Xu, G., Yao, N., Aksau, I. A., Groves, J., *J. Am. Chem. Soc.*, 1998, **120**, 11977.
100. Orme, C. A. Noy, A., Wierzbichi, A., McBrice, M. T., Grantham, M., Teng, H. H., Dove, P. M., DeYoreo, J. J., *Nature*, 2001, **411**, 775.
101. Miyamoto, H., Miyashita, T., Okushima, M., Nakano, S., Morita, T., Matsushiro, A., *Proc. Natl. Acad. Sci. U.S.A.,*1996, **93**, 9657.
102. Aizenberg, J., Hanson, J., Koetzle, T. F., Weiner, S., Addadi, L.. *J. Am. Chem. Soc.*, 1997, **119**, 881.
103. Dobachi, T., Nagata, H., Manabe, Y., Inokuchi, S., *IEEE/ASME Trans. Mechatronics*, 1998, **3**, 106.
104. Nagata, H., Dobashi, T., Manabe, Y., Usami, T., Inokuchi, S., *IEEE Trans. Visualization and Computer Graphics*, 1997, **3**, 307.
105. Mets, G., Wu, X., Smith, S. O., *J. Magn. Reson.,* 1994, **A110**, 219.
106. Bennet, A. E., Rienstra, C. M., Auger, M., Klakshmi, K. V., Griffin, R. G., *J. Chem. Phys.,* 1995, **103**, 6951.
107. Torchia, D. A., *J. Magn. Reson.,*1978, **30**, 613.
108. Ueyama, N., Kozuki H., Doi, M., Yamada, Y., Takahashi, K., Onoda, A., Okamura, T., *Macromolecules*, 2001, **34**, 2607.

< Section 4.4 >

109. Hobbs, C., *Medical Mushrooms*, p.158, Botanical Press, 1991 and the references cited therein.
110. Chihara, G., Maeda, Y., Hamuro, J., Sasaki, T., Fukuoka, F., *Nature*, 1969, **222**, 687.
111. Hamuro, J., Kikuchi, T., Takatsuki, F., Suzuki, M., *Brit. J. Cancer,* 1996, **73**(4), 465.
112. a)Yanaki, T., Ito, W., Tabata, K., Kojima, T., Norisuye,T., Takano, N , Fujita, H., *Biophys. Chem.*, 1983, **17**(4), 337.
b)Tabata, K., Ito, W., Kojima, T., Kawabata, S., Misaki, A., *Carbohydr. Res.*, 1981, **89**(1), 121.
113. Chihara, G., Hamuro, J., Maeda Y., Arai Y., Fukuoka, F., *Cancer Res.*, 1970, **30**, 2776.
114. Shimizu. Y., Chen.J. -T., Hasumi. K., *BIOTHERAPY*, **4**(8), 1990. 1390.
115. Vandamme, E. J., Baets, S. D., Steinbuchel, A. eds., *Biopolymers Polysaccharides*, Chapter 6 in Vol I and Chapters 2 and 3 in Vol II, Wiley-VCH, Weinheim, 2002.
116. Fuhrhip, J. H., Endish, C., *Molecular and Supramolecular Chmemistry of Natural Products and Their Model Compound*, Marcel Dekker, New York, 2000.
117. Rao, V. S. R., Qasba, P. K., Balaji, P. V., Chandrasekaran, R., *Conforamtion of Carbohydrates*, Harwood Academic Press, 1998.
118. Koumoto, K.. Kimura, T., Kobayashi, H., Sakurai, K., Shinkai, S., *Chem. Lett.*, **2001**, 908.
119. Desland, Y., Marchessault, R. H., Sarko, A., *Macromolecules*, 1980, **13**, 1466.
120. a) Norisuye, T., Yanaki, T., Fujita, H., *J. Polym. Sci., Polym. Phys. Ed.*, 1980, **18**(3), 547.
b) T. Sato, T., Norisuye, T., Fujita, H., *Carbohydr. Res.*, 1981, **95**(2), 195.
121. Sato, T., Sakurai, K., Norisuye, T., Fujita, H. *Polym. J.* (Tokyo), 1983, **15**(1), 87.
122. McIntire, T. M., Brant, D. A., *J. Am. Chem. Soc.*, 1998, **120**, 6909.
123. Sakurai, K., Shinkai, S, *J. Am. Chem. Soc.*, 2000, **122**, 4520.
124. Kimura, T., Koumoto, K., Sakurai, K., Shinkai, S., *Chem. Lett.*, **2000**, 1242.

125. Mizu, M., Koumoto, K., Sakurai, K., Shinkai, S., *Chem. Commun.*, **2001**, 429.
126. Sakurai, K., Mizu, M., Shinkai, S., *Biomacromolecules*, 2001, **2**, 641.
127. Sakurai, K., Iguchi, R., Koumoto, K., Kimura, T., Mizu, M., Hisaeda, Y., Shinkai, S., *Biopolymers*, 2002, **65**, 1.
128. a) Koumoto, K., Kimura, T., Mizu, M., Sakurai, K., Shinkai, S. *Chem. Commun.*, **2001**, 1962.
b) Koumoto, K., Kimura, T., Mizu, M., Kunitake, T., Sakurai, K., Shinkai, S., *Perkin Trans 1.*, **2002**, 2477.
129. Saenger, W., *Principles of Nucleic Acid Structure*, Springer-Verlag, New York, 1984.
130. Nakajima, K., Brrova, N., *Circular Dichroism: Principles and Application*, VCH, New York, 1994.
131. Sakurai, K., Mizu, M., Shinkai, S., submitted to *Bioorganic Chemistry*.
132. Sakurai, K., Shinkai, S, *J. Inclusion Phenomena and Macrocyclic Chem.*, 2001, **41**, 173.
133. Filed patent PCT/JP02/02228 and JP 2002-272938, paper in preparation.

**< Section 4.5 >**

134. Xu, M., Lewis, R. V., *Proc. Natl. Acad. Sci. U. S. A.*, 1990, **87**, 7120.
135. Hinman, M. B., Lewis, R. V., *J. Biol. Chem.*, 1992, **267**, 19320.
136. Lazaris, C. A., Arcidiacono, S., Huang, Y., Zhou, J. F., Duguay, F., Chretien, N., Welsh, E. A., Soares, J. W., Karatzas, C. N., *Science*, 2002, **295**, 472.
137. M. W. Denny, *J. Exp. Biol.*, 1976, **65**, 483.
138. Work, R. W., *Text. Res. J.*, 1976, **46**, 485.
139. Work, R. W., Morosoff, N. S., *Text. Res. J.*, 1982, **52**, 349.
140. Gosline, J. M., Denny, M. W., DeMond. M. E., *Nature*, 1984, **309**, 551.
141. Vollrath, F., Edmonds, D., *Nature*, 1989, **340**, 305.
142. Becker, M. A., Mahoney, D. V., Lenhert, P. G., Edy, R. K., Kaplan, D., Adams, W. W., *Silk Polymers*, Kaplan, D., Adams, W. W., Farmer, B., Viney, C. eds., p.185, American Chemical Society, Washington DC, 1995.
143. Cunniff, P. M., Fossey, S. A., Auerbach, M. A., Song, J. W., *Silk Polymers*, Kaplan, D., Adams, W. W., Farmer, B., Viney, C. eds., p.234, American Chemical Society, Washington DC, 1995.
144. Termonia, Y., *Macromolecules*, 1994, **27**, 7378.
145. Brandwood, A., *J. Exp. Biol.*, 1985, **116**, 141.
146. Osaki, S., *Kagaku-no-Ryoiki*, 1982, **36**, 426.
147. Osaki, S., *J. Synth. Org. Chem., Jpn*, 1985, **43**, 828.
148. Osaki, S., *Acta Arachnol.*, 1989, **37**, 69.
149. Osaki, S., *Acta Arachnol.*, 1989, **38**, 21.
150. Osaki, S., *Acta Arachnol.*, 1994, **43**, 1.
151. Osaki, S., *Acta Arachnol.*, 1997, **46**, 1.
152. Osaki, S., *Nature*, 1996, **348**, 419.
153. Osaki, S., *Int. J. Biol. Macromol.*, 1999, **24**, 283.
154. Osaki, S., Ishikawa, R., *Polym. J.*, 2002, **34**, 25.
155. Foelix, R. F., *Biology of Spiders,* p.111, Harvard University Press, Cambridge, 1982.
156. Zemlin, J. C., U. S. Army Natick Laboratories, *Natick MA*, *Technical Report TR* 69-29-CM(AD684333), 1968.
157. McNamee, S. G., Ober, C. K., Jelinski, L. W., Ray, E., Xia, Y., Grubb, D. T., *Silk Polymers*, Kaplan, D., Adams, W. W., Farmer, B., Viner, C. eds., p.176, American Chemical Society, Washington DC, 1995.
158. Osaki, S. ed., *Mystery of Spider's Silk*, Chuokoron-Shinsha, Tokyo, 2000.
159. Osaki, S., *Polym. J.*, 2003, **35**, 261.
160. Nakamae, K., Nishino, T., Hata, K., Matsumoto, T., *Kobunshi Ronbunshu*, 1985, **42**, 211.
161. Osaki, S., Kamachi, M., *Polymer Preprints, Japan*, 1998, **47**, No.4, 710.
162. Osaki, S., *Polymerwerkstoffe 2002*, Halle, Germany, 617, 2002.
163. Osaki, S., *Nature*, 1990, **347**, 132.
164. Jackson, C., O'Brien, J. P., *Macromolecules*, 1995, **28**, 5975.
165. Osaki, S., Yamamoto, K., unpublished data.
166. Osaki, S., 17th Bratislava International Conference on Macromolecules (IUPAC), Bratislava, Slovakia, p.23, 2003.
167. Osaki, S. ed., *Why Don't Spiders Fall Down from Their Lifeline*?, PHP Institute, Tokyo, 2004.s

# Index

## T

## U



## V



## W, X, Y, Z